Groundwater Economics and Policy in South Asia

Groundwater Economics and Policy in South Asia

M. Dinesh Kumar
Executive Director
Institute for Resource Analysis and Policy
Hyderabad, Telangana, India

ELSEVIER

Elsevier
Radarweg 29, PO Box 211, 1000 AE Amsterdam, Netherlands
The Boulevard, Langford Lane, Kidlington, Oxford OX5 1GB, United Kingdom
50 Hampshire Street, 5th Floor, Cambridge, MA 02139, United States

Notices
Knowledge and best practice in this field are constantly changing. As new research and experience broaden our understanding, changes in research methods, professional practices, or medical treatment may become necessary.

Practitioners and researchers must always rely on their own experience and knowledge in evaluating and using any information, methods, compounds, or experiments described herein. In using such information or methods they should be mindful of their own safety and the safety of others, including parties for whom they have a professional responsibility.

To the fullest extent of the law, neither the Publisher nor the authors, contributors, or editors, assume any liability for any injury and/or damage to persons or property as a matter of products liability, negligence or otherwise, or from any use or operation of any methods, products, instructions, or ideas contained in the material herein.

ISBN: 978-0-443-14011-2

For Information on all Elsevier publications visit our website at
https://www.elsevier.com/books-and-journals

Publisher: Candice Janco
Acquisitions Editor: Maria Elekidou
Editorial Project Manager: Teddy A. Lewis
Production Project Manager: Bharatwaj Varatharajan
Cover Designer: Christian J. Bilbow

Working together
to grow libraries in
developing countries
www.elsevier.com • www.bookaid.org

Typeset by Aptara, New Delhi, India

Dedication

This book is dedicated to my beloved Father-in-Law, A. K. Sreedharan

Contents

List of figures

List of tables

Preface

The original idea of this book emerged sometime during 2021 when I wanted to prepare a book manuscript on groundwater economics and policy in India for Sage Publications.

The debates around the management of groundwater resources in South Asian countries, particularly India, Pakistan, and Nepal began in the early 1990s. A major international conference on groundwater titled "**Water Management: India's Groundwater Challenge**" was organized by a reputed Ahmedabad-based NGO, named VIKSAT (Vikram Sarabhai Centre for Development Interaction) with the support of the Ford Foundation in December 1993, and which had scholars, academicians, and groundwater professionals from 22 countries participating, marked the launching of the wide academic and policy debates on the groundwater management issues and challenges not only in India but also the entire South Asian region.

It marked a major departure from the earlier debates around the issues of access equity and efficiency in groundwater use and looked at the issues and challenges of ensuring the sustainability of resource use and thereby sustaining agricultural production in semi-arid regions which was supported by its intensive use, and equity in access to the resource. In the initial years, such debates covered several important topics that many developed countries such as the United States and Spain as well as upper-middle-income countries such as China and Mexico, also debated. The debate was quite rich and comprehensive and covered all crucial aspects of water management—from technical to social to economic to institutional to legal.

Over the years, the debate lost the seriousness that it demanded as no significant initiative came from the government front to address the growing groundwater challenges in a comprehensive manner. Despite being faced with very serious problems of over-exploitation, within South Asia, the governments began to treat the issues with kid gloves. The water bureaucracies at the central as well as the state level, being influenced by a few highly localized experiments of NGOs across the country that were touted as "successful" and that received high media publicity, started advocating decentralized water harvesting and recharge, while also making half-hearted attempts to regulate groundwater abstraction by millions of farmers through legislations which have no teeth. In the process, the debate, which otherwise was academic in nature, got hijacked by the motivated writings of a few influential groups, which seemed to believe that the agenda of sustainable groundwater resource development and management can be fulfilled by a few wealthy farmers and community leaders with some government support through extra energy subsidies and funds for local groundwater recharge schemes.

Playing to the gallery, they advocated some piecemeal approaches. Some of them are "decentralized water harvesting and recharge"; large-scale promotion of micro irrigation systems; electricity supply rationing for the agriculture sector; cash incentives for farmers who restrict power consumption. These approaches lacked any serious understanding of the resource dynamic, particularly the regional variations, and the community behavior, particularly how the farmers respond to the state interventions. The subject which is complex given the nature of the resource and the dynamics of its availability and use, and which demanded serious academic research in disciplines as varied as hydrology and geohydrology, economics, public administration, sociology, and law, thus got trivialized.

The support for these "run-of-the-mill" solutions was built on the narrative that the traditional way of solving groundwater problems using management tools that are tried and tested in the developed countries of the West (like groundwater pricing, energy rationing and "water rights" reforms, registration of wells, providing cheap replacement water to reduce groundwater dependence, etc.) is "mundane" and would not work in the South Asian context. To eclipse the occasional concerns raised about distributional inequity that such quick-fix solutions can pose, very often comparisons are made with public surface irrigation, falsely claiming that access to groundwater is more equitable with the resource largely benefiting the poor small and marginal farmers, and its use more efficient, and any amount of investment in managing groundwater can be justified.

Having realized the need to challenge some of the dominant views about groundwater management alternatives, I prepared a book proposal on groundwater economics and policy in India. It was promptly reviewed and accepted by Sage, New Delhi for publication. Within 3–4 weeks of submitting the full manuscript to the publisher for final review and publication, I heard the news that Sage was closing its book publishing portfolio in India. This was quite disturbing for me as an author.

After a few days of coming out of the shock created by Sage's decision not to proceed with the publication of the manuscript, I decided to contact Elsevier, with which I had already published two of my books and two edited volumes on various themes related to water. The suggestion from Elsevier was that I should work on a larger canvas and focus on the entire South Asia instead of India for them to consider it further. While the idea of writing a book on South Asia's groundwater had been there in my mind for quite some time, I took this challenge as an opportunity. It meant doing a lot of additional work to update the manuscript based on the work that was done during the past decade or so in the rest of South Asia, particularly Pakistan and Bangladesh by various scholars and my own understanding of the region. The reviews of the book proposal from the four anonymous reviewers were highly encouraging and exciting.

What is attempted in the book is infusing some fresh ideas into the minds of the readers on what can work and what cannot for managing a complex resource like groundwater, based on a thorough analysis of the groundwater management approaches that had worked and approaches that had failed in the region and elsewhere.

I sincerely hope this book "**Groundwater Economics and Policy in South Asia**" would elevate the level of the international debate on several topics such as the technologies, economic instruments, institutions, and policies required for achieving the multiple goals of improving the efficiency, equity, and sustainability of groundwater use without compromising on the livelihoods of hundreds of millions of farmers in South Asia region; and the need to improve the rigor in the research on groundwater resource assessment, groundwater development strategies, and groundwater management.

M. Dinesh Kumar
February 22, 2023

Acknowledgment

First of all, I am extremely thankful to the four reviewers, who had gone through the book proposal, and the editorial office of Elsevier Science which immensely helped in framing the "problem" that the volume addresses and enhancing the value of the final manuscript. Thanks are particularly due to Ms Maria Elekidou, Acquisitions Editor: Aquatic Sciences, Elsevier Science, and Mr Theodore Lewis, Editorial Project Manager, Elsevier for their encouragement and immense support throughout the course of this work—starting from conceptualization of the book proposal to preparation and submission of the full manuscript.

I am also extremely thankful to my colleagues, Ms Vaishnavi, Research Officer, Institute for Resource Analysis and Policy (IRAP), for digging out the latest scientific articles and reports on groundwater in the South Asia region and outside. I used them in the several chapters of the book. My sincere thank are also due to Mr Ajath Sanjeev, my Executive Assistant, who meticulously prepared the table of contents, and list of tables, figures, and abbreviations of the book and incorporated the editorial corrections in the manuscript.

I am also extremely grateful to my beloved daughter, Archana Dineshkumar, who also works currently as the Consulting Junior Science Editor for IRAP, for meticulously editing the entire manuscript for language, consistency, and style, and checking the correctness of the citations and references.

I owe profusely to my beloved family, especially my wife Shyma and my daughter who stood by me during the difficult times, provided moral strength, and became a constant source of motivation to do more and more.

Lastly, I proudly dedicate this book to my beloved father-in-law, Shri A. K. Sreedharan, whose love and affection towards me is no less than that of my late father. He has been a great pillar of support for me during times of crisis.

M. Dinesh Kumar
February 22, 2023

List of abbreviations

ADB-	Asian Development Bank
AP-	Andhra Pradesh
APFMGS-	Andhra Pradesh Farmer Management of Groundwater Systems
APGWD-	Andhra Pradesh Ground Water Department
APWALTA-	Andhra Pradesh Water, Land and Trees Act
BCM-	Billion Cubic Meter
BGL-	Below Ground Level
BISAG-	Bhaskaracharya Institute for Space Applications and Geo informatics
CAPART-	Council of Advancement of People's Action and Rural Technology
CCAFS-	Climate Change, Agriculture and Food Security
CEEW-	Council on Energy, Environment and Water
CF-	Consumptive Fraction
CGIAR-	Consultative Group on International Agricultural Research
CGWB-	Central Ground Water Board
COTAS-	*Comite Tecnicos de Aguas Subterraneas*
CWC-	Central Water Commission
DNA-	Deccan News Agency
DTW-	Deep Tube Well
EGP-	Eastern Gangetic Plains
ET-	Evapo transpiration
FAO-	Food and Agriculture Organization
FCI-	Fluoride Contamination Index
FW-	Farmers Welfare
GBM-	Ganga–Brahmaputra Meghna
GCA-	Gross Cropped Area
GDP-	Gross Domestic Product
GEC-	Groundwater Estimation Committee
GMD-	Groundwater Management Districts
GOAP-	Government of Andhra Pradesh
GOI-	Government of India
GRACE-	Gravity Recovery and Climate Experiment
GWS-	Ground Water Storage
Ha-	Hectare
HDPE-	Highdensity Polyethylene
HMZ-	Hydrometeorological Zones
HP-	Horsepower
Hr-	Hour
ICT-	Information and Communication Technologies
IFPRI-	International Food Policy Research Institute

IGP-	Indo-Gangetic Plains
INR-	Indian Rupee
IRMA-	Institute of Rural Management Anand
IRMED-	Institute for Resource Management and Economic Development
IRR-	Internal Rate of Return
ITGE-	Instituto Tecnológico GeoMinero de España
IWMI-	International Water Management Institute
JGY-	Jyotigram Yojana
kg-	Kilograms
kW-	Kilo watt
m-	Meter
MCM-	Million Cubic Meters
MGNREGA-	Mahatma Gandhi National Rural Employment Guarantee Act
MGVCL-	Madhya Gujarat Vij Company Limited
MI-	Microirrigation
mm-	Millimeter
MNRE-	Ministry of New and Renewable Energy
MOA-	Ministry of Agriculture
MP-	Madhya Pradesh
MPR-	Monopoly Price Ratio
MSSRF-	MS Swaminathan Research Foundation
NA-	Not Available
NABARD-	National Bank for Agriculture and Rural Development
NGO-	Non-Governmental Organization
NHM-	National Horticulture Mission
NSSO-	National Sample Survey Office
NWC-	National Water Commission
OW-	Open Well
PET-	Potential Evapotranspiration
PGWM-	Participatory Ground water Management
PRI-	Panchayati Raj Institution
PV-	Photovoltaic
RF-	Rheumatoid Factor
RKVY-	Rashtriya Krishi Vigyan Yojna
Rs-	Rupees
RWH-	Rainwater Harvesting
SAS-	Sahibzada Ajit Singh
SBS-	Shaheed Bhagat Singh
SMS-	Short Message Service
SOLAW-	State of The World's Land and Water Resources for Food and Agriculture
SPICE-	Solar Pump Irrigators' Cooperative Enterprise
SRTT-	Sir Ratan Tata Trust
SSP-	Sardar Sarovar Project
STW-	Shallow Tube Well
SWAP-	Soil–Water–Atmosphere–Plant
TDS-	Total Dissolved Solids
TMC-	Thousand Million Cubic

TN-	Tamil Nadu
TOI-	Times of India
UCLM-	University of Castilla-La Mancha
UNDP-	United Nations Development Programme
UP-	Uttar Pradesh
US-	United States
USA-	United States of America
UT-	Union Territory
VIKSAT-	Vikram Sarabhai Centre for Development Interaction
WARR-	Weighted Average Rate of Return
WB-	West Bengal
WFP-	World Food Programme
WRD-	Water Resources Department
WRIS-	Water Resources Information System
WTP-	Willingness to Pay
WUAs-	Water User Associations

Chapter 1

Introduction

1.1 Changing character of groundwater debate in South Asia

Globally, the sub-continent of South Asia uses the largest amount of groundwater (Qureshi, 2020). Among the South Asian countries, India is the largest user of groundwater not only in the sub-continent but in the entire world, surpassing the most populous country, i.e., China (Jadhav, 2018; Shiferaw, 2021; World Bank, 2010). Pakistan is the second-largest user of groundwater in the sub-continent (60 BCM per annum), with a major part of the country experiencing arid climatic conditions. The third largest user of groundwater in the sub-continent is Bangladesh, with an annual withdrawal of 30 BCM (Subhadra, 2015; Suhag, 2016; Qureshi et al., 2014). With a net irrigation of 48 m.ha from groundwater, these three countries account for 42% of the world's well-irrigated area (Mukherji et al., 2015). Among all the South Asian countries, India is quite unique in the sense that all the climates that are witnessed in the sub-continent are experienced in one or more parts of the country, from hot and hyper-arid to hot and arid to hot and sub-humid to hot and humid to cold and humid to a cold, dry desert. Almost all the geological formations that bear groundwater and that are encountered in other countries in South Asia, such as unconsolidated formations, semi-consolidated formations, and consolidated formations, are encountered in India as well.

With its intensive use for irrigation, rural and urban drinking, and industry, groundwater resources are over-exploited in many semi-arid and arid parts of India, with several manifestations such as alarming drops in water levels in alluvial areas, rapid seasonal depletion in hard rock areas, increasing salinity of groundwater in inland areas and seawater intrusion in coastal regions (FAO, 2012). There are also pockets or regions in India where groundwater resources are plentiful and the water table remains high, but direct access to the resource is limited to a few due to the poor economic conditions of a large proportion of the farmers. Problems of over-exploitation of groundwater are also witnessed in large parts of Pakistan, along with growing problems of salinity in groundwater (Qureshi et al., 2010; Qureshi, 2020). In countries such as Afghanistan and Sri Lanka, groundwater exploitation is still limited due to a variety of reasons. In Bangladesh, in spite of having heavy rainfall and good surface water availability, groundwater draft has become significant in the past 30 years or so. As per a

Groundwater Economics and Policy in South Asia. DOI: https://doi.org/10.1016/B978-0-443-14011-2.00007-3

2014 report, there are 1,523,322 shallow tube wells and 35,322 deep tube wells in Bangladesh to provide water for irrigation. About 79% of the total cultivated area in that country is irrigated by groundwater (Qureshi et al., 2014).

There are no particular groundwater conditions observed elsewhere in South Asia but not experienced in India—whether in terms of quantity or quality. As a result, strategies evolved to tackle groundwater problems in various Indian situations can provide strong clues on the ways to tackle similar problems in other South Asian countries. Even by the official estimates of the Central Ground Water Board (CGWB), whose methodology to estimate groundwater resource availability is still not perfect and tends to over-estimate the resource availability (Kumar & Singh, 2008), nearly 17.5% of the 6881 assessment units in the country are already over-exploited—with abstraction exceeding annual recharge (CGWB, 2019). There are many regions where groundwater suffers from problems of contamination, with excessive levels of salinity, fluorides, chlorides, nitrates, and iron in many parts, and arsenic in some pockets of the eastern region (CGWB, 2010).

Currently, roughly 65% of India's gross irrigation comes from groundwater sources such as open wells, bore wells, and tube wells. In addition, 80% of the rural drinking water supplies and half of the urban water supplies are managed through supplies from wells. Therefore, groundwater depletion and groundwater quality deterioration problems, if left unchecked, can pose a serious threat to agricultural production and food security, economic growth in rural areas, and drinking water security, especially in rural areas and small towns. In many semi-arid areas underlain by alluvium, with a significant long-term decline in groundwater levels, investment for well construction has enormously increased, with direct access to the resource now being limited to only rich farmers. In hard rock areas, the rampant failure of wells and reduction in well yields is making well irrigation unaffordable for resource-poor farmers. Over-exploitation is also resulting in the widespread failure of drinking water wells in villages falling in the hard rock areas.

What is generally seen in any part of the world is that the communities as well as the governments exploit groundwater when good quality surface water is not available locally in adequate quantities to meet local water demands. In all other South Asian countries, groundwater is a source of water for irrigation and domestic and industrial water supply. However, the intensity of use varies from country to country and from region to region depending on the rainfall conditions, surface water availability, climate, and geohydrology, with the first three factors determining the demand for groundwater and the fourth one determining the supply potential of the aquifers. Exceptions are African countries, where the high cost of well drilling, poor rural electrification, and expensive fossil fuels are constraints to exploiting groundwater, and as a result, the intensity of resource use is still very low in most of them. The current thinking about groundwater problems and management solutions in India and a few other South Asian countries such as Pakistan and Bangladesh, and the experience of these countries in dealing with the problems, will be extremely relevant not only for every country in South Asia but also for many countries in Africa.

The debates around the management of groundwater resources in South Asian countries, especially India and Pakistan, began in the early 1990s. It marked a major departure from the earlier debates around issues of access equity and efficiency in groundwater use and looked at the issues and challenges of ensuring the sustainability of resource use and sustaining agricultural production supported by its intensive use in semi-arid regions, as well as equity in access to the resource that was becoming increasingly scarce. In the initial years, such debates covered the following: (1) community awareness for resource protection and conservation; (2) artificial recharge of groundwater in the over-exploited regions using local water resources with the involvement of communities; (3) introducing water conservation practices in irrigated agriculture that accounts for a major share of the total groundwater consumed annually; (4) regulating groundwater use through powerful state legislations; (5) enabling government legislations for the local communities to manage groundwater; (6) large-scale water transfers for conjunctive management of groundwater and surface water; (7) creating a proper water rights system and allocating these rights amongst various users/user groups; (8) building local community institutions for groundwater management at various scales from village to aquifer, with the support of the institutional regime of water rights supported by law; (9) policy interventions to promote equity in access to groundwater; and (10) using economic instruments such as groundwater pricing and pricing of energy used for pumping groundwater. Thus, the debate was quite rich and comprehensive and covered all crucial aspects of water management—from technical to social to economic to institutional to legal (Moench et al., 1993). But such debates never found adequate space in academic literature and course materials of advanced studies in water resources management and water economics and policy.

Over the years, the debate lost the seriousness and depth that it calls for as no significant initiative came from the government front to address growing groundwater challenges in a comprehensive manner. In spite of having serious groundwater depletion problems, in India, which had the highest dependence on groundwater for irrigation among all the South Asian countries, the governments of the country began to treat the problems with kid gloves. The water bureaucracies at the central as well as the state level, being influenced by a few highly localized experiments of NGOs across the country that were touted as "successful" and that received high media publicity, started advocating decentralized water harvesting and recharge (Kumar, 2018; Kumar & Pandit, 2018), while also making half-hearted attempts to regulate groundwater abstraction by millions of farmers through vacuous legislation. In the process, the debate, which was otherwise serious and academic in nature, got hijacked by the motivated writings of a few influential researchers and groups, who seemed to believe that the agenda of sustainable groundwater resource development and management can be finished with the help of a few wealthy farmers and community leaders with some government support through extra energy subsidies and funds for groundwater recharge (see, Mukherji et al., 2009, 2010; Shah, 1993, 2000). They advocated some piecemeal approaches such as "decentralized water harvesting and recharge" (Shah, 2000; Shah et al., 2003); large-scale promotion of micro irrigation systems; electricity supply rationing for the agriculture sector (Shah & Verma, 2008); participatory forms of groundwater management, using "aquifer-based common pool resource" approaches (Kulkarni et al., 2015); cash incentives for farmers who restrict power consumption (Gulati & Pahuja, 2015); and heavily subsidized power connections for farm wells in water-rich regions, and subsidized and unmetered electricity supply to such wells (Mukherji et al., 2009).

These approaches lacked any serious understanding of the resource dynamic, particularly the regional variations in resource availability, and the community behavior, particularly how the farmers respond to the state interventions such as electricity supply rationing for agriculture, provisioning of cash incentives for regulating electricity use by them, tariff subsidy for grid-connected solar power, subsidy for construction of farm ponds, etc. The subject which is otherwise complex given the nature of the resource and the dynamics of its availability and use, and which demanded serious academic research in disciplines as varied as hydrology and geohydrology, economics, public administration, sociology, and law, thus got trivialized.

The support for these quick-fix solutions was built on the narrative that the traditional way of solving groundwater problems using management tools that are tried and tested in the developed countries of the west (like groundwater pricing, energy rationing, and "water rights" reforms, registration of wells, providing cheap replacement water to reduce groundwater dependence, etc.) is "mundane" and would not work in the Indian context. However, there are occasional concerns raised about distributional equity that such quick-fix solutions can pose. To eclipse such concerns, comparisons are made with public surface irrigation, falsely claiming that access to groundwater is more equitable, with the resource largely benefiting the poor small and marginal farmers, and its use more efficient, so any amount of investment in managing groundwater can be justified.

On the teaching and training front, the academic and research institutions running courses in natural resource management and resource economics and policy always were short of materials based on empirical research that could be used for teaching advanced courses for M.A., M. Phil, and Ph.D. This was especially true for water resource management, and water economics and policy. The faculties in such institutions had to depend on books written several decades ago, which were not updated to keep pace with the advanced concepts, theories, and practices in the related physical and social science fields that are essential for policy making. Worse is the case for thematic areas like groundwater, where there are no solid empirical research-based materials available from the developing countries of Asia, including China, for secondary reading in groundwater resource dynamics, resource assessment, socioeconomic aspects of groundwater use, and economic instruments and institutions for groundwater management.

There were a few attempts to apply some of the solutions earlier mentioned by a few state governments to arrest groundwater depletion through new programs, which include the governments of the states of Punjab, Gujarat, Madhya Pradesh, Andhra Pradesh, Maharashtra, and Telangana in India through popular schemes. Sufficient time has lapsed since these programs were first piloted or implemented, and thus their outcomes can be clearly assessed through proper methodologies to see whether the intended outcomes have been realized or not. While the researchers who championed these schemes eulogized them like evangelists through their academic writings, a few empirical studies have happened in the past that critically and objectively evaluated the impacts of these policies and programs from inter-disciplinary and multi-disciplinary perspectives. Interestingly, there were also isolated attempts by the governments in Bangladesh to implement innovative solutions that use advanced concepts to regulate groundwater over-draft. An example is the introduction of pre-paid meters for agricultural groundwater pumping in Bangladesh. Given this context, it is time to revisit the old debates in light of the emerging concepts, theories, and practices with available empirical knowledge of their application on the ground to have a more nuanced and policy-oriented debate on viable alternatives for sustainable groundwater management in South Asia.

Such a critical analysis of the management options, if presented with theoretical frameworks and several empirical examples of their applications, will be of immense value not only for groundwater professionals and policymakers in South Asia but also for the researchers, multi-lateral donors, and government policymakers in countries of Africa as well as other countries of Asia that are on the same development trajectory as that of India, thereby being able to channelize the debate to come up with sound groundwater development and management policies. For the academicians who currently use fringe literature that is based on outdated concepts and theories for teaching courses in "water resources management," and "water economics and policy," this will serve as secondary reading material.

1.2   Purpose and scope of the book

The purpose of the book is to generate a nuanced and informed academic debate on the physical, social, economic, legal, institutional, and policy aspects of groundwater management in South Asia, the subcontinent which constitutes the largest user of groundwater in the world on an annual basis. It does so by presenting the theoretical arguments and empirical evidence about the characteristics of the resource—quantity, quality, and variations across regions and seasons—its use (levels of use, equity, and efficiency), the magnitude of over-exploitation problems, and the various groundwater management solutions initiated by governments and NGOs on the resource availability, and equity in access and control over the resource. The idea is to thoroughly discuss various concepts, theories, and practices in disciplines such as hydrology, economics, and institutions that are discussed and tried so far in the South Asian context to manage groundwater. It is also to help academics handling advanced courses in groundwater resource management and groundwater economics and policy with methodological and analytical approaches, and provide government policymakers, groundwater professionals, and donors with decision-making frameworks.

By first critically analyzing the resource conditions and the over-exploitation problems, and then evaluating the past attempts to manage groundwater on the engineering, social, economic, institutional and legal fronts, using knowledge from disciplines such as geohydrology, hydrology, agronomy, economics, and institutions, the book provides a thorough database and knowledge on what can work and what cannot for managing groundwater in a complex situation like the one present in South Asia. It offers groundwater management solutions for different typologies of socio-ecological settings (alluvium with intensive groundwater use, alluvium with low levels of groundwater use, and hard rocks with intensive groundwater use). It also shows how programs that are ill-conceived but touted as highly successful can produce impacts that are totally undesirable. It thus makes a unique contribution to the field of groundwater economics, management, and policy.

The book offers a thorough analysis of the groundwater availability and use situation in India with a particular focus on regional patterns; provides a critical assessment of how over the years various public policies have impacted the access to and use of groundwater by various segments of the farming communities, particularly the equity in access to groundwater and efficiency in use for agricultural production; critically reviews the resource assessment methodologies adopted by official agencies, highlighting their flaws and shows how such flaws can lead to wrong management decisions; and critically evaluates several schemes implemented in different parts of the sub-continent, some of which are touted as "highly successful" and some controversial, for their effectiveness in checking groundwater depletion using sound principles in water management, and shows how the past analyses of their impacts on the resource condition have been misleading and flawed. Drawing definite clues from these Indian experiences and some of the international experiences, the author presents some viable alternatives for groundwater management for the distinct socio-ecologies that exist in the sub-continent, built on strong conceptual and theoretical foundations, and discusses how they can be implemented. Finally, some policy lessons for other countries, especially those from Africa that are at an earlier stage in the groundwater development trajectory, are provided.

1.3 Contents of the book

The analysis and arguments in this book are presented in eleven chapters. Following the introduction chapter that provides the rationale and purpose of the book and an outline of individual chapters, in Chapter 2, we would provide the macro picture of the characteristics of the resources—particularly its quantity and the regional variations across South Asia, and the structures that are used to abstract the resource and their changing composition and character over time. We would then provide a glimpse of the arguments that shaped the public policies in the groundwater sector. We would then analyze the issue of equity in access to groundwater by looking at the ownership of different types of wells across landholding categories in the region, and the amount of land each category of farmer irrigates. Further, we would analyze the equity impacts of groundwater markets in the South Asian region by providing the available empirical evidence. Finally, we will also look at how groundwater irrigation in India and other South Asian countries has been growing over the years and will confront the question, i.e., whether the groundwater boom will sustain.

The author's view is that many of the past studies that analyzed the socio-economic aspects of groundwater use lacked rigor mainly as they failed to bring in the nuances such as groundwater system characteristics (aquifer types and yields), types of groundwater irrigation structures, source characteristics (well outputs, depth of pumping, water level trends), the cost of maintaining the wells, the area commanded by wells, and the types of crops cultivated in their analysis, in spite of their being critical in analyzing the performance of irrigation systems. Such gaps in the analysis heavily influenced the results vis-à-vis cost-effectiveness, the irrigation potential of the system and the cost per ha of irrigation from wells, the long-term growth potential, and therefore the policy prescriptions they came up with as regards investment choices, institutional financing, and provisioning of subsidies for well construction or electricity.

Neither the professional agencies involved in groundwater development planning nor the mainstream academic books in water resources management and water policy provide a proper definition of groundwater over-exploitation. Chapter 3 deals with the definition of aquifer over-exploitation. A review of the various definitions and criteria for assessing over-exploitation is presented. Subsequently, the existing methodologies in India, the only country in the South Asian region to have developed methodologies for the assessment of groundwater resources, are reviewed to examine: the robustness of the criteria used, and the scientific accuracy of the methodologies and procedures suggested. Finally, the current estimates of groundwater over-development for India are reviewed from the perspective of detailed water balance, geology, hydrodynamics, and negative social, economic, ecological, and ethical consequences. The chapter argues that there are several conceptual issues involved in the assessment of aquifer over-exploitation. One reason is that over-exploitation is linked to various "undesirable consequences" of groundwater use that are physical, social, economic, ecological, environmental, and ethical in nature. The other reason is that there are differences in the way undesirable consequences are perceived by different stakeholders. The principle of inter-generational equity, used in the concept of sustainability, is one of the goals that are built into the standard definitions of aquifer over-exploitation. But defining and assessing over-exploitation is both difficult and complex, and not amenable to simple formulations. We demonstrate through selected illustrative cases how integrating data on complex hydrology, geology, hydrodynamics, and socio-economic, ecological, and ethical aspects of groundwater use with the official statistics could change India's groundwater scenario altogether.

In recent years, a myopic view favoring only private well irrigation in preference to canal irrigation has emerged among some irrigation researchers in the region. They try to weigh surface irrigation against groundwater irrigation. Even in Pakistan, which has the largest canal irrigation network, the importance of canal irrigation in sustaining the yield of wells is often underplayed. This has led to several fallacies, trivializing the serious problems facing the groundwater sector, which are as follows. Well irrigation is superior to canal irrigation. Surface irrigation is becoming increasingly irrelevant in India's irrigation landscape in spite of growing investments, and therefore, future investments in irrigation should be diverted to well irrigation. The growth in well irrigation in semi-arid regions of India can be sustained by using local runoff for recharging the aquifers. Theoretical comparisons of surface water and groundwater available in academics are limited to discussions on the biological quality of water from the point of view of portability, and the patterns and levels of contamination of the two sources by toxic substances. The academic literature on water resources is conspicuous by the absence of a proper assessment of the differential benefits and impacts of irrigation from two sources, based on the ground situation.

Such distorted views had actually trivialized the serious problems facing the groundwater sector and had done a lot of damage to the cause of finding long-term solutions for groundwater depletion problems in the countries of the sub-continent. Given this context, Chapter 4 presents the following arguments. The inherent advantages of surface irrigation system over well irrigation, such as higher dependability of the system and the ability to effectively address spatial mismatch in resource availability and demand, means the second is not a substitute for the first. The use of outdated irrigation management concepts that treat "vertical drainage" as waste leads to an underestimation of the efficiency of surface systems. Sustaining well irrigation in semi-arid and arid regions would need "imported surface water" rather than local runoff for recharging. The use of simple statistics of "area irrigated" to pass judgments about the performance of surface irrigation systems is sheer misuse of statistics.

Chapter 5 critically evaluates two controversial government schemes of recent times in the water sector, viz., the farm pond scheme of Maharashtra and the solar irrigation cooperatives with the feed-in-tariff system from Gujarat, two western provinces in India. The first was launched by the government of Maharashtra with the objective of reducing the pressure on groundwater for irrigation, and the second was launched by the electricity utility of Central Gujarat with the twin objectives of promoting groundwater and energy conservation in collaboration with the International Water Management Institute. The analysis was done for the potential impacts on the ground as well as their welfare effects. The first critique (farm ponds in Maharashtra) is based on the evidence available from the field, and some of the earlier critics of the same. The second critique is based on a synthesis of a series of articles published on the concepts underlying the scheme design and the working of the model by the author himself and others. The questions that the chapter raises are: whether these schemes provide any welfare benefits for the heavy government subsidies they involve, apart from subsidizing the rich, and whether the outcomes and impacts that these schemes produce are as intended.

There is perhaps no region in the world that received as much attention from scholars as Gujarat when it comes to debating groundwater over-exploitation problems and the solutions attempted. While the 80s and 90s discussed the problems of groundwater over-exploitation, the discussions during the next two decades centered on the range of management interventions attempted by the state government and the NGOs. The groundwater build-up in the state in recent years has caught the attention of researchers worldwide. Many researchers have empirically argued that the improvement in the groundwater condition in this region was because of the government-sponsored massive water harvesting work and the introduction of a policy to restrict power supply to the agriculture sector for energizing irrigation wells, which accounts for a large share of groundwater draft in that region.

Though Gujarat is known for several large gravity irrigation systems, the importance of such large water systems for water banking, improving the sustainability of well irrigation, and addressing water-energy nexus issues has not been as much a subject of interest for academicians in the field of groundwater management as water harvesting and artificial recharge. Chapter 6 will analyze the various parameters that can influence groundwater behavior, especially the large-scale import of surface water made possible through the irrigation canal networks of the multi-purpose Sardar Sarovar project, regional rainfall trends, decentralized water harvesting, and policy to restrict electricity supply to the farm sector, and argues that the recent upward trend is due to the large-scale import of surface water from the Sardar Sarovar project for gravity irrigation and an overall upward trend in rainfall regionally.

With 49% of the total irrigation from groundwater, the erstwhile state of Andhra Pradesh accounts for 5.3% of the net groundwater irrigated area in the country. While the state remains one of the largest exporters of rice, with paddy accounting for nearly 70% of the state's total irrigated area, groundwater depletion poses serious challenges to not only agricultural production and rural livelihoods but also the nation's food security.

A major part of the hard rock formations of South Asia lies in India. While most of the scientific literature on groundwater is based on research on alluvial and sedimentary formations, which support a major share of well irrigation not only in India but across the world, Chapter 7 is about groundwater management challenges in erstwhile Andhra Pradesh, which is largely underlain by hard rock formations. It shows that the assessment of groundwater over-exploitation based on simplistic considerations of aggregate abstraction and recharge provides highly misleading outcomes about the condition of the resource in the state. The gravity of the problems can be gauged from the extent of well failures, the sharp decline in the average area irrigated by wells, and the increase in energy consumption for irrigation. The adverse effects of groundwater intensive use on tank irrigation also need due attention. Further, the chapter argues that ideas to improve groundwater management in the state, such as promotion of drip systems, replacement of paddy with dry land crops, rainwater harvesting, community management of groundwater, and rationing of power supply in the farm sector are shallow and not based on any scientific consideration of the key physical, socio-economic, and institutional parameters that determine the success or effectiveness of these interventions.

There is a scant appreciation of the way irrigated paddy influences groundwater balance. The role of drip irrigation in changing agricultural water demand is often over-stated, without proper consideration of the conditions under which it becomes the "best-bet technology." The debate on the scope of rainwater harvesting and artificial recharge often ignores the fact that the regions facing groundwater depletion do not have surplus water. Pricing of electricity in the farm sector appears to be the only option the state is left with to tackle the multiple problems of wasteful expenditure on well drilling, and inefficient energy and water use. In the long run, establishing property water rights in groundwater might be a viable option if one considers the opportunity cost of not having them.

The Indo-Gangetic plain is the "cereal bowl" of South Asia, and has a history of settled agriculture, with irrigation from wells, river diversion systems, storage reservoirs, and canal networks, which are several centuries old. In the past five decades, the region has experienced a virtual explosion in well irrigation, manifested by a manifold increase in the number of wells and pump sets, and irrigated areas. However, the groundwater problems and management challenges are distinctly different. While farmers in the alluvial plains of the Indus experience secular declines in groundwater levels, soil salinization, and the rising cost of well drilling as a result of overdraft for irrigation, their counterparts in the eastern Gangetic plains experience problems of rising energy costs and high inequity in access to groundwater for irrigation, while groundwater contamination affects drinking water supplies.

Chapter 8 first provides an overview of the groundwater issues in alluvial aquifers of South Asia extending from the Indus basin areas in Pakistan to the Brahmaputra basin area in Bangladesh, which form the largest alluvial aquifer basin in entire South Asia, and the governance and management challenges they pose. It then analyses the performance of the existing formal and informal institutions (both local, provincial, and national), concerned with groundwater development for irrigation and water supply and resource management in the two basins, vis-à-vis the resource evaluation and planning methodologies, key public sector interventions for groundwater irrigation development, and the effectiveness of various legal/regulatory instruments for controlling groundwater overdraft and groundwater contamination. The chapter also reviews various policies related to groundwater, energy, and food security in the basins, which have implications for the way groundwater resources are developed, used, and managed. It then goes on to analyze their impact on the sustainability of groundwater use and equity in access to the resources in the region. Finally, the chapter discusses the institutional challenges in achieving proper groundwater governance and sustainable use of groundwater.

In Chapter 9, we argue that the recently-introduced model for direct delivery of energy subsidy to well-irrigators in the Indus basin area of Indian Punjab is just another piecemeal and ad-hoc approach for checking groundwater over-exploitation in the state, and has a weak conceptual footing. The approach, which works on incentivizing farmers who stick to using just the pre-determined quota of electricity by the utility, is done in an arbitrary way based on the connected load of the well and the season. By comparing the income returns per unit of electricity consumed and the cash incentive being offered to the farmers, we argue that the energy quota based on connected load will only lead to the resource-rich appropriation of the subsidy benefits. Reduced use of well water to irrigate paddy will not arrest depletion since a large proportion of that water returns to the pumped aquifer anyway. We suggest some institutional alternatives for bringing about long-term changes in the groundwater balance of the alluvial parts of the Indus basin based on sound water-use hydrology and the use of technology and market forces.

Chapter 10 is about groundwater management institutions. While several researchers in the past three decades have mooted the idea of local, community management of groundwater, the discussion has failed to make much advancement in the absence of any clarity on the management structures and institutional arrangements. It suggests the establishment of tradable private property rights in groundwater as a major institutional reform for local communities to establish rights over the water they manage and to address the issues of efficiency, equity, and sustainability. In the chapter, we argue that well-defined property rights in groundwater would promote efficient markets for water allocation, boosting resource use efficiency and agricultural growth. While suggesting management of groundwater at the local level (village), we have also identified the "externalities" in it. We provide a general framework for the design of groundwater management institutions based on a participatory institutional format, which can internalize these externalities and implement the groundwater management solutions for the aquifers, including defining and establishing groundwater property rights. The viability of the institutional reform options is examined through a review of some international experiences with tradable property rights in water and groundwater management. The last section examines how far an enabling environment for setting up groundwater management institutions exists in South Asia.

Chapter 11 provides a synthesis of the most important findings from the individual chapters from Chapter 2 to Chapter 10, and, based on the key learnings and insights distilled from them, suggests what can actually work to improve the efficiency, equity, and sustainability of groundwater use in the semi-arid and arid regions, and improve the equity in access to groundwater in the groundwater rich areas. It also comes up with a list of management paradigms and solutions that were introduced to South Asia's groundwater sector during the last three decades, whose usefulness has never been proven by scientific evidence but has dominated the discourse at various levels because of constant propaganda and lobbying by certain interest groups. Based on a sound theoretical framework, scientific principles, and available hard evidence, the chapter then suggests key policy measures and actions for the governments to engage in over the next decade or so to address the surmounting challenge of sustainable groundwater management. It also provides a strong methodological and analytical framework for academicians who teach advanced courses in groundwater resource management and groundwater economics and policy.

References

Central Ground Water Board. (2010). *Ground water quality in shallow aquifers of India*. Central Ground Water Board, Ministry of Water Resources.

Central Ground Water Board. (2019). *National compilation on dynamic ground water resources of India 2017*. Central Ground Water Board, Dept. of Water Resources, RD & GR, Ministry of Jal Shakti.

Food and Agriculture Organization of the United Nations. (2012). Coping with water scarcity: an action framework for agriculture and food security, Food and Agriculture Organization of the United Nations, No. 38, FAO Water Reports, Rome.

Gulati, M., & Pahuja, S. (2015). Direct delivery of power subsidy to manage energy–ground water–agriculture nexus. *Aquatic Procedia, 5*, 22–30.

Jadhav, R. (2018, June 9). The world's highest, and most reckless, user of groundwater. *The Times of India*. https://timesofindia.indiatimes.com/india/the-worlds-highest-and-most-reckless-user-of-groundwater/articleshow/64515989.cms.

Kulkarni, H., Shah, M., & Vijay Shankar, P. (2015). Shaping the contours of groundwater governance in India. *Journal of Hydrology: Regional Studies, 4*(A), 172–192. https://doi.org/10.1016/j.ejrh.2014.11.004.

Kumar, D. M. (2018). *Water policy science and politics: An Indian perspective* (1st ed.). Elsevier Science.

Kumar, M. D., & Pandit, C. M. (2018). India's water management debate: Is the 'civil society' making it everlasting? *International Journal of Water Resources Development, 34*(1), 28–41. https://doi.org/10.1080/07900627.2016.1204536.

Kumar, M. D., & Singh, O. P. (2008). How serious are groundwater over-exploitation problems in India? A fresh investigation into an old issue. In M. D. Kumar (Ed.), *Managing water in the face of growing scarcity, inequity and declining returns: Exploring fresh approaches. 7th Annual Partners' meet of IWMI-Tata Water Policy Research Program, ICRISAT, Patancheru, AP, 2–4 April 2008*.

Moench, M., Turnquist, S., & Kumar, M. D. (Eds.). *Water management: India's groundwater.* Proceedings of the workshop Water Management: India's Groundwater Challenge, VIKSAT, Thaltej Tekra, Ahmedabad, 14–16 December 1993.

Mukherjee, A., Sahab, D., Harvey, C. F., Taylor, R. G., Ahmed, K. M., & Bhanja, S. N. (2015). Groundwater systems of the Indian sub-continent. *Journal of Hydrology: Regional Studies, 4*(A), 1–14.

Mukherji, A., Das, B., Majumdar, N., Nayak, N. C., Sethi, R. R., & Sharma, B. R. (2009). Metering of agricultural power supply in West Bengal, India: Who gains and who loses? *Energy Policy, 37*(12), 5530–5539. https://doi.org/10.1016/j.enpol.2009.08.051.

Mukherji, A., Facon, T., Burke, J., de Fraiture, C., Giordano, M., Molden, D., & Shah, T. (2010). *Revitalizing Asia's irrigation: To sustainably meet tomorrow's food needs.* International Water Management Institute Food and Agricultural Organization of the United Nations.

Qureshi, A. S. (2020). Groundwater governance in Pakistan: From colossal development to neglected management. *Water, 12*(11), 3017. https://doi.org/10.3390/w12113017.

Qureshi, A. S., Ahmed, Z., & Krupnik, T. J. (2014). *Groundwater management in Bangladesh: An analysis of problems and opportunities* [Research report no. 2]. Cereal Systems Initiative for South Asia Mechanization and Irrigation (CSISA-MI) Project, CIMMYT.

Qureshi, A. S., McCornick, P. G., Sarwar, A., & Sharma, B. R. (2010). Challenges and prospects of sustainable groundwater management in the Indus Basin, Pakistan. *Water Resources Management, 24*, 1551–1569. https://doi.org/10.1007/s11269-009-9513-3.

Shah, T. (1993). Groundwater markets and irrigation development: political economy and practical policy, Oxford University Press, New Delhi, p. 267.

Shah, T. (2000). Mobilising social energy against environmental challenge: Understanding the groundwater recharge movement in western India. *Natural Resources Forum, 24*(3), 197–209. https://doi.org/10.1111/j.1477-8947.2000.tb00944.x.

Shah, T., Roy, A. D., Qureshi, A. S., & Wang, J. (2003). Sustaining Asia's groundwater boom: An overview of issues and evidence. *Natural Resources Forum, 27*(2), 130–141. https://doi.org/10.1111/1477-8947.00048.

Shah, T., & Verma, S. (2008). Co-management of electricity and groundwater: An assessment of Gujarat's Jyotirgram Scheme. *Economic and Political Weekly, 43*(7), 59–66.

Shiferaw, B. (2021). *Addressing groundwater depletion: Lessons from India, the world's largest user of groundwater.* Independent Evaluation Group, World Bank Group. https://ieg.worldbankgroup.org/tblog/addressing-groundwater-depletion-lessons-india-worlds-largest-user-groundwater.

Subhadra, B. (2015). Water: Halt India's groundwater loss. *Nature, 521*(7552), 289. https://doi.org/10.1038/521289d.

Suhag, R. (2016). *Overview of groundwater in India.* PRS Legislative Research.

World Bank. (2010). *Deep wells and prudence: Towards pragmatic action for addressing groundwater overexploitation in India.* World Bank Group. http://documents.worldbank.org/curated/en/272661468267911138/Deep-wells-and-prudence-towards-pragmatic-action-for-addressing-groundwater-overexploitation-in-India.

Chapter 2

South Asia's groundwater: The resource characteristics, ownership regimes, and access

2.1 Introduction

There is a large body of research studies that look at South Asia's groundwater issues. They cover the following aspects: results of the micro-level geophysical investigation on the occurrence of groundwater, and geohydrological studies to assess aquifer properties; (mathematical) modeling studies to understand the groundwater behavior (flow and levels) in relation to various natural and human-induced processes; large-scale aquifer mapping studies; agro-economic research studies to understand the returns from groundwater irrigation; geochemistry studies to understand the natural quality of groundwater, and groundwater pollution studies from environmental science's perspective; technical studies to assess groundwater problems and identify management solutions; and studies that look at the socioeconomic aspects of groundwater development and use.

Though technical studies by groundwater scientists (geologists, geo-hydrologists, and geophysicists) that looked at groundwater occurrence, characterization of aquifers and groundwater flow dominated the initial years of research on groundwater in South Asia, particularly India, during the past three decades, the groundwater research agenda was set mainly by social scientists, especially development economists. The primary aims of their research were: quantifying the returns from groundwater irrigation; comparing the performance of groundwater irrigation systems against surface irrigation; understanding equity and efficiency in groundwater irrigation; assessing ownership of wells and access to well irrigation; and functioning of pump irrigation (groundwater) markets, especially their equity impacts. For instance, several studies done on groundwater in India and Pakistan looked at the working of informal groundwater markets (Meizen-Dick, 1996; Palmer-Jones, 1994; Shah, 1993; Saleth, 1998).

As regards the performance of the groundwater irrigation sector, the irrigated area has been growing in India since the late 1960s and had overtaken surface irrigation several years ago (Kumar et al., 2009). As per the official statistics, the net area under well irrigation stood at 39.05 m.ha, benefiting hundreds

Groundwater Economics and Policy in South Asia. DOI: https://doi.org/10.1016/B978-0-443-14011-2.00002-4

15

of millions of farmers spread across the country. There has been significant growth in groundwater schemes from 18.5 million in 2000–01 (GoI, 2014) to 19.75 million in 2006–07 to 20.52 million in 2013–14 (GoI, 2017). In Pakistan too, the well-irrigated area made impressive growth from 2.90 m.ha in 1960 to 14.30 m.ha in 2015, with an exponential increase in the number of shallow tube wells (Qureshi, 2020).

In Bangladesh, with the introduction of high-yielding rice varieties in the 1980s and 1990s that responded favorably to irrigation and fertilizer and were suitable for Boro rice, the demand for reliable irrigation increased. Since aquifer conditions were favorable in most parts of the country, the development of groundwater resources received great attention. The installation of deep tube wells (DTWs) started in the late 1960s but gained momentum in the late 1980s. By 1992, there were about 25,500 DTWs in the country (BADC, 2013). Currently, 35,322 DTWs meant for irrigation are working in Bangladesh. However, despite the visible benefits of groundwater irrigation, shallow tube wells (STWs) were not initially adopted due to restrictions on tube well spacing and an embargo on the import of all types of diesel engines (Dey et al., 2013). Things, however, changed following the devastating floods of 1988 and subsequent cyclones in the early 1990s. Realizing the need to kick-start a boost in agricultural production, the government lifted all restrictions and embargoes on the import of irrigation equipment. Consequently, local markets were flooded with irrigation pumps that were easy to operate and inexpensive (<12 HP) from India and China.

The increased availability of equipment led to the maturation of Bangladesh's mechanized agricultural economy. Today, there are 1,523,609 STWs operating in the country (BADC, 2013). The leap in the population of STWs was linked to their suitability given the prevailing socioeconomic conditions of Bangladesh's burgeoning Boro farmers. Currently, about 79% of the total cultivated area is irrigated by groundwater, whereas the remaining is irrigated by surface water (BADC, 2013).

With this impressive growth, several scholars from the rural development sector have turned their attention to conducting research on public policies in the groundwater sector, especially those that can promote equitable and sustainable use of the resource, as it is a widely known fact that irrigation is a powerful tool to reduce rural poverty. Ownership of wells was a central research topic, especially in view of the fact that large amounts of public funds had gone into subsidizing the construction of irrigation wells, installation of pump sets, and electricity and diesel used for groundwater pumping (Kumar, 2007). Researchers had also begun to look at groundwater markets, especially their equity impacts and the kind of public policies that can create efficient groundwater markets which make well irrigation affordable for poor non–well-owning farmers.

Many of these studies that looked at the socioeconomic aspects of groundwater use lacked rigor, mainly because they could not bring in the nuances such as groundwater system characteristics (type of aquifers being tapped, i.e., whether alluvial or semi-consolidated or hard rock; the long-term trends in the groundwater levels), types of groundwater irrigation structures (i.e., dug well or bore well or shallow tube well or deep tube well); the area commanded by wells, and the types of crops cultivated, in their analysis. Such gaps in the analysis heavily influenced the results vis-à-vis cost-effectiveness, the irrigation potential of the system, the cost per ha of irrigation from wells, the long-term growth potential, and therefore the policy prescriptions they came up with as regards investment choices (i.e., whether to invest in private groundwater irrigation or publically owned gravity irrigation), institutional financing (whether to provide loans and whom to provide), and provisioning of subsidies for well construction or electricity.

In this chapter, we will provide a glimpse of the arguments that shaped the public policies in the groundwater sector. We would then analyze the issue of equity in access to groundwater by looking at the ownership of different types of wells across landholding categories, and the amount of land each category of farmers irrigates. Further, we would analyze the equity impacts of groundwater (or pump rental) markets in India by providing the available empirical evidence. Finally, we will also look at how groundwater irrigation in India has been growing over the years and will confront the question, i.e., of whether the groundwater boom will sustain. However, prior to doing that we would provide the macro picture of the characteristics of the resource—particularly their quantity and the regional variations—and the structures that are used to abstract the resource and their changing composition and character over time.

2.2   Geohydrology, dynamic groundwater resources, and variations

South Asia has one of the most complex and fascinating aquifer systems in the world, marked by major spatial variations in the geological, geohydrological, and geohydrochemical conditions. The types of aquifers vary from unconsolidated formations (alluvial deposits) to semi-consolidated formations (sedimentary origin) to consolidated formations of igneous and metamorphic origin. While some of the aquifers extend over a large region (such as the alluvial deposits of the Indo-Gangetic plain), spreading over several hundred thousand square kilometers, some of them are localized formations. The yield characteristics of the aquifers, such as specific yield and transmissivity, also vary widely amongst the formations (Mukherjee et al., 2015). They can also vary within the same type of formation depending on the extent of weathering and density of fractures (in the case of consolidated rocks) and the texture of the soils and compaction (in the case of unconsolidated and semi-consolidated formations). The region has extensive freshwater aquifers and aquifers containing minerals (such as chlorides, fluorides, arsenic, and iron) of geogenic origin. Fig. 2.1 provides a map of the aquifer formations of South Asia.

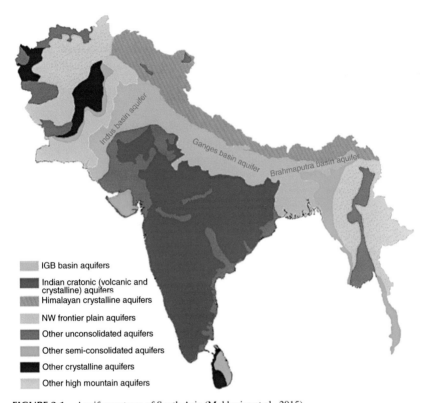

IGB basin aquifers

Indian cratonic (volcanic and crystalline) aquifers

Himalayan crystalline aquifers

NW frontier plain aquifers

Other unconsolidated aquifers

Other semi-consolidated aquifers

Other crystalline aquifers

Other high mountain aquifers

FIGURE 2.1   Aquifer systems of South Asia (Mukherjee et al., 2015).

2.2.1   India

The annual dynamic groundwater resources in India are estimated to be 431.86 BCM (CGWB, 2019). India's groundwater situation is complex because of the presence of different types of geological formations that bear water. India has unconsolidated alluvial formations in the Indo-Gangetic plain, in the Cambay basin in Gujarat, and in coastal areas of Odisha, Andhra Pradesh, Tamil Nadu, and Karnataka. It has semi-consolidated formations of sedimentary origin (in Rajasthan, Gujarat, and Madhya Pradesh). It also has consolidated formations of igneous origin in large parts of the Deccan plateau and central Indian plateau regions. Consolidated (hard) rocks of metamorphic origin are also found all over India, with gneiss and schist in the Himalayas, Assam, West Bengal, Bihar, Odisha, Madhya Pradesh, and Rajasthan; quartzite in Rajasthan, Bihar, Madhya Pradesh, Tamil Nadu and areas surrounding Delhi; marble in Alwar, Ajmer, Jaipur, and Jodhpur districts of Rajasthan and parts of Narmada Valley in Madhya Pradesh; slate in Rewari (Haryana), Kangra (Himachal Pradesh), and parts of Bihar (slate); and graphite in Odisha and Andhra Pradesh. The properties of the aquifers (mainly specific yield and transmissivity) that determine the groundwater storage and flow characteristics vary widely from the alluvial formations to the consolidated formations.

There are deep and extensive alluvial formations in India, and they extend over the entire Indo-Gangetic plain. But, 2/3rd of the geographical area is underlain by consolidated formations or hard rocks (Kulkarni & Vijay Shankar, 2009; Kulkarni et al., 2011). While a well yield of 25 liters per second is quite common in the Gangetic plains, in the hard rock, it is often as low as 1–2 liters per second. While a tube well in the Gangetic plains can irrigate 30–40 ha of land, a bore well in the hard rock areas can hardly irrigate 0.4–1.0 ha. But, the ardent supporters of well irrigation ignore this conveniently while analyzing the crude statistics of groundwater structures (Kumar et al., 2013b). A geohydrological map of India, showing the extent of different types of aquifers (unconsolidated, consolidated, and semi-consolidated), is provided in Fig. 2.2.

It is also well known that in the same locality, the geo-hydrological environment changes remarkably over time. In the alluvial areas, it can change from shallow water table conditions in phreatic aquifers to deep piezometric levels in high-yielding confined aquifers. In hard rock, it can change from shallow water table conditions in large open wells to deep water table conditions in bore wells. Accordingly, the irrigation potential of wells also could change drastically. While the shallow open wells in the alluvium irrigate 1.2–1.6 ha of land, the irrigation potential of deep tube wells could be as high as 30–40 ha, as mentioned above. In the case of hard rock areas, just the opposite usually happens. While the large open wells could irrigate 1.6–2.0 ha, the bore wells will hardly be able to irrigate 0.4–0.80 ha. Also, their life is very short (Kumar et al., 2013b).

Over the years, the proportion of open wells in India's well irrigation landscape has decreased, and the proportion of deep tube wells (including deep bore wells) has increased, as indicated by the data available from the 4th and 5th Minor Irrigation Census (Fig. 2.3). In absolute terms, the number of open wells reduced from 9.2 million to 8.78 million and the number of deep tube wells recorded a significant jump from 1.44 million to 2.61 million over a period of seven years (GoI, 2017). This entire change has happened over a period of just 7 years. In the case of hard rock areas, one major reason for this reduction in open wells is groundwater depletion. Since open well construction is not feasible beyond a certain depth (say, 100–120 feet), farmers go for the construction of bore wells, despite them being low-yielding. In the alluvial areas too, the dominance of open wells has reduced, and is now being replaced by shallow and deep tube wells. The irrigation potential could be different depending on the situation. In the alluvial areas of Punjab and north Gujarat and western Uttar Pradesh, drying up of the shallow aquifers is the reason, whereas, in most parts of the eastern Gangetic plains, the cost of well construction is an important factor. It is also well understood that a tiny fraction of the deep tube wells, whose number has increased over the years, are owned by marginal and small farmers.

Therefore, groundwater hydrology in India is too heterogeneous for any spatial analysis of groundwater development and use to be amenable to any simple

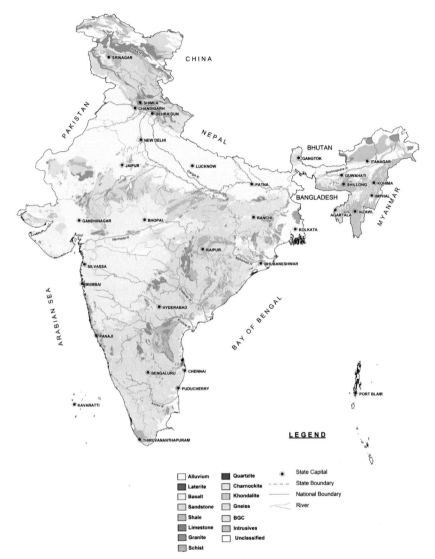

FIGURE 2.2   Major aquifer systems of India. (*From: Central Ground Water Board*).

formulations that use the commonly used criteria such as well numbers, well density, the annual rate of change in water levels, and the stage of groundwater development.

Given the complexity of groundwater systems, the researchers should exercise some caution while drawing conclusions on the basis of trends emulating from data on well numbers, particularly when the data used for comparisons of different time periods do not belong to the same states or geographical areas.

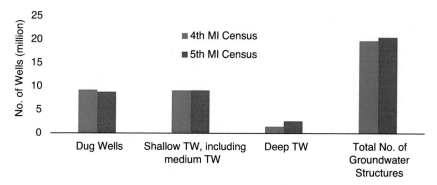

FIGURE 2.3   Changing composition of groundwater abstraction structures in India.

Policies formulated based on conclusions drawn from analysis of insufficient and rudimentary data can have serious unintended consequences.

One such conclusion is that minor irrigation is all about groundwater. To make this point, a comparison is made between the numbers of groundwater abstraction structures against other minor irrigation (MI) structures (see Mukherji et al., 2013). The importance of groundwater in India's irrigation cannot be judged from a mere number of structures. From the point that the surface MI structures account for only 6% of the total MI structures in the country, an argument is made that we can ignore them (This is simply absurd. Many tanks irrigate hundreds of hectares of land, often 50–100 times the land irrigated by wells.). The tanks account for 25% of the net irrigated area. By this standard, we should ignore all large irrigation schemes in the country, as they are only a few thousand of them, against around 20.5 million wells (as reported by the 5th Minor Irrigation Census, GoI, 2017). They are comparing the un-comparable (Kumar et al., 2013b). Doing simple statistical projections (of growth in well numbers) based on past trends without understanding the underlying physical processes is tardy. Many, in fact, assumed that the number of groundwater structures can keep increasing with time with a proportional increase in area irrigated, without being constrained by lack of availability of groundwater (particularly in the semi-arid, hard rock regions), arable land, finance, etc. The outcome of this skewed view of irrigation is that well irrigation is pitched as an alternative to gravity irrigation.

2.2.2   Pakistan

Pakistan is an arid country with an average annual rainfall of just 297 millimeters. Pakistan's Indus Basin is underlain by one of the richest alluvial aquifers in the world (Khan et al., 2018). According to one estimate, the groundwater system underneath Pakistan's flowing rivers in the Indus plains has at least 400 million acre feet (MAF) or 4800 BCM of pristine water. This storage is equivalent to more than three years of the mean annual flow of the Indus (or 1000 days of storage, after excluding polluted areas) (Hussain & Abbas, 2019).

The rivers of the Indus system have carried huge silt loads for millions of years, depositing them in the plains all the way to the delta. The high sediment loads in the Indus river system have created nearly 200,000 square kilometers of flatlands. These flatlands, to a considerable depth, are made up of unconsolidated and granular formations capable of holding large volumes of water. "This reservoir of water is so vast, it ranks among the natural wonders of the world." (Ahmad & Chaudhry, 1988). According to Michel (1967), "The Indus Rivers, describes these alluvial deposits as unconsolidated material, deeper than one mile, forming a large homogeneous groundwater reservoir with a capacity "at least ten times the annual runoff of the Indus River."

A study by Khan et al. (2018) found that the four doabs formed by the five rivers of the Indus river system constitute the largest of the freshwater aquifers in the entire Indus plains. The Thal Doab, Chaj Doab, Rechna Doab and Bari Doab, with an area of 3.5 m.ha, 1.3 m.ha, 3.12 m.ha, and 2.96 m.ha, respectively, are all independent hydrological units. They are overlain by the largest contiguous irrigation system in the world, i.e., the Indus Basin Irrigation System. The total amount of useable water was estimated to be 959 BCM from Thal Doab, and the average useable groundwater in the area was estimated to be 8993 BCM, with an active (dynamic) storage of 2748 BCM. The isotope studies revealed the significance of river plains in the replenishment of groundwater in UIPA. Regional groundwater modeling revealed that the safe yield of UIPA is around 7.5 BCM per m.ha of area, with the highest safe yield in the Thal Doab at 9.0 BCM per m.ha (Khan et al., 2018).

In the Indus delta, good-quality groundwater may occur only within the top hundred feet or less. Away from the main river channels, the water becomes brackish. The existence of deep high-capacity tube wells in previous Salinity Control and Reclamation (SCARP) projects and many deep tube wells in the private sector today have contaminated the shallower fresh water with saline upwelling. In many areas away from the rivers, shallow layers of groundwater have been exploited by irrigation tube wells, and what remains now is brackish to saline groundwater (Hussain & Abbas, 2019).

The map in Fig. 2.4 shows the extent of freshwater aquifers in the Indus plains. It excludes areas where the water quality is saline, brackish, or of marginal quality. Out of 200,000 square kilometers of the Indus plain, the current fresh groundwater occurrence in Pakistan is approximately spread over 88,000 square kilometers.

Balochistan, which covers 44% of the country's land mass and has a 770-kilometer-long coastline, is unique in its geopolitical significance. It is an arid region with very low rainfall, high temporal variability, and an annual potential evaporation ten times higher than the annual rainfall (Ashraf & Majeed, 2006). Balochistan is among the most drought-prone regions of the country, where severe droughts have been recorded in 1967–69, 1971, 1973–75, 1994, 1998–02, and 2009–15 (Ahmad et al., 2016).

Groundwater in Balochistan is present in both confined and unconfined aquifers in all river basins and sub-basins, and groundwater flow largely follows

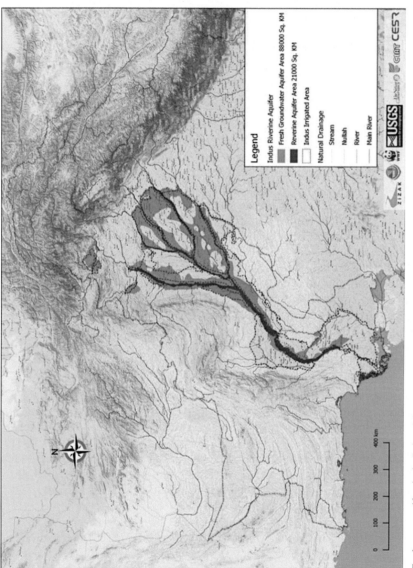

FIGURE 2.4   Freshwater aquifers in the Indus plains and riverine corridors.   *(From: Hussain & Abhas, 2019).*

the general trend of surface drainage and is found in the alluvial fans and pied-mont plains. The Water and Power Development Authority (WAPDA) evaluated these resources from 1976 to 1980 for a number of basins. In 2007, Halcrow Pakistan (Pvt.) Ltd. assessed the groundwater resources of 14 of 18 basins. The total annual potential was estimated to be 1071 million cubic meters per year (van Steenbergen et al., 2015).

2.2.3 Bangladesh

Bangladesh is a riverine country that occupies two-thirds of the delta of Ganga–Brahmaputra basin. The surface geology of Bangladesh is primarily character-ized by unconsolidated sediments (of Halocene origin) that cover nearly 80% of the land surface. Pre-Holocene sediments occur primarily in terraces as well as hills located in the eastern part of Bangladesh, occupying 20% of the land area. The terrace deposits (of Plio-Pleistocene origin), located in northwest (Barind Tract) and north-central (Madhupur Tract) Bangladesh (Alam et al., 1990; Reimann, 1993) occupy 8% of the nation's land surface and feature annual groundwater withdrawals of around 9.7 BCM, which is approximately 30% of the total groundwater withdrawals estimated for the country (Shamsudduha et al., 2009).

Both the quaternary deposits consist of sands, silts, and clays that form aquifers throughout the basin. Groundwater generally occurs at shallow depths (10 m bgl) in the fluvio-deltaic sediments of the Madhupur and Barind Tracts (Ahmed et al., 2004; Shamsudduha et al., 2011). These two primary hydro-geological units feature contrasting permeabilities on the surface, relatively higher and lower, respectively. The unconsolidated sediments are unconfined or semiconfined with high transmissivity, whereas the terrace deposits are confined or semiconfined with low transmissivity (Ravenscroft & Rahman, 2003). Plio-Pleistocene aquifers typically occur below a clay unit, stratigraphically known as the Madhupur Clay Formation that varies in thickness from 8 to 45 m; the un-derlying aquifer is known as Dupi Tila (Ravenscroft et al., 2004; Shamsudduha & Uddin, 2007).

The surface geology of Bangladesh (Fig. 2.5) ranges from alluvial sand, alluvial silt, alluvial sand and clay, deltaic sand, gravely sand, deltaic silt, Madhapur and Barind clay residuum, and tidal delataic deposits, marsh clay and peat and estuarine deposits (in the coastal areas).

The quantification of groundwater recharge using the water table fluctua-tion method (Shamsudduha et al., 2011) estimates an average net groundwater recharge to be around 150 mm per year for the period of 1985–07. However, the regional variation in recharge rates is extremely high, with the lowest recharge rate of 10–50 mm in the southern coastal zone and the highest of 350–600 mm in the north-western and north-central parts (Shamsudduha et al., 2011). Though previous studies have indicated that focused recharge can be an important recharge pathway in the Bengal Basin (Burgess et al., 2011; Zahid et al., 2008),

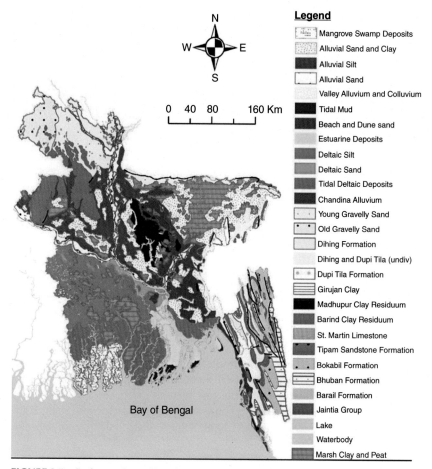

Legend

- Mangrove Swamp Deposits
- Alluvial Sand and Clay
- Alluvial Silt
- Alluvial Sand
- Valley Alluvium and Colluvium
- Tidal Mud
- Beach and Dune sand
- Estuarine Deposits
- Deltaic Silt
- Deltaic Sand
- Tidal Deltaic Deposits
- Chandina Alluvium
- Young Gravelly Sand
- Old Gravelly Sand
- Dihing Formation
- Dihing and Dupi Tila (undiv)
- Dupi Tila Formation
- Girujan Clay
- Madhupur Clay Residuum
- Barind Clay Residuum
- St. Martin Limestone
- Tipam Sandstone Formation
- Bokabil Formation
- Bhuban Formation
- Barail Formation
- Jaintia Group
- Lake
- Waterbody
- Marsh Clay and Peat

FIGURE 2.5 Surface geology of Bangladesh. (*From: Islam et al., 2021*).

locally developed recharge estimation tools and hydrological models (Kirby et al., 2015), as well as large-scale models such as the PCRaster GLOBal Water Balance model (Wada et al., 2010), are restricted to the estimation of diffused recharge that results solely from the direct infiltration of precipitation at the soil surface.

An interesting hydrological phenomenon observed in Bangladesh is the changing groundwater recharge pattern with changing pre-monsoon water table conditions. With increased pumping of groundwater during the summer for Boro rice irrigation, which began in recent decades and has resulted in a lowering of water levels in the shallow aquifer, the rate of infiltration of rainwater and the cumulative recharge has increased. As brought out by Shamsudduha et al. (2011), the net recharge increased in the North-Western region and along the river floodplains of the Brahmaputra and Ganges. The mean annual recharge during 2002–07 was found to be greater than the long-term mean for 1985–07 in the North-Western region and in the floodplains of the Brahmaputra. It increased recently in Jessore, Khulna district, Mymensingh, and Comilla regions. Changes in recharge were found to be limited in the rest of the country. The increase in net recharge between the pre-groundwater irrigation period (1975–80) and the developed groundwater irrigation period (2002–07) was substantial.

There were clear spatial trends in the estimated changes in recharge rates. The recharge increased 5–15 mm/year in the North-Western, Western, and North-Central districts and in Comilla district in the east. This coincides with areas of intensive groundwater-fed irrigation for Boro rice. Recharge decreases by −0.5 to −1 mm/year in the Southern GBM Delta Sylhet depression. Recharge remained unchanged in the rest of Bangladesh (Shamsudduha et al., 2011).

2.3 Arguments that shaped public policies in groundwater and agriculture in South Asia

The arguments that were largely instrumental in shaping public policies in the agricultural groundwater sector in South Asia are: high density of farm wells increases the transaction cost of metering and charging for electricity on a pro-rata basis, as a tool to control groundwater draft (Qureshi et al., 2009; Shah et al., 2004; Scott & Shah, 2004); and groundwater economy is controlled by millions of small and marginal farmers, and that any attempts to regulate it would threaten their livelihoods and therefore are politically sensitive (Debroy & Shah, 2003; Halcrow-ACE, 2003; Qureshi et al., 2009; Scott & Shah, 2004; Veenot & Molle, 2008).

One argument against the shift in power pricing is that the marginal cost of supplying electricity will be high owing to the high transaction cost of metering, which may reduce the net social welfare as a result of reductions in demand for electricity and groundwater and net surpluses that individual farmers generate from farming (Shah, 1993). Another argument is that for the power tariff to be in the responsive (price-elastic) range of the power demand curve, prices have to be so high that they become socially unviable or politically untenable (Saleth, 1997). The third argument is that under pro rata tariffs, the increased cost of pumping groundwater would be transferred to the water buyers (Mukherji et al., 2012).

In the context of Pakistan, some additional arguments against electricity pricing were made by a few scholars on the following grounds (see Qureshi, 2020). For instance, Qureshi (2018) notes that during the 1980s, Pakistan moved from a flat tariff system to consumption-based pricing of electricity and then to a combined flat and consumption-based pricing system. The flat billing system helped in collecting revenue but failed to control groundwater abstraction. The flat tariff policy was also criticized by small farmers due to its "fixed cost" even in cases of minimally or completely non-operational tube wells (Venot & Molle, 2008). According to Venot (2008), these energy pricing policies did not help in controlling groundwater over-draft, and groundwater abstraction kept rising because it was crucial to meet agricultural water demands. One of the primary reasons for this failure, as argued by Zoumides and Zachariadis (2009), was that in Pakistan only 15% of the total tube wells are electrically powered. Therefore, in their opinion, changing energy prices would have only a marginal effect on groundwater abstraction. Further, they pointed out the case of farmers in Pakistan shifting from electric to diesel pumps as a fallout of an increase in the agricultural power tariff, resulting in very little reduction in the actual groundwater abstraction (Zoumides & Zachariadis, 2009).

These arguments were very effectively used to question the effectiveness of the measures such as agricultural metering and pro rata pricing of electricity; and state control of groundwater abstraction by farmers, and push for subsidized electricity for agricultural groundwater pumping (Kumar, 2016).

The "high transaction cost" argument has mainly stemmed from the sheer number of groundwater abstraction structures in India and Pakistan. But the proponents of this argument have paid little attention to the number of wells in regions that actually require management measures. The fact is that the overwhelming numbers of wells in India are in the Indo-Gangetic belt (Scott & Sharma, 2009), which does not experience serious problems of groundwater overdraft (Sharma and Ambili, 2009). Secondly, the scholars from Pakistan, who argued against electricity pricing, did not explore the option of introducing a subsidized metered (pro-rata) tariff for electricity in a way that can encourage efficient use without hurting the farmers in the form of an excessive increase in the cost of inputs and forcing them to shift to diesel engines. Thirdly, the shift from electric pumps to diesel pumps for groundwater pumping itself would have automatically incentivized the farmers in the region to use groundwater efficiently due to the positive marginal cost of pumping, going by the empirical evidence available from India (see Kumar et al., 2010). More importantly, the cost of pumping groundwater using diesel pumps was far greater than that of electric pumps (Qureshi & Akhtar, 2003).

The works that examined the validity of the arguments and claims find serious flaws in them. They pertain to equity in access to groundwater and the question of who controls the groundwater economy (Kumar, 2007; Sarkar, 2012); how electricity and diesel subsidies can be used to promote equity in access to groundwater; the impact of the rise in power tariff on efficiency and sustainability of groundwater use (Badiani et al., 2012; Kumar, 2007) and access equity in groundwater, including the functioning of water markets in semi-arid water-scarce regions (Kumar, 2007; Kumar et al., 2013b). We will examine the following three key aspects: (1) ownership of wells; (2) equity in access to groundwater for irrigation; and (3) impacts of public policies on the functioning of groundwater markets.

2.4 Is access to groundwater equitable in South Asia?

2.4.1 India

In the context of India, some researchers have argued that groundwater is a more democratic resource, with a greater geographical spread of wells, unlike canal irrigation which is concentrated, and therefore promotes access equity (Debroy & Shah, 2003). But, the analysis of data on ownership of wells, obtained from 11 major Indian states, shows that there is skew-ness in ownership of all types of wells towards medium and large farmers. A little more than 20% of large farmers own dug wells; 16.5% of them own shallow tube wells and 0.4% own deep tube wells. Hence, a total of 37% of large farmers own wells. But, as regards marginal farmers, 2.5% own dug wells and 3.5% own shallow tube wells. Ownership of deep tube wells is close to nil in this category of farmers. Hence, only 6% of marginal farmers own wells (see Table 2.1).

Mukherji et al. (2013) wrongly argued on the basis of flawed analysis that wells are mostly owned (66.5%) by small and marginal farmers. Further analysis could have been done by looking at the characteristics of wells owned by each category of farmers. Instead, the total number of wells owned by farmers under each category (in the denominator) against the number of wells was used to arrive at their conclusion (see Kumar et al., 2013b).

What the authors conveniently ignored is that around 80% of the farmers in India belong to the small and marginal category (see Table 2.1) and a very small fraction of them own wells and pump sets. Against this, less than 1% of the farmers are large holders and around 69% of the large farmers own wells and pump sets. So, there exists a high inequity in access to groundwater abstraction structures. They stretch their argument by saying that "groundwater irrigation continues to remain as the only source of irrigation for India's poorest farmers" (Mukherji et al., 2013, p. 117). By stating this, they assume that small and marginal farmers do not have any access to water from public irrigation systems wherein the landowners do not have to contribute towards capital expenditure. Research in Punjab shows that when groundwater is over-exploited, canal water becomes the major source of water for poor, small and marginal farmers to sustain their income from irrigated production as they lose out on investing in technology to chase the water table (Sarkar, 2012).

The authors extend the argument further by stating that small and marginal farmers own a major share of India's groundwater resources, and because the eastern Indian states are not very prosperous, free power connections and high diesel subsidies are pro-poor strategies (Mukherji et al., 2013). It does not take much effort for one to realize that all these benefits would end up with rich farmers who are running wells using diesel pumps now.

The fallacy in the authors' argument can be understood from Table 2.1, which is based on data from the agricultural census (2005–06) and 3rd Minor Irrigation Census (2000–01) on well ownership. Table 2.1 shows that India had nearly 83.6 million marginal and 23.9 million small farmers, while there are

TABLE 2.1 Estimation of well irrigated area for different land holding classes.

Landholding category	Total no. of holders	Total holding (ha)	Average holding (ha)	% of farmers owing wells			Estimated net area under well irrigation (ha) and % of total area (%)
				OW	STW	DTW	
Marginal	83,694,372	32,025,970	0.4	2.5	3.5	0.01	1,921,559 (6.0)
Small	23,929,627	33,100,790	1.4	10.0	12.0	0.02	7,282,174 (22.0)
Medium	20,502,460	74,481,092	2.8	13.4	13.2	0.09	19,886,452 (26.7)
Large	1,095,778	18715131	17.1	20.4	16.5	0.40	6,905,883 (37.3)
Total	135,597,577	158,322,983	1.16				35,996,067 (22.7)

From: Authors' own estimates based on the agricultural census (2005–06) and data from the 3rd Minor Irrigation Census on well ownership in India (Kumar, 2007).

1.05 million large and 20.5 million medium farmers, as per the agricultural census of 2005–06. As per our estimates, the 107.5 million small and marginal farmers together irrigate only 9.1 m.ha of land (net) from wells, against 21.55 million medium and large farmers irrigating 26.8 m.ha of land (net). The average well irrigated area of small and marginal farmers is 0.09 ha, while that of a large farmer is around 6.8 ha. The average area for medium and large farmers is 1.24 ha. So, undoubtedly, India's groundwater economy is controlled by medium and large farmers.

A far more dangerous conclusion emerging from the analysis by Mukherji et al. (2013) is vis-à-vis the changes in access inequity in groundwater over time. They contended that a far greater percentage of the wells were owned by small and marginal farmers in India in 2013, than in 1987. While it is true that the number of wells owned by small and marginal farmers has gone up in the past 25 years or so, the authors do not look at the increase in the number of operational holders in this category, or that with the rising cost of drilling wells and the high risk involved in well construction, wells have become a product of the rich farmers.

Even the most recent data, for 2013–14 (from the 5th Minor Irrigation Census) show that a very small percentage of the wells (6.5%) are owned by marginal farmers, followed by small farmers (21.8%), semi medium farmers (7.19%), medium farmers (27.25%) and large farmers (53%). As one can see, over the 13-year period, there was no increase in the proportion of marginal farmers owning wells, though there was a very sharp increase in the percentage of large farmers owning wells (from 37.3% to 53.0%) (Table 2.2). So far as equity in access to groundwater is concerned, this is dismal performance vis-à-vis when we consider the fact that the real cost of well drilling (per unit depth) has actually come down over the years in all parts of India owing to the deep penetration of the technology.

Some researchers argue that the water markets contribute to more than 60% of India's well irrigated area. For instance, Mukherji (2005) estimated the contribution of pump rental markets to India's irrigated area to be nearly 20 million ha (m ha). For this, the study used the data available from NSSO. Though this is the first attempt of its kind in assessing the size of groundwater markets in India, the assessment suffers from some flaws.

Mukherji (2005) used the data on the percentage of holders who resort to water purchase, and the gross cultivated area within that particular category of holders as the basis for this estimate, a procedure which has two potential sources of error: (1) a water buyer might irrigate only a fraction of the land with purchase water, and figures of cropping intensity would not represent irrigation intensity; and (2) there could be major variations in the gross cropped area amongst farmers belonging to the same land-holding category. Following this methodology would lead to over-estimation of the area covered. If the estimates by Mukherji (2005) are anywhere close to reality, then the actual area irrigated by the water market would be 1/3rd of the net irrigated area.

TABLE 2.2 Ownership of different types of wells among different land holding classes.

Land holding category	No of farmers ('000)	Percentage of farmers belonging to each class owing wells					Total
		Dug wells	Shallow tube well	Medium tube well	Deep tube well		
Marginal (<1 ha)	92,826	2.430	2.742	0.916	0.520		6.6082
Small Farmer (1–2 ha)	24,779	9.358	6.402	3.809	2.275		21.845
Semi-Medium (2–4 ha)	13,896	10.003	7.618	5.377	4.194		27.192
Medium (4–10 ha)	5,875	9.635	5.347	5.925	6.346		27.253
Large (>10 ha)	973	20.270	14.871	7.781	10.094		53.016

From: Author's own analysis based on data from the 5th Minor Irrigation Census (GoI, 2017) and Pocket Book of Agricultural Statistics (MOA & FW, 2017).

If we treat the area irrigated through lifting from wells/tube wells as nearly 60% of the total net irrigated area, then pump rental markets should account for nearly 12 m.ha. But the following statistics expose the fallacy of this claim. The total number of land holdings of marginal farmers as per the 1991 census is 61.9 million, and the total size of the holdings is 24.5 m.ha. According to NSSO figures, out of the 25 million households that reportedly hired pump rental services in 1997–98, 75% (18.5 million) are marginal farmers. Going by the estimates provided by Mukherji (2005), about 15.2 m.ha (75% of the 20.3 million) of area (net) must have been irrigated by marginal farmers through water purchase, assuming that area irrigated through rental services is more or less the same across land-holding categories. The land owned by marginal farmers, who hire pumps, can be estimated at 7.33 m.ha. If we assume that the marginal farmers do not own wells, the net irrigated area becomes more than the total land owned. These comparisons show that irrigation through pump renting is over-estimated.

There are also issues with the way pump rental markets are viewed. It is assumed that pump rental markets operate between well owners, and non–well-owning, small and marginal farmers, who lack the capital to invest in wells and pump sets. Such a view leads to the interpretation that many non–well-owning, small and marginal farmers are served by pump rental markets. But the evidence available from water-abundant eastern India throws contradicting views. In Bihar, many farmers who buy water from their neighbors are also well owners, having land parcels elsewhere that are irrigated by their own wells (Kishore, 2004; Kumar, 2016).

Underlying the question of "who controls the groundwater economy" is the issue of access equity. Lesser the monopoly price of water, higher would be the degree of access equity through groundwater markets (Kumar et al., 2013a; (Palmer-Jones, 1994). But diametrically opposite views exist about the role of water markets in promoting access equity in groundwater (see, for example, Palmer-Jones, 1994; Shah, 1993). While Shah (1993) argues that groundwater markets are "oligopolies," (Palmer-Jones, 1994) showed that the electricity pricing policies and lack of institutional regime governing the use of water increase the monopoly of well owners.

Rights to groundwater in India are attached to land ownership rights, and are governed by the "Law of Dominant Heritage." Hence, every land owner has the right to access groundwater (Saleth, 1996). But there are no limits to the volume of groundwater that a land owner can abstract. This peculiar situation gives a strategic advantage to resource-rich farmers to maximize the outputs and profits under two conditions: regions where the physical availability of groundwater resources is limited; and regions where the risk involved in investing in well development is high (Kumar, 2016). This comparative advantage is leading to natural monopolies. The power pricing policies followed in most of the groundwater-scarce states (Karnataka, Tamil Nadu, Telangana, Madhya Pradesh, and Rajasthan) only increase the opportunities for the resource-rich, large well owners as they do not have to pay for electricity in proportion to the [water] production volume.

Hence, the analysis of equity impacts of water markets and the informal groundwater economy from a pure market perspective is distorted. The returns from irrigated production that the water buyers get when compared to what the well owners get is very important from an equity perspective. This is much less for water buyers (Deepak et al., 2005). Under drought conditions, when the well yields go down, the poor water-buyer farmers are badly hit as the well owners would either deny them irrigation services or raise water charges exorbitantly. To sum up, as resources become more and more scarce, the water markets would give greater opportunities to resource-rich farmers to earn extra income, and put the resource-poor farmers in a highly disadvantageous position. The current institutional regime governing the rights to access groundwater and the power pricing policies increases the monopoly power of resource-rich well owners (Kumar, 2018).

2.4.2 Pakistan

Groundwater development in Pakistan is not uniform. It is heavily skewed towards Punjab. The canal system contributes to more than 80% of the total groundwater recharge in that province (Tariq et al., 2020). The shallow depth and better quality of groundwater owing to its replenishment from surface irrigation and rivers favored the massive development of private tube wells in Punjab. In Punjab, only 23% of the area has poor groundwater quality, whereas in Sindh, nearly 78% of the groundwater suffers from high levels of salinity (Bakshi & Trivedi, 2011). For this reason, the development of private tube wells in Sindh remains limited. In Balochistan, groundwater levels are deep, and turbine/submersible pumps are needed to run these tube wells. The average installation cost of a deep electric tube well (>20 m) is US $ 10,000, compared to US $ 1000 for a shallow tube well (Qureshi, 2020).

Research studies conducted till the mid-1980s had shown conclusively that it is the wealthier farmers in Pakistan's Punjab who have had the greatest access to private tube wells. Every study has shown that the distribution of tube wells among farmers with different sized holdings is skewed towards large-scale landholders. There are a number of reasons for this. First, large-scale farmers have had a comparative advantage in procuring assistance from past government programs. For example, the Agricultural Development Bank of Pakistan requires a farmer seeking credit for installing a private well to have at least 2.024 ha (5 acres) that he can mortgage. In its early years, the diesel subsidy program required that applicants have at least 10.12 ha (25 acres) in compact holdings. With the Water and Power Development Authority's electrical connection subsidy, farmers are usually expected to pay the entire cost of connection, including the cost of the connecting line and transformer. These costs were usually much higher than the subsidy of Rs 15,000, discouraging its utilization by small-scale farmers. Even without formal program restrictions on small-scale farmers' access to public programs, local political influence wielded by large-scale farmers has often been enough to guarantee access to scarce public resources (Johnson, 1989).

The second reason for the concentration of private wells in large landowners' hands is that the technologies available were financially unattractive when holding sizes fell below a minimum level. While water selling in Punjab was common, the Water and Power Development Authority (1980) claimed that the amount of water sold in Pakistan compared to the amounts pumped was very small, and therefore the primary source of returns to a farmer installing a private well must come from his own land and not from the sale of water. Also, while fractional cusec technologies were available, costs were relatively inelastic when compared to capacities. Hence, cutting a well's capacity in half does not reduce its installation cost by a corresponding amount. Finally, the wells owned jointly by several farmers are a relatively small fraction of the total number of private wells. The estimate of the number of wells, owned by groups, ranges from 11% (Yasin, 1975) to 25% (Water and Power Development Authority, 1980). Again, this is partly due to the fact that the public programs have done little to encourage joint ownership of private wells; in fact, some discourage such arrangements (Johnson, 1989).

From the available data, it can be inferred that such inequities still exist in Pakistan. For instance, in Punjab province, which has a shallow water table, groundwater abstraction is the least expensive. Yet only one out of four farmers there owns a shallow well (Leghari et al., 2012). In Pakistan, the cost of pumping 1000 cubic meters of water from a shallow tube well is US $ 4.5, compared to US $ 15 from a deep tube well (Basharat, 2015). Those who do not have wells purchase water through informal groundwater markets (Leghari et al., 2012). A recent study in the Punjab province of Pakistan looking at the participation in groundwater markets showed that 99.4% of those farmers who were buying water were marginal and small farmers, and all those who were selling water were large farmers. Moreover, nearly 46% of the large farmers were found to be selling water (Razzaq et al., 2022, p. 11).

In the Balochistan province, where the cost of drilling a well is very high due to deep water table conditions, electricity is subsidized to make groundwater pumping affordable for smallholder farmers. However, the large landholders own deeper wells. During 2015–16, the subsidy provided by federal and provincial governments on electricity was on the order of PKR 28 billion or US $280 million (UNDP, 2016). According to the Agricultural Statistics of Pakistan in 2014–15, there were 30,387 private electric tube wells in Balochistan. Therefore, the subsidy per tube well comes to approximately PKR 0.92 million (US $9200) per year (Ashraf & Sheikh, 2017). Most of the large farmers own more than one tube well. Thus, the subsidy benefits also go to the large farmers (Qureshi, 2014). In fact, less than 0.3% of the population receives direct benefits from this huge subsidy (Ashraf & Sheikh, 2017).

2.4.3   Bangladesh

Over the past four decades, one big transformation that has happened in Bangladesh's agriculture is the rapid increase in the number of shallow tube wells for irrigation, which almost revolutionized the agricultural economy of that country and increased more than 15 times from around 100,000 in 1982 to 1.523 million in 2013. These wells, which irrigate 2–4 ha of land, are relatively inexpensive when compared to the deep tube wells, which irrigate around 26 ha and are owned by small and marginal farmers (BADC, 2013). Against this astronomical growth, the number of deep tube wells recorded a very nominal increase of around 10,000 over a period of three decades beginning in 1992. When compared to the expansion in area irrigated by DTWs that are owned by very rich landowners, the expansion in area irrigated by STWs during the past few decades is far greater. Therefore, it can be inferred that access to groundwater is more equitable in Bangladesh than in India and Pakistan.

2.5   Impacts of public policies on the functioning of groundwater markets

With groundwater becoming very scarce in many arid and semi-arid regions and the cost of accessing it becoming prohibitively high in those regions, it is increasingly coming under the control of a few resource-rich people, with unclear legal ownership rights over it (Kumar et al., 2013b, for India; Qureshi, 2014, for Pakistan). The rich well farmers who have the wherewithal to drill deep and install high-capacity pumps keep chasing the falling water table in regions where surface water supplies for irrigation are limited. They keep offering irrigation services to the neighbouring farmers, who do not own wells, at prohibitive price—be in Kolar in Karnataka, Coimbatore in TN, North Gujarat or Madhya Pradesh, or the Punjab province in Pakistan. They enjoy high monopoly power and in fact, the price of water and the monopoly price ratio keeps increasing as the number of potential water buyers against the sellers keeps increasing (Kumar, 2018).

No one really knows how old the phenomenon of groundwater markets in South Asia is. One should suspect that water sharing arrangements between well owners and farmers without wells are as old as the practice of well irrigation itself. But, the research on informal groundwater markets is at least 40 years old (Kumar, 2018). That said, the last two decades have seen remarkable debate on the impact of public policies relating to irrigation on the access equity in groundwater, particularly the economic impact of well irrigation on different classes of farmers in eastern India (see Shah, 1993; Saleth, 1997; Mukherji et al., 2012; Shah, 2016) which has the largest concentration of poor people in the country. A few researchers have argued that groundwater development and cheap well irrigation could trigger agricultural growth and economic prosperity (Shah, 2001; Pant 2004; Mukherji et al., 2012 Shah, 2016), and sought appropriate public policies for promoting well irrigation. The focus has been on policy instruments for promoting efficient groundwater markets (Kumar, 2007). Most of the earlier research on groundwater markets viewed it as a case of "natural oligopoly," but one which can be converted into a powerful instrument to promote equity in access to groundwater and reduce poverty, through the right kind of energy pricing policies which create an incentive for well owners to increase the production volume and sell the surplus to his neighbours at lowered prices (Mukherjee & Biswas, 2016; Shah, 1993; Qureshi et al., 2014). Such arguments, however, had no empirical basis.

For instance, in the context of Bangladesh, Qureshi et al. (2014) note: "…, the owners of STWs also provide irrigation services to their neighbors for a fixed seasonal fee in cash or through payment in produce." These "informal" water markets are quite mature and provide one of the most promising institutional mechanisms for increasing access to irrigation from groundwater, particularly for tenants and small farmers. With the expansion of water markets in the private sector, irrigation sellers have adapted their pricing systems to suit farmers' varying circumstances. Water fees vary from one area to another depending on the type of well and the depth of water availability. However, such descriptions of water markets are devoid of any details, such as the irrigation charges paid by the water buyers per hour of irrigation and the cost incurred by the well owners for pumping water per hour.

Ideally, the question that the researchers should have posed is this: "How much would it cost if the farmer invested his own money to drill a well and pump water to water his farm as opposed to buying water from a neighbouring well owner?" A simpler approach would have been to look at the monopoly price ratio for the water-selling farmers. Obviously, the argument that the *groundwater market promotes equity in access to groundwater and therefore helps many millions of poor, small and marginal farmers* will be tenable only if the cost in the former case (farmer investing in his own well and pump) is more or less the same as the price at which water is being bought, or if the monopoly price of water is very low or close to one. However, such an analysis was never attempted.

As Kumar (2018) pointed out: "The fact is that the equity impact of groundwater markets (as they operate today in different parts of our country) in India is a billion dollar question if we consider the following facts: (1) the amount of money government spends in the form of electricity subsidy for well owners, which is close to eight billion dollars (approximately 8 billion dollars per annum); (2) a large share of the subsidy benefit go to the large and medium farmers (as studies by the World Bank in Punjab and erstwhile Andhra Pradesh have shown); and (3) groundwater markets cover nearly 15% of the total well irrigated area in the country (source: estimates by Saleth (1998))."

The exponents of groundwater markets stretched their argument further to build a strong case for offering electricity to this privileged group of well-owning farmers based on a flat rate on the pretext that they could produce groundwater at a very low cost and thus offer water to their neighbours at competitive prices, thereby passing on a share of the subsidy benefit to the non-well owning water buyers (Mukherji et al., 2009). The groundwater markets in India were romanticized due to the sheer scale of irrigation it could produce and the hundreds of millions of farmers it could *prima facie* benefit. Such an argument was palatable for many, including some state electricity utilities. A flat rate system for electricity was introduced for agriculture by these utilities, considering the fact that metering electricity consumption in the farm sector was getting more and more difficult with the explosion in well numbers in the rural areas (Kumar, 2018). Such tendencies to glorify water markets are seen in Pakistan as well. A recent study by a group of researchers, which looked at the equity and productivity impacts of groundwater markets in the Punjab province, argued that the water markets promoted equity in access to groundwater, in spite of the data showing lower yield and income obtained by the water buyers as compared to the water sellers, owing to poorer access to irrigation (Razzaq et al., 2022).

Obviously, its impact on the sustainability of resource use and intra-generational equity in access to the resource was hardly analyzed empirically. Some social scientists, who were not so convinced about the arguments of "positive impacts of private water markets," did question this superficial way of looking at access equity in groundwater, which treats the groundwater market as a natural oligopoly (Palmer-Jones, 1994; Saleth, 1994), and raised concerns about the negative ecological, equity and efficiency effects of water markets. They argued for more fundamental changes in the legal and institutional frame-work governing the use of groundwater (Saleth, 1994; Kumar, 2000). In Pakistan, the flat tariff policy was also criticized by small farmers due to its "fixed cost" even in cases of minimally or completely non-operational tube wells (Venot & Molle, 2008).

In fact, recent empirical research (Uttar Pradesh, Bihar, and Gujarat) clearly shows that the monopoly power of well owners, which determines the price at which water is sold by them against the cost they incur, has nothing to do with the electricity pricing policies that determine the cost of production of water. Instead, it is influenced by the market conditions (number of buyers against the sellers, and overall demand-supply situation). In a given area, irrespective of the differences in the cost of production of water (between diesel well owners and electric well owners), the selling price of water was found to be more or less the same. Many electric well owners, whose marginal cost of pumping water was almost zero, sold water at as high a price as the diesel well owners, who incurred a very high cost of pumping water due to the high fuel cost (Kumar et al., 2013a). Hence, the argument that a large portion of the benefits from the heavily subsidized electricity supplied to well owning farmers would eventually be passed on to poor water-buying farmers is nothing short of a fallacy.

On the contrary, under similar market conditions, under the flat rate (pricing based on connected load), the large well owners, whose implicit cost of irrigation is very low and return from crop production high, may enjoy monopoly power and decide the price at which water should be sold in the market. Conversely, the smallholders owning wells, whose implicit cost of irrigating their own farm is high, are left without much choice but to look for buyers to whom they could sell water to earn extra income, at a price decided by the market, as they will have limited bargaining power. Such prices obviously offer high profit margins for the large farmers, but not for the small farmers.

It was found that the monopoly price ratio (MPR—the ratio of the price at which a commodity is sold and the actual cost of production—charged by electric well owners, who incur zero marginal cost and a very low implicit cost of pumping water, is far higher than that charged by the diesel well owners who incur a high marginal cost of pumping (Kumar et al., 2013a). Studies from eastern Uttar Pradesh, West Bengal, and south Bihar show that these markets can be highly monopolistic in the sense that the price that well owners charge the buyers is much higher than the actual cost incurred by them for abstracting water, depending on supply and demand in the market (Kumar et al., 2013a, 2014). Electric well owners who typically incur very low costs for abstracting water, owing to the very low cost of energy, charge as much as a diesel pump owner, who incurs much higher energy costs. Therefore, the monopoly price of electric well owners was found to be very high (Kumar et al., 2013a). Over a period of three decades or so, the monopoly prices charged by these well owners have however dropped, with a greater proportion of farmers owing wells and pump sets (Kishore, 2004).

Metering and consumption-based pricing create a level-playing field for all landholding classes, while improving the efficiency of the use of water and electricity, and is unlikely to cause an upward trend in the monopoly price of water, and thus promote equity in access to groundwater. If electricity is charged on a pro rata basis, both small and large farmers will incur the same unit cost of pumping water, and the large farmers will not enjoy a comparative advantage over the small farmers. In view of the fact that there are no fixed costs for keeping pump sets, the small landholders are not under pressure to sell water at a very low price to stay in the market. Hence, both large and small farmers would have equal incentives to invest in wells and sell water and have equal opportunities to make profits in the shallow groundwater areas. This will lower the price at which water would be sold in the market. Empirical evidence from West Bengal substantiates this view[1] (Kumar et al., 2014).

West Bengal introduced metering and consumption-based pricing of electricity in agriculture way back in 2006. Today, interestingly, 85% of the agricultural connections in Gujarat—a state that had long been stereotyped as a difficult

---

1. Analysis of primary data from villages in West Bengal (Source: based on Mukherji (2008)) shows that the monopoly power enjoyed by diesel well owners (estimated in terms of monopoly price ratio, MPR = 1.90), who are confronted with a positive marginal cost of pumping groundwater, was much lower than that of electric well owners under the flat rate system of pricing electricity (MPR = 16.70).

turf for anything like this—are now metered. The impacts of such measures on groundwater use efficiency and equity can now be studied using a large sample. Unfortunately, it is common to find researchers manipulating data to present a distorted picture of the scenario that serves certain interests, and policy prescriptions are dished out.

The argument that the zero marginal cost of pumping water would create the incentive to pump more water from the well is by and large correct, but that it would create competitive water markets is fallacious as shown by field evidence. A survey in West Bengal showed that the diesel well owners were selling water more aggressively to make profits than those who had electric submersible pumps and paying for electricity on the basis of the connected load. Against an average area of 22.8 ha irrigated by diesel well owners, the total area of water buyers was 19.2 ha, i.e., nearly 84% of the pumping was done to provide irrigation service to neighbouring farmers. Conversely, in the case of electric well owners, against a total area of 27.0 ha irrigated by the tube well, the area irrigated by water buyers was 22.3 ha (82%) (Mukherji, 2008).

Further, extending the findings of studies from eastern India to water-scarce regions such as western India, central India and the South Indian peninsula would be untenable. The reason is that in water-scarce regions, the price at which water is traded does reflect the scarcity value of the resource (Kumar, 2007)[2]. Nevertheless, in such a situation, one can only reduce the cost of pumping water and not the scarcity value of the resource. But this reduction would be just marginal as the cost of energy is a small fraction of the total well irrigation cost. While the only leverage is to manipulate the electricity prices, all the State Electricity Boards in the western, central and South Indian peninsula are following subsidized flat rates, barring the state of Gujarat (Kumar et al., 2013a). Hence, the benefits due to the reduction in the implicit cost of pumping through the introduction of subsidized flat rate charges are least likely to be transferred to water buyers.

Some researchers paraded such arguments for nearly 25 years (Mukherji et al., 2009; Mukherjee & Biswas, 2016; Shah, 1993) without producing an iota of empirical evidence to support their claim or even providing counter-intuitive evidence. As a matter of fact, the empirical analysis provided by Mukherjee and Biswas (2016) for Madhya Pradesh shows that groundwater markets are monopolistic, while they argued conversely. The result is that we have a blinkered view of groundwater markets. In this blinkered view, the village water trader emerges as the "saviour of the poor"!

Yet, the silent revolution created by the profound equity impact of large irrigation systems has hardly captured the imagination of social science

---

2. The higher value of water comes from the high gap between demand and supply of water and the high capital investment required to secure groundwater through drilling, or the poor chances of hitting groundwater again in situations of well failure.

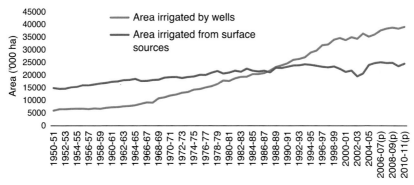

FIGURE 2.6   Net area irrigated by wells and surface water sources.

researchers in India until recently. The ongoing Sardar Sarovar Narmada project, which is irrigating nearly 0.9–1.0 m.ha of land in Gujarat (around 25% of the gross irrigated area in that state), has made a remarkable dent on groundwater markets, which thrived for many decades in the region. Well owners are forced to lower the water prices owing to an overall reduction in demand for irrigation water. Gravity irrigation is rejuvenating millions of wells in the areas that are now receiving canal water. Field evidence suggests that while the area directly irrigated by "purchased water" had declined sharply, the energy consumption per unit of well irrigated area had also declined owing to the rising water table. This effect is seen in the electricity consumption in the agriculture sector, with a steep decline at the state level (Kumar, 2018). In Punjab, there was greater equity in the distribution of agricultural income among farmers in villages receiving canal water as compared to those in villages with tube well irrigation but facing problems of groundwater over-exploitation (Sarkar, 2012).

2.6   Growth of well irrigation in South Asia: will it sustain?

Well irrigated areas have been growing steadily in South Asian countries, particularly in India, Pakistan, and Bangladesh (source: based on Kumar et al., 2013b; Qureshi et al., 2014; Qureshi, 2020). However, the growth patterns are not the same across countries.

In India, well irrigated areas have grown remarkably over the past five decades, beginning in the late 1960s. Though the net area under well irrigation has been steadily growing since 1950–51 (from a mere 5.98 m.ha in 1950–51 to 39.0 m.ha in 2010–11) (Fig. 2.6), the annual growth rate suddenly increased after 1967–68 and continued till 1999–00 for nearly 32 years, at an average rate of 7.97 lac ha per year. By the late 1980s, it had overtaken surface irrigation in terms of net area commanded (see Fig. 2.3). However, the growth rate started declining thereafter, and the average growth rate for the period from 1999–00 to 2010–11 is only 4.00 lac ha, which is a mere 50% of the growth rate recorded during the previous period. This is in spite of the fact that there is still a lot of unutilized groundwater in the country at the aggregate level as per the latest estimates from the Central Ground Water Board (as of 2017)[3], though the report does show that nearly 17.2% of the assessment units in the country witness over-exploitation and another 18.6% belong to the critical and sub-critical stage of exploitation (CGWB, 2019).

Two factors might have surely contributed to this decline in growth rate. First: in many areas where agricultural land is still lying un-irrigated, there is no scope for getting extra water from wells due to groundwater depletion (they include many parts of Punjab, Rajasthan, Gujarat, parts of Andhra Pradesh and Telangana, Tamil Nadu, and Karnataka). Second: in areas where there is good groundwater potential, the arable land for expanding irrigation is either not available in plenty (eastern Uttar Pradesh, Bihar, West Bengal, and Assam) (Kumar et al., 2012). Yet they are not sufficient to explain this drastic decline in the growth rate of well irrigated areas. So we will have to find out whether there are other reasons too. That said, there are large areas that according to the official estimates (of the CGWB), are not "over-exploited" or critical, and where arable land is not a constraint for further utilization of groundwater. They include large parts of Telangana, Karnataka, Maharashtra, Odisha, Tamil Nadu, Madhya Pradesh, Chhattisgarh, and Jharkhand (CGWB, 2019). Most of these areas are underlain by hard rocks and there is hardly any groundwater stock available here. The problem of over-exploitation is actually quite serious in these areas and all the water available from annual recharge does get used up for agriculture and other uses, but the methodology for assessing groundwater development is not robust enough to capture it, as it tends to over-estimate the dynamic utilizable groundwater resources available from precipitation (Kumar & Singh, 2008). These methodological issues will be discussed in the next chapter.

That said, we have already seen that groundwater irrigation is highly in-equitable in India, with a very small fraction of the marginal farmers (who constitute a lion's share of the operational holders) owning wells as per both the 3rd and 5th Minor Irrigation Census. The proportion goes down further when it comes to heavy duty wells (medium deep tube wells and deep tube wells) that have high irrigation potential. The area irrigated by the wells owned by marginal and small farmers is only 25% of the total (net) well irrigated area, though they hold 44.5% of the total agricultural land. Further, only 14% (9.2 m.ha) of their total agricultural land (65 m.ha) is irrigated by the wells owned by them. Hence, the impact of well irrigation on small and marginal farmers in terms of production of farm surplus is much less in comparison to the medium and large farmers. In the absence of well-defined water rights, groundwater

---

3. As per the CGWB estimates (as of 2017), the total groundwater draft is only 248.69 BCM against a net utilizable recharge of 392.7 BCM. The un-utilized groundwater in the country, therefore, is 144 BCM (CGWB, 2019).

markets are also highly inequitable, leaving very few opportunities for the water buyers to generate a surplus from the use of purchased water when compared to the well owners who have unlimited access to the resource, but without being confronted with any opportunity cost of using it. The energy pricing policies that are pervasive and that result in a zero marginal cost of using electricity perpetuate the inequity, leaving little space even for the small well-owning farmers. Under this circumstance, the problems of groundwater over-exploitation can put a brake on the prospects of a large majority of the marginal and small farmers who do not own wells in the future, posing new challenges of inter-generational equity.

In Pakistan, the area irrigated by wells, which started increasing in the early 1960s, continues to rise. Because of the presence of deep alluvial aquifers and arable land, and shrinkage in areas fully irrigated by canals, it is possible to increase the well-irrigated crop production in response to growing agricultural growth and rural economic needs. During the last 50 years, agriculture in Pakistan has changed from surface water to largely groundwater-fed irrigation. The increasing trend of groundwater use has played a pivotal role in enhancing the irrigated area and improving crop yields. Between 1960 and 2015, the area irrigated by groundwater has increased by 390% (from 2.9 m.ha to 14.3 m.ha). This includes areas where groundwater is used either in isolation or in conjunction with canal water (Qureshi, 2020).

The growth in well-irrigated area noticed in Pakistan is much sharper than that of India. While well irrigation is increasing in the tail reaches of canal networks, which receive limited and erratic canal water, the same is causing fast depletion of groundwater because the recharge from gravity irrigation is very limited in those areas. Further, the groundwater, which is saline in the tail reaches, will have to be blended with canal water for dilution to make it usable for irrigation (Qureshi et al., 2009; Qureshi, 2020).

However, continued expansion of well irrigation faces serious constraints induced by the deteriorating quality of groundwater in terms of the rise in salinity levels in the water pumped from deeper layers of the alluvial plains of Punjab and Sindh. Unless sufficient surface water is made available for blending with groundwater, an increase in groundwater use for expanding well irrigation will not be possible. In Balochistan, the situation is even more serious, as water levels are declining at a rate of 2–3 m annually. In the absence of effective management measures, the smallholder farmers in around 20% of the area in Balochistan may lose access to groundwater in the near future (Tariq et al., 2020).

As regards Bangladesh, there are eight distinct hydrological zones. They are: north western, north central, south western, south central, river and estuarine, eastern Himalayas, north eastern and south eastern. Groundwater irrigation is intensive in the north western and north central zones and in nearly half of the south western and north eastern zones. In the rest of the zones, the extent of well irrigation is less. There is limited surface irrigation in those zones, and surface irrigation is greater in the north eastern zone. Against 4.17 m.ha of well irrigation, surface irrigation is only 1.02 m.ha. There was rapid growth in the well irrigated area from the early 1980s until 2005, followed by a slowdown in growth. However, the surface irrigation growth, which was quite low during that period (1980–05), picked up after 2005 owing to the easy availability of low lift pumps (BADC, 2013).

Over-exploitation of groundwater in some regions has resulted in substantial drawdowns in water levels (Jahan et al., 2010; Shahid, 2011). Using data from the Bangladesh Water Development Board (BWDB), Qureshi et al. (2014) have shown that in areas with water tables less than 8 m in depth, decline has increased significantly over time. Between 1998 and 2002, the area in the country experiencing groundwater overdraft (with a decline in water levels) went up from 4% of the country's total geographical area (in 1998–02) to 11% in 2008 and 14% in 2012. The most significantly affected areas lie in the north-west (e.g., Barind Tract) and north-central (i.e., Madhupur Tract) regions. These are areas of intensive *Boro* (rice) cultivation and exhibit declining long-term groundwater trends (Shamsudduha et al., 2009). In the north western region, water tables are declining steadily but more slowly (0.1–0.5 m/year), making the use of STWs tapping shallow aquifers unsustainable for intensive Boro irrigation (Dey et al., 2013). At the same time, water levels are rising in southern Bangladesh due to seawater intrusion and tidal waves creating waterlogging conditions (Brammer, 2014).

In Bangladesh, the biggest constraint on future expansion in groundwater irrigation is the availability of arable land. The total amount of un-utilized land that is suitable for cultivation is quite low. Qureshi et al. (2014) notes that in the north central and north western regions of Bangladesh, where summer cultivation (Boro rice) is already intensive, groundwater levels are depleting fast. Chaudhary (2010) notes that in the coastal region, where groundwater levels in the shallow aquifer are rising due to seawater ingress and intrusion, the extent of dry season cropping is limited mainly due to a lack of irrigation water and soil salinity. According to the author, there are over 700,000 ha of under-utilized medium highlands that can potentially be brought under intensified cultivation during the dry season if irrigation water is available. While the groundwater in the deep aquifers of the coastal zone is free from salinity, it is largely unexploited due to the prohibitively high cost of DTW installation (Qureshi et al., 2014). The country is water-rich, with plenty of surface water available in ponds, especially in the south-central, south-western, and river estuarine zones. If groundwater over-exploitation continues, the increase in the cost of pumping water might force farmers to shift to surface irrigation wherever surface water sources are available. However, the summer cultivation will suffer when these surface water bodies dry up. In any case, expansion of groundwater irrigation at the country level is very unlikely in the current scenario.

2.7   Conclusion

The research agenda for groundwater in South Asia during the past three decades was set by social scientists, especially development economists. The primary aims of their research were to: quantify the returns from groundwater irrigation;

compare the performance of groundwater irrigation systems against surface irrigation; understand the equity and efficiency aspects of groundwater irrigation; assess the ownership of wells and access to well irrigation; and understand the functioning of groundwater markets, especially their equity impacts. Many of these studies lacked rigor mainly because their analysis could not capture the nuances such as groundwater system characteristics, types of groundwater irrigation structures, the area commanded by different types of wells and the types of crops typically cultivated by groundwater-based sources. Such gaps in the analysis heavily influenced the results vis-à-vis cost-effectiveness of the investments for well construction, the irrigation potential of the system and the cost per ha of irrigation from wells, the long-term growth potential, and therefore the policy prescriptions that followed with regard to investment choices, institutional financing, and provisioning of subsidies for well construction or electricity.

So far as India is concerned, the fact remains that the country's groundwater is characterized by a high degree of heterogeneity—with groundwater-rich areas underlain by deep allium and groundwater-scarce areas underlain by hard rock formations; areas with shallow water table and areas with deep water table conditions in both alluvial and hard rock areas; and high yielding deep tube wells in the alluvial areas (of the Indo Gangetic plain of India and Bangladesh) to low-yielding deep bore wells in the hard rocks (of the peninsular). Therefore, the spatial analysis of groundwater development and use across regions is not amenable to simple formulations using commonly used criteria such as well numbers, well density, the annual rate of change in water levels and the stage of groundwater development.

Over a period of seven years (from 2006–07 to 2013–14), with changes in the geohydrological environment caused by resource depletion, the composition of groundwater abstraction structures in India's irrigation landscape has changed, with a reduction in the number of open wells (from 9.2 million to 8.78 million) and a big rise in the number of deep tube wells (from 1.44 million to 2.61 million). The arguments that were very instrumental in shaping public policies in the groundwater irrigation sector in India are: high density of farm wells increases the transaction cost of metering and charging for electricity on a pro-rata basis as a tool to control groundwater draft; the groundwater economy is controlled by millions of small and marginal farmers, and any attempts to regulate it would threaten their livelihoods and therefore are politically sensitive.

In Pakistan and Bangladesh, given the unique geological and geohydrological settings, the number of shallow tube wells has increased dramatically over the past 4–5 decades. In Pakistan, it saw a steep jump from around 100,000 tube wells (diesel + electric) in 1970 to around 6,50,000 (diesel + electric) tube wells in 2002 (Qureshi et al., 2008), and by 2018, Punjab alone had around 1.0 million tube wells (Qureshi, 2020). In Bangladesh, it increased from 500,000 ha in 1981–82 to 4.20 m.ha in 2012–13 (BADC, 2013). A lot of this expansion has happened due to the dramatic increase in the number of shallow tube wells, which went up from around 100,000 in 1982 to 1.523 million in 2012/13 (BADC, 2013).

The analysis presented in this chapter concerning India using data from the 5th Minor Irrigation Census (2013–14) shows that groundwater irrigation, which has expanded remarkably over the past 50 plus years (since 1967–68) from 9.1 m.ha to 39 m.ha, is highly inequitable in India, with a very small fraction of marginal farmers (6.6%) owning wells against 53% of large farmers. But the marginal farmers constitute a lion's share of the operational holders in the country—numbering 92.8 million. The proportion goes down further when it comes to heavy duty wells (medium deep tube wells and deep tube wells) that have high irrigation potential. Further, over a 13-year time period from 2000–01 to 2013–14, the proportion of large farmers owning wells has increased from 37.3% to 53%. In Pakistan too, groundwater access for irrigation is still highly inequitable, especially given the high cost of energy for pumping groundwater in areas facing over-exploitation, especially in Punjab and Balochistan. The degree of inequity in access can be said to be less in Bangladesh given the fact that around 80% of the area irrigated by groundwater is through shallow tube wells.

Under the current institutional regime, groundwater markets are also highly inequitable, leaving very few opportunities for the water buyers to generate a surplus from the use of purchased water, when compared to the well owners who have unlimited access to the resource, but without being confronted with any opportunity cost of using it in the absence of well-defined water rights. The energy pricing policies followed by most Indian states that are pervasive and that result in a zero marginal cost of using electricity, perpetuate this inequity. This, coupled with resource over-exploitation problems in many areas, which are dampening the growth in well irrigation in those areas despite having a sufficient amount of arable land, might pose new challenges of inter-generational equity in access to groundwater.

In the case of Pakistan and Bangladesh, the energy cost for pumping ground-water is far greater than that of India, not only in relation to the total cost of cultivation but also in aggregate terms (source: based on Kumar, 2018; Qureshi, 2020; Qureshi et al., 2014). This is because diesel well owners (who account for a large proportion of the well-owning farmers in these two countries) incur a high marginal cost for pumping groundwater, and the electric well owners, who are fewer in number, also pay for electricity partly on the basis of consumption in most cases. Hence, groundwater markets are expected to be less inequitable in these countries when compared to India.

References

Ahmad, N., & Chaudhry, G. R. (1988). *Irrigated agriculture of Pakistan*. S. Nazir.

Ahmad, K., Shahid, S., Sobri, H., & Wang, X. (2016). Characterization of seasonal droughts in Balochistan Province, Pakistan. *Stochastic Environmental Research and Risk Assessment, 30*(2), 747–762.

Ahmed, K., Bhattacharya, P., Hasan, M., Akhter, S., Alam, S., Bhuyian, M., Imam, M., Khan, A. A., & Sracek, O. (2004). Arsenic enrichment in groundwater of the alluvial aquifers

in Bangladesh: An overview. *Applied Geochemistry,* 19(2), 181–200. https://doi.org/10.1016/
j.apgeochem.2003.09.006.

Alam, M. K., Shahidul, H. A. K., Khan, M. R., & Whitney, J. W. (1990). *Geological map of
Bangladesh.* Dhaka, Bangladesh: Geological Survey of Bangladesh.

Ashraf, M., & Majeed, A. (2006). *Water requirements of major crops for different agro-climatic
zones of Balochistan.* IUCN (The World Conservation Union), Water Programme, Balochistan
Programme Office.

Ashraf, M., & Sheikh, A. A. (2017). *Sustainable groundwater management in Balochistan.* Pakistan Council of Research in Water Resources (PCRWR). https://pcrwr.gov.pk/wp-content/
uploads/2020/Water-Management-Reports/sustainable-grndwtr-in-balochistan.pdf.

Badiani, R., Jessoe, K., & Plant, S. (2012). Development and the environment: The implications of
agricultural electricity subsidies in India. *The Journal of Environment and Development,* 21(2),
244–262.

Bakshi, G., Trivedi, S. The Indus Equation; Strategic Foresight Group: Mumbai, India, 2011.

Bangladesh Agricultural Development Corporation (BADC). (2013). *Minor irrigation survey report
2012-2013.* Bangladesh Agricultural Development Corporation, Ministry of Agriculture,

Basharat, M. (2015). *Groundwater management in Indus plain and integrated water resources
management approach.* International Waterlogging and Salinity Research Institute (IWASRI).

Brammer, H. (2014). Bangladesh's dynamic coastal regions and sea-level risc. *Climate Risk Management,* 1, 51–62. https://doi.org/10.1016/j.crm.2013.10.001.

Burgess, W. G., Hasan, M. K., Rihani, E., Ahmed, K. M., Hoque, M. A., & Darling, W. G. (2011).
Groundwater quality trends in the Dupi Tila aquifer of Dhaka, Bangladesh: Sources of contam-
ination evaluated using modelling and environmental isotopes. *International Journal of Urban
Sustainable Development,* 3(1), 56–76. https://doi.org/10.1080/19463138.2011.554662.

Central Ground Water Board. (2019). *Dynamic ground water resources of India.* Central Ground
Water Board, Ministry of Jal Shakti, Government of India.

Chaudhary, N. T. (2010). Water management in Bangladesh: an analytical review. *Water Policy,* 12,
32–51.

Deb Roy, A., & Shah, T (2003). Socio-ecology of groundwater irrigation in India. In R. Llamas, &
E. Custodio (Eds.), *Intensive use of groundwater: Challenges and opportunities* (pp. 307–335).
Swets and Zetlinger Publishing Co.

Deepak, S. C., Chandrakanth, M. G., & Nagaraj, N. (2005). *Water demand management: A strategy
to deal with water scarcity* [Paper presentation]. In *4th Annual Partners' Meet of IWMI-Tata
Water Policy Research Programme, Anand, Gujarat.*

Dey, N. C., Bala, S. K., Saiful Islam, A. K. M., & Rashid, M. A. (2013). *Sustainability of groundwater
use for irrigation in northwest Bangladesh* p. 89. Dhaka, Bangladesh: Policy Report prepared
under the National Food Policy Capacity Strengthening Programme (NFPCSP).

Government of India (GoI). (2014). *Report of 4th census of minor irrigation schemes.* Minor Irrigation Statistics Wing, Ministry of Water Resources, River Development and Ganga Rejuvenation,
Government of India.

Government of India (GoI). (2017). *Report of 5th census of minor irrigation schemes.* Minor Irrigation Statistics Wing, Ministry of Water Resources, River Development and Ganga Rejuvenation,
Government of India.

Halcrow-ACE. (2003). *Exploitation and regulation of fresh groundwater: Main report.* ACE-
Halcrow JV Consultants.

Hussain, A., Abbas, H., Hussain, A., & Abbas, H. (2019, September 27). To save Pakistan, look
under its rivers. *The Third Pole.* https://www.thethirdpole.net/en/climate/pakistans-riverine-
aquifers-may-save-its-future/

Islam, Md. S., Shijan, Md. H. H., Saif, Md. Samin, Biswas, P. K., & Faruk, M. O. (2021). Petrophysical and petrographic characteristics of Barail Sandstone of the Surma Basin, Bangladesh. *Journal of Petroleum Exploration and Production, 11*, 3149–3161. https://doi.org/10.1007/s13202-021-01196-0.

Jahan, C. S., Mazumder, Q. H., Akter, N., Adham, M. I., & Zaman, M. A. (2010). Hydrogeological environment and groundwater occurrences in the Plio-Pleistocene aquifer in Barind area, Northwest Bangladesh. *Bangladesh Geoscience Journal, 16*, 23–37.

Johnson, R. (1989). *Private tube well development in Pakistan's Punjab: Review of past public programs/policies and relevant research*. IWMI Books, International Water Management Institute.

Khan, A. D., Ashraf, M., Ghumman, A. R., & Iqbal, N (2018). Groundwater resource of Indus Plain aquifer of Pakistan investigations, evaluation and management. *International Journal of Water Resources and Arid Environments, 7*(1), 52–69.

Kirby, J., Ahmad, M., Mainuddin, M., Palash, W., Quadir, M., Shah-Newaz, S., & Hossain, M. (2015). The impact of irrigation development on regional groundwater resources in Bangladesh. *Agricultural Water Management, 159*, 264–276. https://doi.org/10.1016/j.agwat.2015.05.026.

Kishore, A. (2004). Understanding agrarian impasse in Bihar. *Economic and Political Weekly, 39*(31), 3484–3491.

Kulkarni, H., & Shankar, P. S. V. (2009). Groundwater: Towards an aquifer management framework. *Economic and Political Weekly, 44*(6), 13–17. https://www.jstor.org/stable/pdfplus/40278472.

Kulkarni, H., Vijayshankar, & P. S., & Krishnan, S. (2011). India's groundwater challenge and the way forward. *Economic and Political Weekly, 46*(2), 37–45.

Kumar, M. D. (2000). Institutional framework for managing groundwater: A case study of community organisations in Gujarat, India. *Water Policy, 2*(6), 423–432.

Kumar, M. D. (2007). *Groundwater management in India: Physical, institutional and policy alternatives*. Sage Publications.

Kumar, M. D. (2016). Distressed elephants: Policy initiatives for sustainable groundwater management in India. *IIM Kozhikode Society & Management Review, 5*(1), 51–62.

Kumar, D. M. (2018). *Water policy science and politics: An Indian perspective* (1st ed.). Elsevier Science.

Kumar, M. D., Bassi, N., Sivamohan, M. V. K., & Venkatachalam, L. (2014). Breaking the agrarian crisis in eastern India. In M. D. Kumar, N. Bassi, A. Narayanamoorthy, & M. V. K. Sivamohan (Eds.), *Water, energy and food security nexus: Lessons from India for development* (pp. 143–159). Routledge.

Kumar, M. D., Narayanamoorthy, A., & Singh, O. P. (2009). Groundwater irrigation versus surface irrigation. *Economic and Political Weekly, 44*(50), 72–73.

Kumar, M. D., & Singh, O. P. (2008). How serious are groundwater over-exploitation problems in India? A fresh investigation into an old issue. In M. D. Kumar (Ed.), *Managing water in the face of growing scarcity, inequity and declining returns: Exploring fresh approaches. 7th Annual Partners' meet of IWMI-Tata Water Policy Research Program, ICRISAT, Patancheru, AP, 2–4 April 2008*.

Kumar, M. D., Singh, O. P., & Sivamohan, M. (2010). Have diesel price hikes actually led to farmer distress in India? *Water International, 35*(3), 270–284.

Kumar, M. D., Sivamohan, M. V. K., & Narayanamoorthy, A. (2012). The food security challenge of the food-land-water nexus. *Food Security Journal, 4*(4), 539–556.

Kumar, M. D., Scott, C. A., & Singh, O. (2013a). Can India raise agricultural productivity while reducing groundwater and energy use? *International Journal of Water Resources Development, 29*(4), 557–573.

Kumar, M. D., Reddy, V. R., Narayanamoorthy, A., & Sivamohan, M. V. K. (2013b). Analysis of India's minor irrigation statistics: Faulty analysis, wrong inferences. *Economic and Political Weekly*, 48, 76–78.

Leghari, A. N., Vanham, D., & Rauch, W. (2012). The Indus Basin in the framework of current and future water resources management. *Hydrology and Earth Systems Science. Discussion, 8*, 2263–2288.

Meinzen-Dick, R. (1996). *Groundwater markets in Pakistan: Participation and productivity*. International Food Policy Research Institute.

Michel, A. A. (1967). *The Indus Rivers: A Study of the Effects of Partition*. New York, USA: Yale University Press.

Ministry of Agriculture & Farmers Welfare. (2017). *Pocket book of agricultural statistics 2017*. Economics and Statistics Division, Dept. of Agriculture, Cooperation and Farmers' Welfare, Ministry of Agriculture & Farmer Welfare, Government of India. https://agricoop.nic.in/sites/default/files/pocketbook_0.pdf.

Mukherji, A. (2005). *The spread and extent of irrigation rental market in India, 1976–77 to 1997–98: What does the National sample survey data reveal?* Water Policy Research Highlight. IWMI-Tata Water Policy Program.

Mukherji, A. (2008). The paradox of groundwater scarcity amidst plenty and its implications for food security and poverty alleviation in West Bengal, India: What can be done to ameliorate the crisis? In *Paper presented at the 9th Annual Global Development Network Conference, Brisbane, Australia*.

Mukherjee, S., & Biswas, D. (2016). An enquiry into equity impact of groundwater markets in the context of subsidised energy pricing: A case study. *IIM Kozhikode Society & Management Review, 5*(1), 63–73.

Mukherji, A., Das, B., Majumdar, N., Nayak, N. C., Sethi, R. R., & Sharma, B. R. (2009). Metering of agricultural power supply in West Bengal, India: Who gains and who loses? *Energy Policy, 37*(12), 5530–5539.

Mukherji, A., Shah, T., & Banerjee, P. S. (2012). Kick-starting a second green revolution in Bengal. *Economic and Political Weekly, 47*(18), 27–30.

Mukherji, A., Rawat, S., & Shah, T. (2013). Major insights from India's minor irrigation censuses: 1986–87 to 2006–07. *Economic and Political Weekly, 48*(26/27), 115–124.

Mukherjee, A., Saha, D., Harvey, C. F., Taylor, R. G., Ahmed, K. M., & Bhanja, S. M. (2015). Groundwater system of the Indian sub-continent. *Journal of Hydrology: Regional Studies, 4*(Part A), 1–14.

Palmer-Jones, R. (1994). Groundwater markets in South Asia: A discussion of theory and evidence. In M. Moench (Ed.), *Selling water: Conceptual and policy debates over groundwater markets in India*. VIKSAT-Pacific Institute-Natural Heritage Institute.

Pant, N. (2004). Trends in Groundwater Irrigation in Eastern and Western UP. *Economic and Political Weekly, 39*(31), 3463–3468.

Qureshi, A. S. (2014). Reducing carbon emissions through improved irrigation management: A case study from Pakistan. *Irrigation and Drainage, 63*(1), 132–138.

Qureshi, A. S. (2018). Managing surface water for irrigation. In T. Oweis (Ed.), *Water management for sustainable agriculture*. Burleigh Dodds Science Publishing Limited.

Qureshi, A. S. (2020). Groundwater governance in Pakistan: From colossal development to neglected management. *Water, 12*(11), 3017. https://doi.org/10.3390/w12113017.

Qureshi, A. S., Ahmed, Z., & Krupnik, T. J. (2014). *Groundwater management in Bangladesh: An analysis of problems and opportunities* [Research report no. 2]. *Cereal Systems Initiative for South Asia Mechanization and Irrigation (CSISA-MI) Project*. CIMMYT.

Qureshi, A. S., Akhtar, M., & Sarwar, A. (2003). Effect of electricity pricing policies on groundwater management in Pakistan. *Pakistan Journal of Water Resources,* 7(2), 1–9.

Qureshi, A. S., McCornick, P., Qadir, M., & Aslam, Z. (2008). Managing salinity and waterlogging in the Indus Basin of Pakistan. *Agricultural Water Management,* 95(1), 1–10. https://doi.org/10.1016/j.agwat.2007.09.014.

Qureshi, A. S., McCornick, P. G., Sarwar, A., & Sharma, B. R. (2009). Challenges and prospects of sustainable groundwater management in the Indus Basin, Pakistan. *Water Resources Management,* 24(8), 1551–1569. https://doi.org/10.1007/s11269-009-9513-3.

Ravenscroft, P., Burgess, W. G., Ahmed, K. M., Burren, M., & Perrin, J. (2004). Arsenic in groundwater of the Bengal Basin, Bangladesh: Distribution, field relations, and hydrogeological setting. *Hydrogeology Journal,* 13(5–6), 727–751. https://doi.org/10.1007/s10040-003-0314-0.

Ravenscroft, P., & Rahman, A. A. (2003). Groundwater resources and development in Bangladesh background to the arsenic crisis. In: Agriculture Hydrogeology Journal (2020) 28:2917–2932 2931. *Potential and the Environment.* Dhaka, Bangladesh: The University Press Limited.

Razzaq, A., Xiao, M., Zhou, Y., Liu, H., Abbas, A., Liang, W., & Naseer, M. A. U. R. (2022). Impact of participation in groundwater market on farmland, income, and water access: Evidence from Pakistan. *Water,* 14(12), 1832. https://doi.org/10.3390/w14121832.

Reimann, K. U. (1993). *Geology of Bangladesh.* Stuttgart, Germany: Schweizerbart.

Saleth, R. M. (1994). Groundwater markets in India: A legal and institutional perspective. *Indian Economic Review,* 29(2), 157–176.

Saleth, R. M. (1996). Water institutions in India: Economics, law, and policy. Commonwealth Publishers.

Saleth, R. M. (1997). Power tariff policy for groundwater regulation: Efficiency, equity and sustainability. *Artha Vijnana,* 39(3), 312–322.

Saleth, R. M. (1998). Water markets in India: Economic and institutional aspects. In K. W. Easter, M. W Rosegrant, & A. Dinar (Eds.), *Markets for water: Potential and performance* (pp. 187–205). Kluwer.

Sarkar, A. (2012). Equity in access to irrigation water: A comparative analysis of tube-well irrigation system and conjunctive irrigation system. In M. Kumar (Ed.), *Problems, perspectives and challenges of agricultural water management.* Intech Publishers.

Scott, C. A., & Shah, T. (2004). Groundwater overdraft reduction through agricultural energy policy: Insights from India and Mexico. *International Journal of Water Resources Development,* 20(2), 149–164.

Scott, C. A., & Sharma, B. (2009). Energy supply and the expansion of groundwater irrigation in the Indus-Ganges basin. *International Journal of River Basin Management,* 7(2), 119–124.

Shah, M. (2016). Eliminating poverty in Bihar: Paradoxes, bottlenecks and solutions. *Economic and Political Weekly,* 51(6), 56–65.

Shah, T. (1993). *Groundwater markets and irrigation development: Political economy and practical policy.* Oxford University Press.

Shah, T., Scott, C. A., Kishore, A., & Sharma, A. (2004). *Energy irrigation nexus in South Asia: Improving groundwater conservation and power sector viability* [Research report no. 70]. International Water Management Institute.

Shah, T. (2001). *Wells and Welfare in Ganga Basin: Public Policy and Private Initiative in Eastern Uttar Pradesh, India (Research Report 54).* Colombo, Sri Lanka: International Water Management Institute.

Shahid, S. (2011). Impact of climate change on irrigation water demand of dry season Boro rice in northwest Bangladesh. *Climatic Change,* 105(3–4), 433–453.

Shamsudduha, M., Chandler, R. E., Taylor, R. G., & Ahmed, K. M. (2009). Recent trends in groundwater levels in a highly seasonal hydrological system: The Ganges-Brahmaputra-Meghna delta. *Hydrology and Earth System Sciences, 13*(12), 2373–2385.

Shamsudduha, M., Taylor, R. G., Ahmed, K. M., & Zahid, A. (2011). The impact of intensive groundwater abstraction on recharge to a shallow regional aquifer system: Evidence from Bangladesh. *Hydrogeology Journal, 19*(4), 901–916. https://doi.org/10.1007/s10040-011-0723-4.

Shamsudduha, M., & Uddin, A. (2007). Quaternary shoreline shifting and hydrogeologic influence on the distribution of groundwater arsenic in aquifers of the Bengal Basin. *Journal of Asian Earth Sciences, 31*(2), 177–194. https://doi.org/10.1016/j.jseaes.2007.07.001.

Sharma, B. R., & Ambili, G. K. (2009). Hydro-geology and water resources of Indus-Gangetic basin: comparative analysis of issues and opportunities. *Annals of Arid Zone, 48*(3&4), 187–218.

Tariq, M., van de Giesen, N., Janjua, S., Shahid, M., & Farooq, R. (2020). An engineering perspective of water sharing issues in Pakistan. *Water, 12*(2), 477. https://doi.org/10.3390/w12020477.

United Nations Development Programme. (2016). *Development advocate Pakistan: Water security in Pakistan: Issues and challenges* (Vol. 3, Issue 4). https://www.undp.org/pakistan/publications/development-advocate-pakistan-volume-3-issue-4.

van Steenbergen, F., Kaisarani, A. B., Khan, N. U., & Gohar, M. S. (2015). A case of groundwater depletion in Balochistan, Pakistan: Enter into the void. *Journal of Hydrology: Regional Studies, 4*, 36–47.

Venot, J. P., & Molle, F. (2008). Groundwater depletion in the Jordan highlands: Can pricing policies regulate irrigation water use? *Water Resources Management, 22*(12), 1925–1941. https://doi.org/10.1007/s11269-008-9260-x.

Wada, Y., van Beek, L. P. H., van Kempen, C. M., Reckman, J. W. T. M., Vasak, S., & Bierkens, M. F. P. (2010). Global depletion of groundwater resources. *Geophysical Research Letters, 37*(20). https://doi.org/10.1029/2010gl044571.

Water and Power Development Authority. (1980). *Private tube wells and factors affecting current rate of Investment Economic Research Section, Planning and Development Division.* Lahore, Pakistan: Water and Power Development Authority.

Yasin, M. G. (1975). *Private tubewells in the Punjab.* Lahore, Pakistan: Publication no. 159. The Punjab Board of Economic Inquiry, Punjab.

Zachariadis, T., & Zoumides, C. (2009). Irrigation water pricing in Southern Europe and Cyprus: The effects of the EU Common Agricultural Policy and the Water Framework Directive. *Cyprus Economic Policy Review, 3*(1), 99–122.

Zahid, A., Haque, A., Islam, M. S., Hasan, M. A. F. M. R., & Hassan, M. Q. (2008). Groundwater resources potential in the deltaic flood-plain area of Madaripur District, southern Bangladesh. *Journal of Applied Irrigation Science, 43*, 41–56.

Chapter 3

Application of the complex concept of groundwater over-exploitation

3.1 Introduction

Groundwater resources play a major role in India's irrigation and rural economy (Kumar, 2007; Shekri, 2014; World Bank, 2010), and are crucial for meeting the water supply needs of both rural and urban areas (Kumar, 2007, 2018; World Bank, 2010). India's ability to manage its future water needs would depend so much on a proper understanding of the availability of groundwater, and the nature and magnitude of groundwater problems. There are ever-increasing evidence of aquifer over-exploitation in many localities, in terms of negative consequences such as drinking water shortages, enormous increase in the cost of water abstraction from wells, frequent well failures, reduced command area of wells increasing inequity in access to well water for irrigation, and ecological degradation such as increasing groundwater and soil salinity (Kumar, 2007). While the concerns over the future of groundwater use in India are growing (GoI, 2007; Jadhav, 2018; Shiferaw, 2021; World Bank, 2010), the official statistics continue to paint a rosier picture of groundwater status in the country (CGWB, 2019). At the root of the public concern is the need to arrive at a working defini-tion and comprehensive criteria for assessing aquifer over-exploitation that inte-grate various concerns such as shortage of water for basic survival needs, poor economics of groundwater use for irrigation, growing inequity in access to water, eco-system and environmental degradation, and unethical water use practices.

While there is a rich discourse around the concept of groundwater over-exploitation (See Custodio, 2000; Llamas, 1992a, 1992b; Margrat, 1992), mainly originating from the west, neither the professional agencies involved in ground-water development planning nor the mainstream academics in water resources management and water policy have attempted to come up with a proper definition of groundwater over-exploitation.

Aquifer over-exploitation mainly deals with negative aspects of ground-water development (Custodio, 1992, 2000; Delgado, 1992; ITGE, 1991;

Groundwater Economics and Policy in South Asia. DOI: https://doi.org/10.1016/B978-0-443-14011-2.00005-X

Margrat, 1992[1]). Scholars have argued that the concept of "groundwater over-development" or aquifer over-exploitation is not simple, merely linked to recharge and extraction balance, but is rather complex and linked to various "undesirable consequences" that are physical, social, economic, ecological, environmental, and ethical in nature. Again, what these undesirable consequences are, also changes with perception (Custodio, 2000). Hence, defining and assessing groundwater over-development is both difficult and complex and not amenable to simple formulations. Still, the perceptions of official agencies concerned with groundwater development and management are characterized by aggregate views based on simple hydrological considerations of recharge and abstraction (Kumar & Singh, 2001, 2008).

While water-level trends inform us about the hydrologic stresses acting on aquifers and how these stresses affect groundwater recharge, storage, and discharge (Taylor & Alley, 2001; Todd and Mays, 2005) which in turn can change the long-term trends in water levels (Kumar, 2010), such complex processes are not considered in the interpretation of water level data collected from observation wells by the official agencies. Such shortcomings are serious because the hydrological (CWC, 2017), geological, geohydrological (CGWB, 2019), and geomorphological conditions influencing the natural processes such as natural recharge and natural discharge vary widely across regions in India.

Nevertheless, there have been some recent changes in the official perceptions of groundwater over-development, as a result of the recognition of the need to integrate economic and social considerations in assessing the degree of exploitation. This is also reflected in the methodology proposed by the Ground Water Estimation Committee of 1997 (NABARD, 2006). But, how far such concerns are integrated into actual assessment is, however, open to question.

In this chapter, we will first discuss some of the conceptual issues in defining groundwater over-exploitation; and present some of the accepted definitions of aquifer over-exploitations. It then critiques some of the methodologies used in India for assessing groundwater resources and stages of groundwater development. Finally, the chapter demonstrates through illustrative cases in India how integrating some of the complex considerations, such as detailed water balance, geology, hydrodynamics, and negative socio-economic, ecological, and ethical consequences of over-exploitation, with the official methodologies can yield an altogether different scenario of groundwater than what the official statistics provide.

While groundwater problems are growing in other South Asian countries, especially Pakistan and Bangladesh, there are no standard criteria and

---

1. Such consequences may include: large and continuous drops in groundwater levels over long time periods; large seasonal drops in water levels in wells and the drying up of wells in the summer season; increase in the salinity of seawater; land subsidence; an enormous increase in cost of groundwater extraction; and a reduction of groundwater-dependent vegetation, springs and seepage.

methodologies adopted by the official agencies in those countries to empirically assess groundwater development and ascertain the extent of over-exploitation spatially and temporally (source: based on Khan et al., 2018 for Pakistan; Qureshi et al., 2014 for Bangladesh). The current assessment of groundwater depletion is based mainly on data on water levels and groundwater quality available from observation wells (Quresh et al., 2014 for Bangladesh; Tariq et al., 2020 for Pakistan). The refinement of the official methodologies used in India to integrate some of the physical, socioeconomic, ecological, and ethical considerations would help the other countries of South Asia better evaluate their resource conditions, which is crucial for evolving management solutions.

3.2   Assessment of groundwater over-exploitation

This section deals with the conceptual issues involved in defining groundwater over-development. The discussion will not touch upon the methodological issues involved in assessing groundwater recharge and extraction, but will identify the complex considerations involved in assessing the degree of "groundwater over-development" or "aquifer over-exploitation". It will then present some of the most common definitions of "groundwater over-development" that use some of these considerations, so as to provide a comprehensive framework for assessing the same.

3.2.1   Conceptual issues in defining over-exploitation

Terms such as groundwater over-exploitation, over-draft, over-development, overuse and unsustainable use have been commonly used in discussions on hydro-geology and groundwater resources since the 1970s (Custodio, 2000). Such phenomena are predominantly applied in arid and semi-arid regions where large volumes of groundwater are abstracted to irrigate extensive areas in situations when the natural recharge to aquifers is limited due to several reasons such as low rainfall, and un-favorable topographic and geo-hydrological environments. They are applied to aquifer conditions in other regions when exploitation leads to undesirable consequences.

The concept of groundwater over-exploitation predominantly deals with negative aspects of groundwater development (Custodio, 1992, 2000; Delgado, 1992; ITGE, 1991; Margat, 1992). Such consequences may include: (1) large and continuous drops in groundwater levels over long time periods; (2) large seasonal drops in water levels in wells and the drying up of wells in the summer season; (3) an increase in the salinity of groundwater; (4) land subsidence; (5) enormous increase in the cost of groundwater extraction; and (6) a reduction of groundwater-dependent vegetation, springs, and seepage (Kumar & Singh, 2008).

However, the concept of "groundwater over-development" or "aquifer over-exploitation" does not appear to be simple, merely linked to recharge and extraction balance, but is rather complex, linked to various "undesirable consequences" that are physical, social, economic, ecological, environmental, and ethical in nature. Therefore, an assessment of groundwater over-development involves complex considerations and concerns such as fundamental rights, basic survival needs, health, and economic, ecological, and ethical issues, and hence it is not possible to capture its essence with simple definitions. Custodio (2000), for example, contends that while undesirable consequences appear when abstraction exceeds recharge, there is frequently no clear proof that the same is the cause of these undesirable consequences. This is true in the case of Gujarat (in western India) and West Bengal (in eastern India). In Gujarat, the increasing incidence of fluoride in groundwater is a major problem whose causes are not clearly known. Fluoride content in groundwater can rise due to the leaching of fluoride-containing minerals present in geological formations with groundwater, a phenomenon not directly linked to groundwater over-extraction. Similarly, in West Bengal, there were widespread incidences of high levels of arsenic in groundwater, threatening drinking water supplies and public health (McArthur et al., 2001). Though there are many competing theories[2], it is seldom attributed to over-exploitation.

It is nevertheless important to mention here that there are fundamental differences in the way these "undesirable consequences" are perceived by various scholars. For instance, according to Custodio (2000), it is predominantly the point of view of overly concerned conservationists and people suffering from real or assumed damage, and not always of well-informed people.

Such a logical framework for analyzing the various viewpoints does not hold well in several situations, including ours. First of all, the framework assumes that there are conservationists and those who are suffering from the damage, "which is real or assumed", are different from the developers. This is not true. In many situations, including one under consideration, both are the same. It has been the rural communities, especially the farmers, who are mostly engaged in groundwater development for irrigation, and the consequences or damage are also primarily borne by them in terms of increased extraction costs, reduced well yields, and quality deterioration. Therefore, the argument that the concerns about "over-exploitation" are an unconscious over-reaction to a given situation or the result of deeply entrenched hydromyths is itself questionable.

On the contrary, more systematic debates about groundwater "*over-development,*" mainly initiated by researchers and scholars, including those from official agencies and NGOs, were driven by concerns about maintaining

---

2. Three mechanisms were used to explain the release of arsenic to groundwater, and are as follows: (1) reductive dissolution of FeOOH and release of sorbed arsenic; (2) oxidation of arsenic pyrite; and (3) anion exchange of sorbed arsenic with phosphate from fertilizer. However, Mc Arthur et al. (2001) postulated another hypothesis, which challenged the oxidation and anion exchange theories, namely that the distribution of arsenic pollution is controlled by microbial degradation of buried peat deposits, rather than the distribution of arsenic in aquifer formations, and the former drives reduction of FeOOH.

sustainable water use in the drinking water sector and agriculture. While official agencies mainly looked at farmers as the main culprits behind the uncontrolled exploitation of groundwater, researchers and scholars from development circles blamed government policies and the institutional framework. Several researchers from India have pointed to the need to integrate the concerns of intra-generational equity (Saleth, 1994), social development, fundamental rights, economic efficiency (Moench, 1995) and the economics of well irrigation (Kumar et al., 2001) in assessing the over-development of groundwater.

In India, the official versions of over-development were primarily based on estimates of recharge and extraction. Therefore, they continued to treat areas where recharge exceeded extraction as "areas suitable for further exploitation" without worrying much about the consequent side effects. Those areas, where the average annual extraction figures exceeded the annual recharge figures, were treated as "over-exploited areas" without giving due consideration to factors such as the absence or presence of static groundwater storage. For instance, the entire Indo-Gangetic plain is underlain by deep alluvial aquifers, with the quantum of static resources exceeding the annual replenishment many times (CGWB, 2019; GoI, 1999). So, treating an area whose annual abstraction has exceeded the annual recharge for a few years as "over-exploited" may not be appropriate from a resource management perspective. Nonetheless, there have been some recent shifts in official perceptions of groundwater overdevelopment as a result of the recognition of the need to integrate economic and social factors in determining the degree of exploitation. This has come out of observation of field realities. For instance, in certain cases, the regions that are declared "safe" are facing acute drinking water scarcity. Similarly, in certain other cases, such regions are facing long-term declines in water levels (GoI, 2005). This new recognition is also reflected in the methodology proposed by the Ground Water Estimation Committee of 1997. But, how far such concerns are integrated into the actual assessment is open to question. We would take up this issue for further discussion in the subsequent section.

Though the groundwater scientists had the need for maintaining safe yields and sustainable levels of extraction to promote development with minimum negative ecological, economic and social consequences, the manifestations of over-development appear much earlier in certain areas. Thus, such concepts have really not found any place in the practical and policy debates. Part of the reason is the realization that ownership rights in groundwater are not well-defined, well development is highly decentralized under private initiatives, and the government does not have any control over the amount of groundwater that farmers could pump out. In sum, both the estimates based on field manifestations and official data (of recharge and extraction) are static and short-term interpretations of the situation. They do not capture the complex physical characteristics and behavior of aquifer systems, including large static groundwater storage, long-term effects, salinity and water quality issues, leakage from aquitards, system recharge and discharge changes, and uncertainty.

3.2.2 Definitions of groundwater over-exploitation

Several researchers have tried to define groundwater over-exploitation and evolve criteria for assessing degrees of over-development that integrate some of the concerns or considerations discussed early (Kumar & Singh, 2008). The 1986 Regulations of the Public Water Domain of the Spanish Water Act (1985) define overexploitation by its effects: an aquifer is considered over-exploited or at risk of over-exploitation, when the sustainability of existing uses is threatened as a consequence of abstraction being greater than one, or close to the annual mean volume of renewable resources, or when they may produce a serious water quality deterioration problem (Custodio, 2000). Young (1992) defined over-exploitation from an economic point of view, de-linking pumping rates from mean recharge values, as non-optimal exploitation.

Llamas (1992b) introduced the notion of strict over-exploitation leaving room for definitions with a broader scope, such as groundwater abstraction producing effects whose final balance is negative for present and future generations, taking into account physical, chemical, economic, ecological and social aspects.

The Bruntland's (Bruntland et al., 1987) concept of sustainability, which is based on the principle of intergenerational equity, is also used to define groundwater over-exploitation[3] (Custodio, 2000). However, Georgescu-Reogen (1971) and Custodio (2000) have argued that the concept is too broad and cannot be applied to local-specific situations, as it does not take into account the impossibility of complete recycling of matter. Another point of contention of Custodio (2000) is that if one strictly follows the principle of sustainable development, as proposed by the Commission, the non-renewable resources like the large and deep confined aquifers of arid regions yield no benefit to anyone. Thus, there is a need for improving or extending the definition of sustainable development for it to be applicable to aquifers.

Finally, the way over-exploitation is perceived depends on the points of view of different stake-holders involved, such as farmers, water development administrators, ecologists, conservationists, mass media, naturists, and citizens, and professionals such as engineers, scientists, economists, management specialists, environmentalists, lawyers, sociologists and politicians (Custodio, 2000).

For instance, one of the dominant perceptions of the farmers about the consequences of over-development—falling water levels, drying up of wells, etc.—is that it happens due to frequent failure of monsoons and the long-term sharp declines in annual rainfalls, sharply affecting the natural recharge rates. In fact, "declining rainfalls" is a *hydromyth* existing among millions of farmers

---

3. The two major principles of sustainable development are: (1) the rate at which renewable natural resources are exploited should be less than the rate of regeneration, and (2) the waste flow into the natural environment should be kept less than its assimilative capacity.

in the region[4]. Nevertheless, the farmers seem not to see well proliferation and increased groundwater draft as the major factors leading to over-development. On the contrary, they see droughts as a major cause of depletion. Farmers fail to recognize the fact that droughts are not a recent phenomenon but a cyclic phenomenon.

On the other hand, the official agencies claim, with the support of their data of recharge and extraction, that there are no reductions in the quantum of recharge over time[5]. However, here we do not rule out the chances of "bias" in the estimates, as they are often influenced by strong political interests. The direction of such a "bias" could change depending on the kind of vested interest. If the "vested interests" are for drilling more wells, the attempt will be to show lower rates of groundwater level drops and over-estimate the recharge figures. If the "vested interest" is in a large surface irrigation project in an area that has considerable well irrigation, the attempt made would be to overplay the signs of over-development and the unsustainable nature of the present use of groundwater. Custodio (2000) has also mentioned this "bias" and manipulation as an important factor influencing the perception of over-exploitation.

The official perceptions of over-development are driven by "aggregate views". They tend to compare figures of recharge and extraction rates for administrative boundaries or natural boundaries of aquifers. In the process, they miss out on several hidden phenomena, such as excessive draw-downs in water levels due to large well-fields, groundwater pollution, and an excessive rise in water levels causing water-logging, which are often localized. Economists' perception of over-exploitation is often based on consideration of the cost of abstraction of groundwater, including the investment for hitting groundwater and the number of attempts farmers have to make to hit the water table. Whereas, the politician perceives scarcity of groundwater for meeting the basic water needs of the communities as a sign of over-exploitation, and this does not have much to do with the level of groundwater draft against recharge.

In sum, defining and assessing groundwater over-development is both difficult and complex and not amenable to simple formulations (Custodio, 2000).

3.3   Review of existing methodologies for groundwater assessment

During the past four decades, four committees were constituted to propose scientific methodologies for assessment of groundwater development, which

---

4. The arguments about long-term declining trends in rainfall were also contested in the case of Gujarat (Bhatia, 1992). However, a detailed analysis of the time series data on the magnitude and pattern of rainfalls—including the number of rainy days, duration, and intensity—is absent, making it difficult to evaluate the impact of rainfall on groundwater recharge.

5. In fact, the official data for the Sabarmati Basin shows that the recharge had gone up during 1992–97 as compared to the period 1987-91 (Source: based on Kumar et al., 2001; GoG, 1999).

were proposed twice by the Central Ground Water Board of the Ministry of Water Resources. The Groundwater Over-exploitation Committee (1979) was the first committee to be formed. The second committee was constituted in 1984 and named the Ground Water Estimation Committee (GEC-84), and the third one was in 1997 and named the Ground Water Resource Estimation Committee 1997 (GWREC-1997). For our discussions, we would consider the methodologies proposed by GEC-1984 and GEC-1997 only.

3.3.1 GEC-1984 criteria and methodology for assessing groundwater development

GEC 1984 proposed a simple criterion for assessing groundwater development that is based on net groundwater draft against gross groundwater recharge. It proposed two methodologies for assessing groundwater resources for administrative units such as blocks and districts. The first is the "water level fluctuation approach." This is suggested when sufficient numbers of observation wells for monitoring water levels are available within the given administrative unit in question. In this approach, the average annual recharge from precipitation is calculated by the following equation:

$$R_e = \left(A_s * W_f * S_y\right) + D_w \tag{3.1}$$

where S_y is the specific yield of the aquifer, W_f the average water level fluctuation during monsoon, A_s the area of the aquifer, and D_w the pumping during monsoon.

The five-year average of the annual fluctuations in groundwater levels between pre- and post-monsoon time, multiplied by the specific yield values and the geographical area of the aquifer gives the total quantum of recharge.

A major limitation of the GEC-1984 is in the criterion used for assessing groundwater development. At best, it works for simple aquifer units and cannot capture the groundwater dynamics in complex aquifer systems. Water level fluctuation is the net result of recharge, discharge, return-flows, leakage from across the system, lateral inflows and outflows, etc. But, the water level fluctuation approach to estimating recharge—which often uses values of fluctuations in water levels within one or two layers of the aquifer system—does not allow any discounting for the contribution from the existing storage from other layers of the aquifer, which could be very significant in the case of deep alluvial aquifer systems with several layers.

In addition, groundwater contribution to streamflows in the form of base flow is significant in many basins and constitutes the river's lean season flows (Winter et al., 1998). For instance, Kumar (2010) found in the case of the Narmada river basin that in spite of an increase in groundwater draft, the annual rate of decline in groundwater levels had decreased over time. This could be explained by a significant reduction in groundwater outflows into surface streams, resulting from the lowering of water levels. Outflows are losses from the aquifer and reduce the effective annual replenishable groundwater. But GEC-1984 neither did include base flow as a determining factor nor did it suggest a procedure for estimating it. These omissions can lead to an over-estimation of the utilizable groundwater, implying negative consequences for stream flows.

Further, the criterion used in GEC-84 for assessing groundwater development uses aggregate figures for recharge and extraction. However, recharge is often confined to certain layers within the aquifer system—most commonly the upper shallow aquifer. So far as abstraction is concerned, there could be layers of the aquifer system that are tapped but do not get natural replenishment either from rainfall or leakage. As a result, different layers of the aquifer can undergo different degrees of exploitation. It would be different from what the aggregate figures for recharge and abstraction show, and would actually be reflected in the water level fluctuations in the respective aquifer layer. Since the existing methodology treats the entire aquifer system as a single aquifer, it fails to assess the degree of exploitation in the different aquifers under consideration.

Further, a simplified criterion can lead to large errors in the estimation of groundwater recharge. For instance, the approach to estimating recharge also considers the abstraction during the monsoon period (see Eq. 3.1). Though the abstraction could come from more than one layer, the entire amount is attributed to a single recharged aquifer, whose water level fluctuation data are available. The error in the estimation of recharge will be inversely proportional to the contribution of the recharged aquifer in the total abstraction during the monsoon period.

One of the outcomes of using such a simple criterion is that the recharge-abstraction balance of the aquifer often does not correlate with water level trends, a variable that groundwater managers and users are equally concerned with. Maintaining abstraction levels far below the annual recharge does not mean that the draft is within safe limits. There could be a continuous outflow of water into natural drainage systems, due to which water levels can decline. This was seen in the Narmada river basin in Madhya Pradesh. The increase in annual groundwater draft over the years did not result in an increase in the rate of decline in water levels regionally, precisely because with the lowering of water levels below a threshold, the natural discharge of water from the aquifer into the streams was drastically reduced. This was also indicated by the reduction in the stream-flows in the basin without any change in the rainfall. From 1988 to 93 onwards, the average annual charge in regional water levels was positive, unlike during the previous period (Fig. 3.1, Source: Kumar, 2010). On the other hand, a steady recharge-abstraction imbalance does not mean a decline in water levels in the aquifer. The aquifer under study might receive the entire recharge from lateral inflows, as well as from the top, while several overlying aquifers might be contributing to the abstraction from the system. But the criteria used in GEC-1984 are too simplistic to capture the complex hydrological considerations and are therefore not realistic.

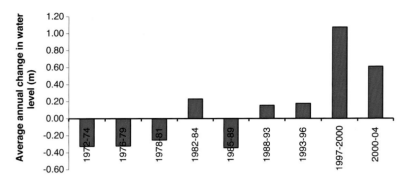

FIGURE 3.1 Average annual groundwater fluctuation at different time periods in Narmada river basin. (*From: Kumar, 2010*).

In the second approach of GEC-1984, the use of ad hoc norms is suggested for the following: (1) recharge from rainfall; (2) recharge due to seepage from unlined canals; (3) return flow from irrigated fields; (4) seepage from tanks; and (5) influent seepage from rivers and streams. Separate norms are used for estimating rainfall recharge for different types of geological formations, such as alluvium, semi-consolidated rocks, and hard rocks.

Return flow from irrigated fields is estimated using the norm of 35% of the irrigation dosage for surface water, 40% for paddy fields irrigated by surface water; 30% of the water delivered at the outlet for well irrigation, except for paddy; and 35% for paddy fields irrigated by well water (NABARD, 2006). The use of such ad hoc norms can invite many sources of error. First: recharge from rainfall is a function of not only the formation geology but also rainfall pattern, soil type, vegetation cover, geo-hydrological environment, and the hydraulic conductivity of the soil in the root zone and below. Again, many regions in India face extreme variability in rainfall and rainy days, and recharge from rainfall is not a linear function of rainfall magnitude. As a result, using normal values of rainfall for recharge estimation can lead to significant errors. Further, for a given crop, return flow from irrigation is a complex function of the total quantum of irrigation water dosage; the irrigation schedule; and agro-hydrological variables that actually determine the return flows from irrigated fields, which are determined by soil hydraulic properties; drainage conditions; agro-meteorology; and crop characteristics (van Dam et al., 2008).

3.3.2 GEC-1997 criteria and methodology for assessing groundwater development

The GEC-1997's criterion for assessing groundwater development is far more realistic and has a better scientific basis than that of GEC 1984. First of all, it proposes an assessment of recharge for monsoon and non-monsoon periods separately. Also, the methodology proposes an analytical approach for the estimation of specific yield using groundwater balance for the non-monsoon period. It also proposes a detailed analytical approach for the estimation of recharge during monsoon using the water level fluctuation approach automatically considering various components of groundwater balance such as storage change, the return flows from irrigation to groundwater, base flows from groundwater into streams and recharge from streams into groundwater, net lateral groundwater inflow into the area, and groundwater draft.

The methodology for estimation of base flow and lateral flows for administrative units, proposed by GEC-1997, however, is not robust. As one would expect, arriving at reasonably accurate figures for these two variables is essential to deduce figures for monsoon recharge. The reason for this is that pre-post monsoon water level fluctuation, which the methodology banks on to estimate monsoon recharge, is caused by storage changes in the groundwater system caused by many inflows and outflows. They include rainfall recharge, net lateral inflows, contribution of stream-flows into the groundwater system, return flows from irrigation, groundwater draft, and base flow into streams.

If the assessment unit is a watershed, a stream gauge station can provide data for the calculation of base flows, and hence the challenges are less. However, only a few states use the watershed as the unit for groundwater assessment, and even in these cases, reliable data on stream-flows are not available, as many lower-order streams are not gauged.

While base flow during the lean season for administrative units is estimated on the basis of groundwater draft during the season, the water level fluctuation and specific yield values, its reliability would depend heavily on the accuracy of the estimation of groundwater draft figures. But these figures are normally estimated using certain ad hoc norms. If data on specific yield is not readily available, it can be estimated using groundwater balance by taking watershed as the unit, which would again involve the use of groundwater draft estimates for the watershed. In a nutshell, the estimation of base flow would involve a lot of errors. This is evident from the groundwater resource assessment for Madhya Pradesh provided by the Central Ground Water Board (GoI, 2005) for the year 2005. It shows that the total groundwater outflow during the lean season is only 1860 MCM. This is a sheer underestimation, when we look at the total amount of lean season flow (from December to May) in just one of the many river basins of Madhya Pradesh, i.e., Narmada alone is 1653.22 MCM (Kumar, 2010), and many perennial rivers originate from the region.

Again, the figures of recharge so obtained include recharge from irrigation, water harvesting structures, and return flows as well. Here again, no scientific methodology is employed for estimating recharge from irrigation return flows. Instead, some modified versions of the earlier norms of 1984 are used. The norm of return flow as a percentage of irrigation dosage changes according to the depth to groundwater table. For areas with a water table higher than 25 metres, the norm is 5% of irrigation dosage; for water table depth between 10 m and 25 m, the norm is 10% of irrigation dosage, and for depth to water table less than 5 m, is taken as the recharge from irrigation return flows, against 40% and 35% considered in the earlier methodology (NABARD, 2006). In intensively canal-irrigated areas of Punjab, Haryana, Uttar Pradesh, Andhra Pradesh, and Maharashtra, the use of such methodologies can lead to highly erroneous estimates of not only return flows but also net natural recharge.

For estimating recharge from water harvesting structures, a uniform rate of 1.4 mm/day is assumed for tanks and ponds, and the total recharge is estimated on the basis of the average area of the pond (NABARD, 2006). But, in hard rock areas of Peninsular India with a large number of tanks and ponds, such assumptions can be unrealistic and can lead to an over-estimation of recharge from recharge structures. The reason is that the sustainability of recharge from tanks and ponds depends on the hydraulic diffusivity of the aquifers, which is very poor in hard rock areas. The creation of a recharge mount in the aquifer underlying the recharge structure can prevent further percolation of water (Muralidharan & Athawale, 1998).

The criterion suggested by GEC-1997 for assessing the stage of groundwater development involves gross groundwater draft against net recharge. The net groundwater recharge takes into account the losses from the groundwater system and net gains from lateral flows. This is a major departure from the earlier methodology that considered net draft against the gross recharge. As Kumar and Singh (2001) note, such an approach had led to over-estimation of recharge and under-estimation of draft, as recharge from irrigation return flows is double counted. Hence, the new methodology reduced the anomalies caused by this. But, when it comes to estimating the net groundwater recharge, neither the state groundwater departments nor the Central Ground Water Board considers the groundwater losses while estimating the net recharge in lieu of the fact that these hydrological variables are difficult to quantify. This was well-acknowledged in a review of the existing methodologies for groundwater assessment carried out by NABARD (NABARD, 2006). Nevertheless, it is not complex enough to realistically assess groundwater development in multi-aquifer systems, where aquifers that get replenished and aquifers that are subject to hydrological stresses could be different.

GEC-1997, however, recommended that hydrological data on recharge and abstraction estimated for the administrative units be integrated with data on mid-term and long-term trends in water levels to make the final assessment of the stage of groundwater development in areas that are showing a continuous decline in groundwater levels. But this is hardly done in actual practice by central and state agencies.

3.3.3 Deficiencies of GEC-84 and GEC-97

The biggest challenge posed by the methodologies proposed under both GEC-84 and GEC-97 is in estimating the specific yield values of aquifers, to which the recharge estimates are highly sensitive. The issue is very crucial for hard rock areas, as specific yield values could vary widely within small geographical areas

(NABARD, 2006). While groundwater balance during the non-monsoon period can be used to estimate specific yield, this would require realistic estimates of groundwater draft during the season. Lack of reliable data on groundwater draft is a major factor affecting the reliability of the entire exercise of assessing the stage of groundwater development, as the inaccuracy in the estimation of this parameter increases the inaccuracy in the estimation of both the denominator and the numerator.

Both committees proposed three methods for estimating groundwater draft: (1) using the well census and the norm of annual draft for different types of wells; (2) using electric power consumption, and the estimate of the quantity of water pumped per unit power consumed; and (3) using groundwater—irrigated area under different crops and the water requirement for each one. All three of these approaches suffer from inadequacies. As regards the first one, it is hard to get the exact number of operational wells in a region at a given point in time, especially in hard rock regions, due to the increasing incidence of well failures (Kumar et al., 2012) and farmers owning many wells at a time. Further, when it comes to quantifying the amount of water abstracted, wide variations in well outputs are seen within regions and over the seasons in hard rock areas (Kumar, 2007). In the case of the second method, it does not take into account the water abstraction by diesel wells that are in operation in many shallow groundwater areas in Odisha, Uttar Pradesh, Bihar, Assam, West Bengal, Madhya Pradesh, and some parts of Gujarat, Kerala, and Andhra Pradesh. As regards the third one, the challenge is in getting reliable data on the groundwater-irrigated area under different crops, as data on source-wise gross irrigated area is not compiled by most state governments.

In a nutshell, both GEC-84 and GEC-97 suffer from problems in the criteria used for assessing the stage of groundwater development. The criteria used in GEC-1984 are only physical, involving only simple hydrological variables such as groundwater recharge and abstraction. Whereas in GEC-1997, the criteria used are a little more complex, with the inclusion of base flows and lateral flows and the replacement of gross recharge by net recharge and net draft by gross draft. However, none of them involve complex hydro-dynamic, economic, social and ecological variables that help determine the negative consequences of groundwater over-exploitation. Some of them are long-term trends in water levels, depth to water table, cost of abstraction of groundwater, and availability of water in wells during the lean season. This aside, there are issues of reliability in the estimation of groundwater recharge and draft.

3.4   Are groundwater over-exploitation problems serious in India?

The first set of alarms about groundwater over-exploitation were raised almost three decades ago based on observations for selected locations in India, including Mehsana in north Gujarat, coastal Saurashtra and Kachchh, Coimbatore in Tamil Nadu, Kolar in Karnataka, and Jaipur in Rajasthan. Several scholars

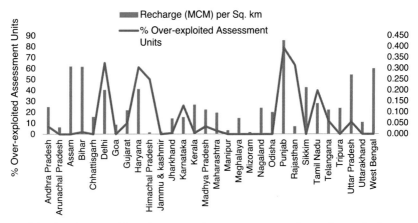

FIGURE 3.2   Net renewable recharge per unit area versus stage of groundwater development (% over-exploited assessment units).

had looked at the problem of groundwater depletion from many disciplinary angles (see Dhawan, 1997; Janakarajan, 1994; Moench, 1995; Phadtare, 1988). Dewas district, located in the Narmada valley in Madhya Pradesh, was another region about which a lot had been written (see, for instance, Shah et al., 1998). Over the years, several new regions have been classified as falling under the "over-exploited" category. Punjab is one such region—where many blocks were shown as experiencing falling water table conditions. There has been a lot of whistleblowing about the impending groundwater crisis in many arid and semi-arid regions based on anecdotal evidence from some of these regions on groundwater level trends. Over the years, the number of safe assessment units in the country had reduced, as per the official estimates, from 71.2% in 2004 to 63.5% in 2017 (CGWB, 2019, p. 2).

But, if one goes by the official estimates of groundwater development in 2017 from the CGWB, 248.7 BCM out of the 392 BCM of extractable groundwater in the country is currently utilized (CGWB, 2019). Again, if one goes by the most recent disaggregated data, only 17.2% of the groundwater basins in the country are over-exploited; 4.5% are critically exploited, and 14.1% are in the semi-critical stage. Nearly 64.2% of the groundwater basins are still "safe" for further exploitation (CGWB, 2019). Interestingly, as per the official statistics, Punjab and Delhi, which are rich in groundwater (in terms of utilizable recharge per unit area), are two of the states where over-exploitation, according to official estimates, is most serious. They are followed by Rajasthan, Haryana and Tamil Nadu. Strangely, the number of over-exploited assessment units is low in many states with scarce groundwater (such as Telangana, Karnataka, Maharashtra, and Andhra Pradesh) and a high incidence of well failures (see Fig. 3.2). Such a situation is quite unlikely given the fact that these states (viz., Karnataka, Andhra Pradesh, Maharashtra and Telangana) have excessively high agricultural water demands, and the pressure on groundwater to meet this demand is very high. Hence, these are big anomalies that need to be rectified through refinements in the assessment methodology.

Nevertheless, the doomsday prophecies were not based on a rational assessment of the scenario using data on hydrological changes and hydrodynamics. Collin and Margat (1993) have argued that this is an unconscious or incited over-reaction to a given situation, while Custodio and Llamas (1997) and Llamas (1992a) assert that this is the result of deeply entrenched "*hydromyths*." Custodio (2000) further opines that the groundwater developers take the opposite position, which focuses on "beneficial use" and uses the concepts of safe yield rational exploitation, and the economics side of sustainable development, to present their viewpoints. This is not to say that groundwater over-exploitation is not a cause for concern in India. The moot point here is that such anomalies in the assessment, which are probably due to problems with the assessment criteria and methodologies, were hardly ever raised by researchers and scholars. In the subsequent section, we will examine how far these "doomsday prophecies" are correct.

3.4.1   What do water level trends really mean?

Groundwater level trends are a net effect of several changes taking place in the resource conditions, owing to recharge from precipitation, return flows from irrigated fields, seepage from water carriers (canals, channels, etc.), abstraction or groundwater draft, lateral flows (either inflow or outflow) or outflows into the natural streams (Taylor & Alley, 2001; Todd and Mays, 2005, pp. 218–229).

Water-level measurements can only provide information about the hydrologic stresses acting on aquifers and how these stresses affect groundwater recharge, storage, and discharge (Taylor & Alley, 2001). In a region where groundwater pumping on a long term basis remains less than the average annual recharge, the groundwater levels can experience short-term declining trends as a result of a drastic increase in groundwater pumping owing to monsoon failure. But such a phenomenon does not represent long-term trends. It is important to note here that semi-arid regions in our country also experience significant inter-annual variability in rainfall (source: based on Pisharoty, 1990; Kumar, 2010).

In a region with static groundwater resources, even during the monsoon, the aquifer can witness depletion, depending on how recharge and abstraction are balanced—a phenomenon that does not occur in hard rock areas experiencing over-exploitation, owing to the absence of groundwater stock and unutilized recharge from the previous year. The point is that in such situations, abstraction during the monsoon cannot exceed the annual recharge.

A comparative analysis of average long-term water level trends in Ludhiana district in Punjab and Ahmednagar district in Maharashtra illustrates this point. In the case of Ludhiana, Punjab, which is underlain by deep alluvial formations, in many years, the (spatial) average water level fluctuation during monsoon

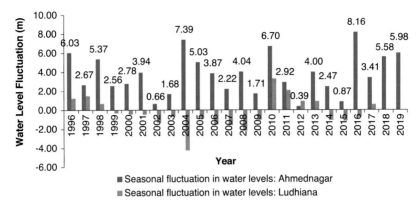

FIGURE 3.3   Water level fluctuation during monsoon in Ahmednagar and Ludhiana.

was negative, whereas in Ahmednagar, which has basaltic terrain, in none of the years, was the monsoon water level trend negative (see Fig. 3.3). In the case of Punjab, in low rainfall years, as the irrigation demand increased owing to heavy pumping during the monsoon season exceeding the limited monsoon recharge, water levels dropped. The additional water came from the stock in the aquifer. Long-term and consistent decline in water levels is encountered in such regions. On the contrary, in Aurangabad, even during drought years, the water level fluctuation was positive.

The mean value of monsoon water level fluctuation for Ludhiana was −0.29 m for the 22-year period, whereas in the case of Ahmednagar, the average was 3.77 m for the 24-year period. Though the water level fluctuation during monsoon indicates the net storage change (monsoon recharge-withdrawal), the relatively higher fluctuation in water levels in Ahmednagar (in high rainfall years when pumping during the monsoon season is unlikely) is because of the low specific yield of the aquifer (water level fluctuation = net storage/specific yield).

Further, it is not correct to attribute all changes in groundwater conditions to hydrological stress induced by human action. In a region where groundwater out-flows into the surface streams are quite large due to the peculiar geo-hydrological environment (Winter et al., 1998), even if the net annual groundwater draft is far less than the net recharge, water levels can decline on an annual basis, as illustrated by the study of surface-water groundwater interactions in the Narmada river basin. In such situations, increasing draft over time can actually reduce the rate of decline in water levels on a long time horizon (Kumar, 2010). As pointed out by Taylor and Alley (2001) of the US Geological Survey, the velocity of flow of water in the aquifer (natural discharge and lateral flows) along with the recharge rates should be two of the factors determining the frequency of monitoring of observation wells (Taylor & Alley, 2001, p. 12). In fact, this is the situation prevailing in many river basins of Central India, such as the Mahi, Tapi, Krishna, Mahanadi and Godavari. Such situations also prevail in the Western

Ghats and north-eastern hilly regions. This means that integrating environmental considerations, such as maintaining lean season flows in rivers, would limit the safe abstraction rates, to levels much lower than what is permissible on the basis of renewable recharge in such areas. Hence, in such regions, estimating the base flows would be very crucial in arriving at the net utilizable recharge and therefore the actual stage of development of groundwater. We have already seen that the groundwater outflows are not properly accounted for in the estimates of the net recharge. Due to this reason, the estimates would show a much lower stage of development than what the region is experiencing.

3.4.2   How do we look at groundwater balance for assessing over-draft?

Ideally, in a region where lateral flows and outflows from groundwater systems are insignificant, groundwater "over-draft" can take place if the total evapotranspiration (ET) demand for water per unit area is more than the total effective rainfall, i.e., the portion of the rainfall remaining in situ after runoff losses, and the amount of water imported from outside perunit area. In many semi-arid to arid regions of India, cropping is intensive, demanding irrigation water during the winter and summer months. The ET demands for crops are much higher in comparison to the effective rainfall (Kumar & Singh, 2008).

Alluvial Indian Punjab, which is under very intensive cropping with paddy and wheat, is a good example of this (Kumar et al., 2021). The deficit has to be met either from local or imported surface water or groundwater pumping (Kumar & Singh, 2008). As a matter of fact, in Punjab, the imported water to meet the deficit is to the tune of approximately 17,000 MCM per annum (Kumar et al., 2021). Hence, the reduction in groundwater storage would be the imbalance between the total recharge from rainfall and return flows from irrigation and groundwater draft. In semi-arid and arid regions, natural recharge from precipitation is generally very low. In an area with intensive surface irrigation, a negative balance in groundwater indicates high levels of over-draft or a deficit in effective rainfall for meeting the ET requirements.

The semi-arid to arid climatic conditions and the intensive cropping make ET per unit area very high in Punjab. Again, since the rains last for very few months, the effective rainfall is low. The water levels are falling throughout Punjab at a rate of 0.3 metres per annum (Hira & Khera, 2000). Let us examine the groundwater balance in an ideal situation like in Punjab. The change in groundwater storage (Δ_s) could be written as:

$$R_{rech} + RF_1 - NGD$$

But,

$$NGD = ET + \Delta_{dep} - (S_1 + P_e - RF_1)$$

Therefore,

$$\Delta_s = R_{rech} + RF_1 - \left\{ ET = \Delta_{dep} - [S_1 + P_e - RF_1] \right\}$$

$$\Delta_s = R_{rech} + S_1 + P_e - ET - \Delta_{dep}$$

where R_{rech} is rainfall recharge; RF_1 is irrigation return flow; NGD is the net groundwater draft; Δ_{dep} is the total of water depleted from the soil during the fallow period and the water stored in the soil profile below the root zone; S_1 is the surface irrigation water applied; and P_e is effective rainfall.

Going by the above groundwater balance equation, if S_1 (i.e., surface water input) is removed, then the change in groundwater storage would become negative if the entire land is cultivated, which is the condition almost throughout Punjab. This is because rainfall (P) is less than ET requirement, and as a result, P_e+ Recharge is also less. This is because "P_e+ Recharge" would always be less than the total rainfall (P). Hence, surface irrigation's role in maintaining groundwater balance is greater than that of the return flows from it and equals the actual amount of surface water applied. This also means that if water levels are falling even with canal irrigation inputs, then the storage depletion and drop in water levels without exogenous water inputs would be much larger (source: Kumar, 2007).

3.4.3 How do geological settings matter?

The geological conditions under which drops in water levels occur are also important in assessing the extent of groundwater over-draft conditions. Here, the distinction between a hard rock area or area underlain by consolidated formations and an area underlain by an unconsolidated (alluvial) or semi consolidated (sedimentary) formation is important (Kumar & Singh, 2008). Many semi-arid and arid areas in the country fall under hard rock conditions. Examples are Peninsular India (except the Western Ghats), Saurashtra in Gujarat, western parts of MP, almost the entire Maharashtra and most parts of Odisha (CGWB, 2019). In these regions, the specific yield of aquifers is very small—from 0.01 to 0.03. Large seasonal drops in water levels are a widespread phenomenon in these areas. During monsoon, a sharp rise in water levels is observed and after the monsoon rains, water levels start receding. Many open wells get dried up during summer. Often, the drop in water levels between pre- and post-monsoon is in the range of 5–6 m. So, one should make a clear distinction between seasonal depletion and annual depletion.

Further, in hard rock areas, a unit volume of groundwater pumped from the aquifer results in up to 12–13 times the annual drawdown that occurs in alluvial areas for the same amount of over-draft (Kumar et al., 2022). A fall in water level of one metre in alluvial Punjab should be a cause for much greater concern than a one metre fall in water levels in hard rock areas of Tamil Nadu, or Saurashtra or Karnataka given the fact that the specific yield of alluvium in Punjab is in

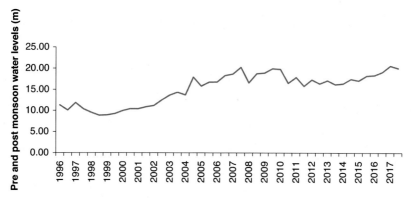

FIGURE 3.4   Long-term trends in water levels in Ludhiana. (*From: India-WRIS*).

the range of 0.13–0.20. As Fig. 3.4 indicates, the long-term decline in Ludhiana district of Punjab is not very high (less than 10 m over a period of 24 years), in spite of the fact that the district has remained in the category of "over-exploited" district for many decades. The average annual rate of decline is only around 0.40 m, which corresponds to a net negative groundwater balance of 0.052 m to 0.08 m (i.e., 50 to 80 mm).

This is not to say that the magnitude of the water level drop is not important. In fact, a sharp fall in water level would also have serious implications for the investment required for pumping groundwater and also the efficiency with which groundwater could be abstracted. Hence, what is more important is at what rate water levels fall on a long-term basis.

3.4.4   Integrating negative consequences of over-exploitation in assessing groundwater development

As Custodio (2000) notes, there are many complex considerations involved in assessing groundwater over-exploitation in terms of various undesirable consequences. They are hydrological, hydro-dynamic, economic, social, and ethical in nature. However, some of the most important ones are: the groundwater stock available in a region; water level trends; net groundwater outflows against inflows; the economics of groundwater intensive use, particularly irrigation, which take the lion's share of the groundwater in most semi-arid and arid areas; the criticality of groundwater in the regional hydro-ecological regime; and the ethical aspects and social impacts of groundwater use. Let us examine how the use of these complex considerations in assessing groundwater over-draft would change the groundwater scenario in India.

First of all, as regards the groundwater stock, a region with a huge amount of static groundwater resources may experience over-draft conditions, with a resultant steady decline in water levels. The region that can be cited is the alluvial

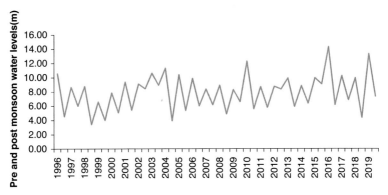

FIGURE 3.5 Long-term water level trends of Ahmednagar, Maharashtra. (*From: India-WRIS*).

plains of the Ganges, whose groundwater stock is many times more than the average annual replenishment (source: based on GoI, 1999). In such regions, assessing over-draft conditions, purely in terms of average annual pumping and recharge, may not make sense. In such regions, the long-term sustainability goal in groundwater use can be realized even if one decides to deplete certain portions of the static groundwater resources along with the renewable portion annually (Custodio, 2000). Limiting groundwater use to renewable resources with the aim of benefiting future generations can mean foregoing large present benefits.

As regards the influence of water level trends, a region may not experience over-draft when pumping is compared against recharge. But, partial well failures could be an area of concern due to the seasonal drops in water levels. Such steep seasonal drops in water levels are characteristic of hard rock areas, as reported by Kumar and Singh (2008) based on data from the hard rock areas of Maharashtra. The most recent data from Ahmednagar district in Maharashtra state also shows similar trends. The water level data of open observation wells obtained from the India-WRIS were analyzed to understand the dynamics of water level in the shallow aquifers (Fig. 3.5).

Fig. 3.5 shows the graphical representation of the long-term trends in average (depth to) water levels covering pre- and post-monsoon conditions in Ahmednagar. As it is clear from Fig. 3.5, there was no significant long-term trend in water levels. But, there was a very sharp seasonal fluctuation in water levels, with water levels rising during the monsoon, followed by a sharp fall after the monsoon. The largest rise in water level during monsoon was 8.16 m, and the lowest was 0.39 m (Fig. 3.5). There was not a single year in which the water level fluctuation during monsoon was negative.

As noted by Kumar et al. (2001), as per official estimates, many such regions are still categorized as "white and grey," though these areas face severe groundwater scarcity during summer. Fig. 3.6 shows the data on wells that have permanently failed and wells that have temporarily not been in use, as available from the 5th minor irrigation census of 2013–14. The total number of failed

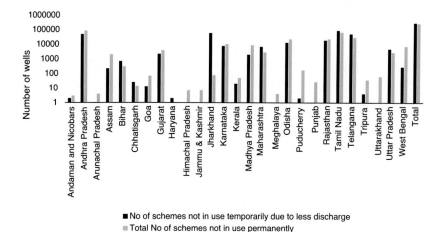

FIGURE 3.6   Number of wells that have failed, and number of wells that are temporarily out of use in different Indian states. (*From: Minor Irrigation Census 2013–14*).

wells includes both wells that have "permanently gone dry" and wells that are "temporarily not in use". The second category essentially refers to wells that are seasonal due to seasonal depletion of groundwater.

The data shows that in the states that are mostly underlain by hard rock formations, both the percentage of wells that have failed and those that are not in use are high. For instance, in Andhra Pradesh, even as per 2017 official data, the stage of groundwater development was only 44.15% (CGWB, 2019). But a large number of wells were reported to have failed (84511) in the state as per the 5th minor irrigation census. Similar trends are found in Tamil Nadu and Rajasthan as well. In Tamil Nadu, the total number of failed wells was 65,718, while the overall groundwater development was only 80.9%. As per the census, a total of 263,554 wells have permanently gone out of use in the country, either due to drying up or due to an increase in salinity or due to industrial pollution, though the drying up of wells due to lowering water levels is the primary cause of well failures reported (GoI, 2017).

Similarly, the current district-wise assessment of groundwater development does not take into account the long-term trends, as the latest methodology suggests. A region might have experienced a long-term decline or rise in water levels, but a few years of abnormal precipitation (either drought years or wet years) may change the trends in the short run. Hence, assessment of over-draft conditions should integrate hydro-dynamics, i.e., the way groundwater levels behave.

Another dimension of groundwater over-exploitation is economic. The cost of production of water should not exceed the benefits derived from its use or the cost of the provision of water from alternative sources. Drops in water levels beyond a certain limit cause negative economic consequences by raising the cost of abstraction of a unit volume of water, not only in irrigation but also

TABLE 3.1 Reduction in average command area of wells over time in Narmada Basin, Madhya Pradesh.

Name of district falling in Narmada Basin	Average area irrigated by a well in Ha					
	1974–75	1980–81	1985–86	1991–92	1995–96	2000–01
Balaghat	4.50	2.25	2.35	2.57	1.73	1.96
Chhindwara	4.56	2.58	2.26	1.42	1.50	1.75
Shahdol	2.04	0.18	0.50	0.70	0.99	0.47
Jhabua	2.93	1.87	0.89	1.20	1.26	0.57
Betul	6.97	3.37	3.02	1.98	2.06	2.18

From: Authors' own estimates based on primary data as provided by Kumar (2007).

in other sectors like municipal uses. Though there could be plenty of water in the aquifers, the fixed cost and variable costs of abstraction of water could be prohibitively high. In alluvial north and central Gujarat and arid Rajasthan, groundwater irrigation is viable due to heavy electricity subsidies. An analysis by Kumar et al. (2001) in the Sabarmati river basin of north-central Gujarat showed that groundwater irrigation would be economically unviable if the full cost of energy used for pumping groundwater was borne by the farmers.

In many hard rock areas underlain by basalt and granite, the highly weathered zones in the geological formations, which yield water, have small vertical extents-up to 30 m. When the regional groundwater level drops below this zone, farmers would be forced to dig bore wells tapping the zone with poor weathering. The reason is that tapping groundwater from strata below this depth using open wells would be not only technically infeasible but also economically unviable. These bore wells have poor yields. For instance, Fig. 3.6 shows that as many as 86,760 failed borewells exist in the case of Tamil Nadu, which is largely underlain by crystalline hard rock formation. A little more than 58,000 wells in Jharkhand suffer from poor yield. The figure is 48,503 wells (that are in use) in AP.

Withdrawal of groundwater from these bore wells creates excessive draw-downs as the specific yield and transmissivity values of these hard rock formations are very low. Due to excessive draw-downs and high well interference, well failures become widespread. Therefore, before a farmer hits water in a successful bore well, he/she would have sunk money in many failed bore wells. Due to this reason, the actual cost of abstraction of groundwater becomes extremely high. The command area of wells is also on a downward trend. For instance, in the case of five districts falling in the basaltic area of the Narmada river basin in Madhya Pradesh, the average command area of energized wells was found to be declining almost consistently from 1974 until 2000 (see Table 3.1). In Betul district, the average area irrigated by a well decreased from 6.97 ha to 2.18 ha during the 26 year period. In Chhindwara, it decreased from 4.56 ha to 2.75 ha.

So the investment for well construction, compounded by the reduction in command area reduces the overall economics of well irrigation. However, this aspect has not been captured in the criteria for assessment of over-exploitation. As per the official data, these five districts are still in the safe category and are safe for further exploitation (CGWB, 2019).

Interestingly, the economics of groundwater use is not a function of depth to water table alone, as is often perceived. As pointed out by Kumar and Singh (2008), even in areas with shallow water table conditions, the cost of abstraction could be very high due to the high cost of energy. In Bihar, due to poor rural electrification (GoI, 2017), farmers are forced to use diesel and kerosene pumps for lifting water from wells. Though the depth to water table is nearly 15–20 feet, it costs them Rs. 50 per hour for pumping water with an output of nearly 15 liters per second. The unit variable cost comes to Rs. $1/m^3$ of water. This is higher than the variable cost farmers incur in north Gujarat (Rs. $0.50/m^3$) for pumping out water from a depth of 400–500 feet.

The economics of groundwater use is not static. The economic viability of groundwater abstraction can change under two circumstances: (1) opportunities for using pumped water for more productive uses emerge with changing times; and (2) the cost of abstraction of groundwater changes due to improvements in pumping technologies or changing cost of energy for pumping groundwater. With massive rural electrification, the cost of groundwater abstraction in Bihar could come down to negligibly low levels. On the other hand, adoption of new high-yielding varieties or high-valued crops can increase the gross returns from farming.

The social consequences of groundwater use are equally important. One serious issue associated with groundwater intensive use is that it excludes resource-poor farmers from directly accessing the resource when water levels begin to fall. Equity in access to the resource (aquifer) should be an important consideration in assessing the degree of over-exploitation of aquifers. In many areas, it is only the rich farmers who are able to pump groundwater, owing to the astronomical rise in the cost of digging/drilling wells, and they enjoy unlimited access to the resource. While the well owners of Mehsana incur an implicit cost of nearly Rs. $0.5/m^3$ of water[6], they charge the buyers to the tune of Rs. $1.5/m^3$ to Rs. $2/m^3$ (Kumar & Singh, 2008). Even when the farmers of all land holding classes are able to continue accessing water from wells in spite of resource degradation, the cost burden is not uniform across them. The direct and indirect costs borne by small and marginal farmers were much higher than those borne by the large farmers, as shown by a study in the hard rock areas of erstwhile Andhra Pradesh (Reddy, 2005, p. 552). In many areas, groundwater intensive use leads to water quality

---

6. This is based on the capital investment of 10 lac rupees amortized over the life of the tube well (12 years), the annual operation and maintenance costs of Rs. 50,000, and the average volume of 2.5 lac cubic metres of water pumped during a year at a rate of 100 m^3/hour.

deterioration, causing scarcity of safe water for drinking. In such situations, the draft does not necessarily exceed the recharge. Examples are the Saurashtra and Chennai coasts, alluvial north and central Gujarat, and the Gangetic alluvium of West Bengal. While the issue is of salinity in coastal Saurashtra and Chennai, it is arsenic content in deep aquifers in West Bengal (CGWB, 2010).

Groundwater over-use, like the use of other natural resources, involves ethical considerations (Custodio, 2000). The ethical considerations concerning water use mainly revolve around the distribution of benefits and costs of water use and the risks associated with it (Llamas & Priscoli, 2000). The extent to which wasteful use practices are involved in major sectors of water use and the degree to which water abstraction practices reduce the opportunities of other users— neighboring farmers, the individual himself, and others—are the major issues to be investigated (Kumar et al., 2001). In a water-scarce region, physically and economically inefficient uses should be discouraged. But in reality, even in regions where acute scarcity of groundwater exists, farmers use wasteful methods of irrigation and allocate water to economically inefficient uses in terms of the value of the crop outputs per unit volume of water (source: based on Kumar, 2005; Deepak et al., 2005). In hard rock areas, competitive drilling by powerful farmers causes a reduction in the yield of neighboring wells due to well interference, depriving resource-poor farmers (Janakarajan, 2002; Deepak et al., 2005). In sum, the current method of assessment of groundwater over-exploitation does not give a clear picture of the actual degree of over-exploitation in both absolute and relative terms. It tends to underestimate the problem in India, which needs to be assessed from the points of view of negative social, economic and ecological consequences.

3.5 Summary and conclusion

There are many conceptual issues in defining groundwater over-development (Custodio, 2000). Similarly, there are complex considerations involved in the analysis of long-term water level data (Taylor & Alley, 2001). First of all, the concept of "groundwater over-development" or "aquifer over-exploitation" is complex and linked to various "undesirable consequences", which are physical, social, economic, ecological, environmental, and ethical in nature. Further, there are varying perceptions of the undesirable consequences. The reasons are as follows: varying perceptions of people concerned; the terms used to define over-exploitation vary with space and time; difficulty in calculating aquifer recharge and integrating water quality with quantity; difficulty in assessing long-term trends in recharge rates that are very important; importance of localized effects in the overall picture; changing social perceptions and priorities; improvements in water use technology; the need to consider the net socio-economic ben-efits; the complex nature of cost-benefit calculations; and the use of scarce, poor, and inappropriate data to define over-development (Kumar & Singh, 2008).

The criteria adopted by the CGWB for assessment of groundwater over-development are inadequate for complex aquifer conditions and at best give aggregate scenarios of recharge and abstraction for simple aquifer conditions. Such simple conditions exist with homogeneous alluvial aquifers under flat terrains (like in central Punjab). But there were some improvements in the criteria and methodologies proposed by various expert committees since 1984. The most significant improvement in the criteria is the inclusion of base flows and lateral flows to determine the net groundwater recharge. The second significant improvement is in the criteria for assessing the stage of groundwater development. GEC-1997 suggests the use of gross groundwater draft against the net recharge instead of the net draft against the gross recharge. It had recommended the inclusion of all hydrological variables in the estimation.

But the methodology proposed for the estimation of these variables becomes deficient when the assessment is to be made for administrative units. This is because of the following: (1) absence of an analytical procedure for estimation of base flows during the monsoon season; and (2) questionable reliability of the estimates of lean season base flows. Even the CGWB's own assessment of groundwater development takes into account only base flows during the lean season (estimated). So, all these can induce major errors in the estimation of recharge. But, the challenge becomes mounting when data on the specific yield of aquifers is not available. When an assessment is to be made for watersheds, the gauge data of stream flows can be used to estimate the base flows during monsoon. Also, the lean season flow can be directly measured.

The issue of getting reliable estimates of recharge and abstraction notwithstanding, the criteria for assessing aquifer over-exploitation in India are too simplistic, involving "net recharge" and "gross draft." They do not take into account the complex hydrological, geological, hydro-dynamic, economic, social and ecological variables that determine the physical, social, economic, ecological and ethical consequences, such as the safe yield of the aquifer, drinking water scarcity during the lean season, poor economics of groundwater use, water quality deterioration, equity in access to water, and the efficiency with which water is used.

Using selected illustrative cases, we have demonstrated how combining official statistics of the status of groundwater development in the country with information on detailed water balance, geology, water level fluctuations, and socio-economic, ecological and ethical aspects would cast an altogether different scenario of the degree of over-exploitation problems. The available assessments of groundwater over-exploitation provide a highly misleading picture of the groundwater exploitation scenario in India. As per the most recent official estimates of the Central Ground Water Board (CGWB, 2019), many hard rock areas where wells temporarily go out of use or wells permanently fail (as reported by the 5th Minor Irrigation Census: GoI, 2017) are shown as "safe" areas. Many areas in central India, which are facing problems of yield decline, are still categorized as under-developed areas. Such anomalies between official estimates and the ground situation need to be rectified through refinements in the assessment methodologies.

To get an assessment of aquifer over-exploitation that reflects the concerns of the stakeholders, two important steps are to be taken. The first step is improving the reliability of groundwater recharge and withdrawal estimates. The most important challenge is the accurate estimation of groundwater draft and natural outflows from and lateral flows in the groundwater system. With the easy availability of good quality data on water level fluctuations from the observation wells for considerably long time periods, the accuracy in the estimation of groundwater recharge would depend much on the availability of reliable data on specific yield values for the aquifer under consideration. The second step is broadening the criteria for assessing aquifer over-exploitation to capture the complex hydrologic, hydro-dynamic, economic, social and ecological variables that reflect the negative consequences of over-development. For this, a lot of data on socio-economic and ecological aspects of water use needs to be generated and combined with the official data.

Besides this, as our analyses of data from Ahmednagar and Ludhiana districts and the Narmada river basin in Madhya Pradesh have shown, the interpretation of long-term data on water levels collected by the CGWB from the many thousands of observation wells needs to factor in the differences in characteristics of the aquifers (i.e., whether consolidated or unconsolidated or semi consolidated, whether there is static ground water or not, and whether the terrain is undulating or plain). The reason is that water-level trends are the net effect of the hydrologic stresses acting on aquifers (draft and infiltration of rainwater or snow melt) and their effect on the processes of groundwater recharge, storage, and discharge, and the conditions that affect these processes change from region to region.

While many other countries in South Asia also extensively and intensively use groundwater, there are no standard criteria or methodologies used by the official agencies in those countries to assess the status of groundwater development. In Pakistan and Bangladesh, where groundwater use for irrigation is very large, the official agencies currently use the trends in water levels and water quality based on data from the observation wells to assess the nature and magnitude of groundwater problems (such as groundwater depletion, excessive rise in water levels and an increase in groundwater salinity). Thus, the criteria and indicators used are purely technical. They are also narrow in the sense that such data are inadequate to capture the complete picture with regard to changes in groundwater quantity and quality in different aquifer layers and the factors driving those changes. More importantly, the socioeconomic and ecological impacts of groundwater overuse are not considered in the assessments.

For instance, as is evident from the discussions in Chapter 2, the impact of lowering water levels on the cost of well irrigation is quite significant in both Pakistan and Bangladesh. In the same way, an increase in salinity of groundwater resulting from reducing recharge of shallow aquifers due to limited surface supplies for irrigation in the tail reaches of canal networks is a cause for serious concern in Pakistan. This is because the farmers use poor quality groundwater for irrigating their crops in the absence of good quality surface water, causing soil salinity. There is also a problem of increasing groundwater salinity in the lower parts of the Indus basin due to seawater intrusion caused by excessive withdrawal for irrigation (Qureshi, 2020). There is an imminent need to develop comprehensive criteria for assessment of groundwater in these countries that are capable of capturing the effect of water level decline on the cost of well irrigation (in both Bangladesh and Pakistan) and the ecological effects of groundwater irrigation (mainly in Pakistan).

References

Bhatia, B. (1992). Lush Fields and Parched Throats: Political Economy of Groundwater in Gujarat. *Economic and Political Weekly, XXVII* (51–52): A 12–170.

Bruntland, G. M., et al. (1987). Our Common Future. *World Commission on Environment and Development*. Oxford University Press.

Central Ground Water Board. (2010). *Ground water quality in shallow aquifers of India*. Central Ground Water Board, Ministry of Water Resources.

Central Ground Water Board. (2019). *National compilation on dynamic ground water resources of India 2017*. Central Ground Water Board, Dept. of Water Resources, RD & GR, Ministry of Jal Shakti.

Central Water Commission. (2017). *Reassessment of water availability in India using space inputs*. Basin Planning and Management Organization, Central Water Commission.

Collin, J. J., & Margat, J. (1993). Overexploitation of water resources: Overreaction or an economic reality? *Hydroplus, 36*, 26–37.

Custodio, E. (1992). *Hydrological and hydrochemical aspects of aquifer overexploitation*. Selected papers on aquifer overexploitation. *International Association of Hydrogeologists, 3*, 3–28.

Custodio, E. (2000). *The complex concept of over-exploited aquifer (Secunda Edicion)*. Uso Intensivo de Las Agua Subterráneas.

Custodio, E., & Llamas, M. R. (1997). Consideraciones sobre la nergiz y evoluión de ciertos "hidromitos" en España. *En Defensa de la Libertad: Homenaje a Victor Mendoza* (pp. 167–179). Instituto de Estudios Económicos.

Deepak, S. C., Chandrakanth, M. G., & Nagaraj, N. (2005). *Water demand management: A strategy to deal with water scarcity* [Paper presentation]. In 4th Annual Partners' Meet of IWMI-Tata Water Policy Research Programme, Institute of Rural Management, Anand, India.

Delgado, S. (1992). Sobreexplotación de acuíferos: una aproximación conceptual. *Hidrogeología y Recursos Hidráulicos. Asoc. Española de Hidrología Subterránea, 25*, 469–476.

Dhawan, B. D. (1997). *Studies in minor irrigation with special reference to groundwater*. Commonwealth Publishers.

Georgescu-Reogen, N. (1971). *The entropy law and the economic process*. Harvard University Press.

Government of Gujarat. (1999). *Report of the Committee on Estimation of Ground Water Resource and Irrigation Potential in Gujarat State: GWRE –1997*. Gandhinagar: Narmada & Water Resources Department.

Government of India (GoI). (1999). *Integrated water resource development a plan for action*. National Commission for Integrated Water Resource Management, Ministry of Water Resources, Government of India.

Government of India (GoI). (2005). *Dynamic ground water resources of India* August. Central Ground Water Board, Ministry of Water Resources, Government of India.

Government of India (GoI). (2007). *Groundwater management and ownership in India*. Planning Commission, Government of India.

Government of India (GoI). (2017). *Report of 5th census of minor irrigation schemes*. Minor Irrigation Statistics Wing, Ministry of Water Resources, River Development and Ganga Rejuvenation, Government of India.

Hira, G. S., & Khera, K. L. (2000). *Water resource management in Punjab under rice-wheat production system* [Research bulletin no. 2/2000]. Department of Soils, Punjab Agricultural University.

ITGE. (1991). *Sobreexplotación de acuíferos: Análisis Conceptual*. Instituto Techológico Geominero de España.

Jadhav, R. (2018, June 9). The world's highest, and most reckless, user of groundwater. *The Times of India*. https://timesofindia.indiatimes.com/india/the-worlds-highest-and-most-reckless-user-of-groundwater/articleshow/64515989.cms.

Janakarajan, S. (1994). Trading in groundwater: A source of power and accumulation. In M. Moench (Ed.), *Selling water: Conceptual and policy debates over groundwater markets in India*. VIKSAT, Pacific Institute for Studies in Environment, Development and Security, Natural Heritage Institute.

Janakarajan, S. (2002). *Wells and illfare: An overview of groundwater use and abuse in Tamil Nadu, South India* [Paper presentation]. In 1st IWMI-Tata Annual Partners' Meet, IRMA, Anand.

Khan, A. D., Ashraf, M., Ghumman, A. R., & Iqbal, N (2018). Groundwater resource of Indus plain aquifer of Pakistan investigations, evaluation and management. *International Journal of Water Resources and Arid Environments, 7*(1), 52–69.

Kumar, M. (2005). Impact of electricity prices and volumetric water allocation on energy and groundwater demand management. *Energy Policy, 33*(1), 39–51. https://doi.org/10.1016/s0301-4215(03)00196-4.

Kumar, M. D. (2007). *Groundwater management in India: Physical, institutional and policy alternatives*. Sage Publications Pvt. Ltd.

Kumar, M. D. (2010). *Managing water in river basins: Hydrology, economics and institutions*. Oxford University Press.

Kumar, M. D. (2018). *Water Policy, Science and Politics: An Indian Perspective*. Amsterdam: Elsevier Science.

Kumar, M. D., Singh, O. P., & Singh, K. (2001). *Groundwater depletion and its socioeconomic and ecological consequences in Sabarmati River Basin* [Monograph # 2]. India Natural Resources Economics and Management Foundation.

Kumar, M. D., & Singh, O. P. (2008). How serious are groundwater over-exploitation problems in India? A fresh investigation into an old issue. In M. D. Kumar (Ed.), *Managing water in the face of growing scarcity, inequity and declining returns: Exploring fresh approaches. 7th Annual Partners' meet of IWMI-Tata Water Policy Research Program, ICRISAT, Patancheru, AP, 2–4 April 2008.*

Kumar, M. D., Sivamohan, M. V. K., & Narayanamoorthy, A. (2012). The food security challenge of the food-land-water nexus. *Food Security Journal, 4*(4), 539–556.

Kumar, M. D., Bassi, N., & Verma, M. S. (2021). Direct delivery of electricity subsidy to farmers in Punjab: Will it help conserve groundwater? *International Journal of Water Resources Development, 38*(2), 306–321. https://doi.org/10.1080/07900627.2021.1899900.

Kumar, M. D., Kumar, S., & Bassi, N (2022). Factors influencing groundwater behaviour and performance of groundwater-based water supply schemes in rural India. *International Journal of Water Resources Development*, 1–21. https://doi.org/10.1080/07900627.2021.2021866.

Llamas, M. R. (1992a). La sobreexploitation de aguas subterraneas bendicion, maldicion on entele-quio? *Tecnologia del aqua Barcelona, 91*, 54–68.

Llamas, M. R. (1992b). La surexploitation des nergiza: Aspects Techniques et Institutionnels. *Hydrogeologie, Orleans, 4*, 139–144.

Llamas, M. R., & Priscoli, J. D. (2000). *Report of the UNESCO working group on ethics of freshwater use*. Water and Ethics, Special Issue. Papeles del Proyecto Aguas Subterraneas, Uso Intensivo de Las Agua Subterraneas.

Margat, J. (1992), *The overexploitation of Aquifers* (No. 3, pp. 29–40). Selected Papers on Aquifer Overexploitation. International Association of Hydro-geologists.

McArthur, J. M., Ravenscroft, P., Safiulla, S., & Thirlwall, M. F. (2001). Arsenic in groundwater: Testing pollution mechanisms for sedimentary aquifers in Bangladesh. *Water Resource Research, 37*(1), 109–117.

Moench, M. (1995). When good water becomes scarce: Objective and criteria for assessing over-development in groundwater resources. In M. Moench (Ed.), *Groundwater availability and pollution: The growing debate over resource condition in India* [Monograph]. VIKSAT-Pacific Institute for Studies in Environment, Development and Security.

Muralidharan, D., & Athawale, R. N. (1998). *Artificial recharge in India*. Base paper prepared for Rajiv Gandhi National Drinking Water Mission, Ministry of Rural Areas and Development, National Geophysical Research Institute.

National Bank for Agriculture and Rural Development. (2006). *Review of methodologies for estimation of ground water resources in India*. National Bank for Agriculture and Rural Development.

Phadtare, P. N. (1988). *Geo-hydrology of Gujarat* [Unpublished report]. Central Ground Water Board, Western Region.

Pisharoty, P. R. (1990). *Characteristics of Indian rainfall* [Monograph]. Physical Research Laboratory.

Qureshi, A. S. (2020). Groundwater governance in Pakistan: From colossal development to neglected management. *Water, 12*(11), 3017. https://doi.org/10.3390/w12113017.

Qureshi, A. S., Ahmed, Z., & Krupnik, T. J. (2014). *Groundwater management in Bangladesh: An analysis of problems and opportunities* [Research report no. 2]. Cereal Systems Initiative for South Asia Mechanization and Irrigation (CSISA-MI) Project, CIMMYT.

Reddy, V. R. (2005). Costs of resource depletion externalities: A study of groundwater overex-ploitation in Andhra Pradesh, India. *Environment and Development Economics, 10*(4), 533–556. https://doi.org/10.1017/s1355770x05002329.

Saleth, R. M. (1994). Groundwater markets in India: A legal and institutional perspective. *Indian Economic Review, 29*(2), 157–176.

Sekhri, S. (2014). Wells, water, and welfare: The impact of access to groundwater on rural poverty and conflict. *American Economic Journal: Applied Economics, 6*(3), 76–102. https://doi.org/10.1257/app.6.3.76.

Shah, M., Banerji, D., Vijayshankar, P. S., & Ambasta, P. (1998). *India's drylands: Tribal societies and development through environmental regeneration*. Oxford University Press.

Shiferaw, B. (2021). *Addressing groundwater depletion: Lessons from India, the world's largest user of groundwater*. Independent Evaluation Group, World Bank Group. https://ieg.worldbankgroup.org/blog/addressing-groundwater-depletion-lessons-india-worlds-largest-user-groundwater.

Tariq, M., van de Giesen, N., Janjua, S., Shahid, M., & Farooq, R. (2020). An engineering perspective of water sharing issues in Pakistan. *Water, 12*(2), 477. https://doi.org/10.3390/w12020477.

Taylor, C. J., & Alley, W. M. (2001). *Ground-water-level monitoring and the importance of long-term water-level data.* U.S. Geological Survey Circular, 1217.

Todd, D. K., & Mays, L. W. (2005). *Groundwater hydrology* (third edition). USA: John Wiley & Sons, Inc.

van Dam, J., Groenendijk, P., Hendriks, Rob F. A., & Kroes, J. (2008). Advances of Modeling Water Flow in Variably Saturated Soils with SWAP. *Vadoze Zone Journal, 7*(2), 640–653.

Winter, T. C., Harvey, J. W., Franke, O. L., & Alley, W. M. (1998). *Ground water and surface water a single resource.* US Geological Survey Circular, 1139.

World Bank. (2010). *Deep wells and prudence: Towards pragmatic action for addressing groundwater overexploitation in India.* World Bank.

Young, R. A. (1992). Managing aquifer over-exploitation: Economics and policies. Selected papers on aquifer over-exploitation. *International Association of Hydro-geologists, 3*, 199–222.

Chapter 4

Comparing well irrigation with gravity irrigation

4.1 Introduction

India has the world's second-largest irrigated area. It is known that major irrigation projects have contributed to expanding irrigated areas in the country over the years. A few researchers have in the recent past documented the larger socio-economic (Bhalla & Mookerjee, 2001) and welfare impacts (Jagadeesan & Kumar, 2015; Kumar et al., 2014; Perry, 2001; Rotiroti et al., 2019; Shah & Kumar, 2008) of large surface irrigation projects. The private well irrigation system has witnessed rapid growth over the last three decades, surpassing flow irrigation in its contribution to the net irrigated area (Deb Roy & Shah, 2003; Kumar et al., 2009). This was because of massive rural electrification, heavy electricity subsidies, and institutional financing for wells and pump sets (GoI, 2017; Kumar, 2007).

The theoretical comparisons between surface water and groundwater available in academics are limited to discussions on the biological quality of water from the point of view of portability, as well as the patterns and levels of contamination from toxic substances. The academic literature on water resources is conspicuous by the absence of a proper assessment of the comparative advantages of the use of water from two sources for socioeconomic production purposes, especially crop production, based on the situation on the ground.

In recent decades, a myopic view favoring only private well irrigation in preference to canal irrigation is emerging among a few irrigation scholars in India (see for instance, IWMI, 2007; Shah, 2009, 2016). This distorted thinking of considering one system superior to the other came because of a poor understanding of the determinants of irrigation growth; the fundamental difference between well and surface irrigation; and the basic concepts in hydrology and water management. This has led to several fallacies in the irrigation sector. They are four of them: (1) Future growth in India's irrigation would come from groundwater (Amarasinghe et al., 2008; Shah, 2009). (2) Surface irrigation systems are highly inefficient (Shah, 2016). (3) Of late, returns on investments in surface irrigation systems have become negative. (4) Local water harvesting

Groundwater Economics and Policy in South Asia. DOI: https://doi.org/10.1016/B978-0-443-14011-2.00008-5

and recharge can help sustain well irrigation in semi-arid and arid regions (Shah et al., 2003; Shah, 2009).

The key questions being investigated in this chapter are: (1) Can well irrigation alone sustain expansion in India's irrigated area or, in India's current water resource-water demand scenario, is canal irrigation substitutable by well irrigation? (2) Is surface irrigation truly inefficient, or are we failing to understand the water use hydrology of surface irrigation systems? (3) Does the declining area under canal irrigation suggest negative returns on investments in surface irrigation systems, or are there significant positive externalities that surface irrigation produces but that remain out of sight of most researchers? (4) Can local rainwater harvesting and recharge arrest groundwater depletion and sustain well irrigation economies in the arid and semi-arid regions? (5) What role does canal irrigation play in sustaining well irrigation and in the management of groundwater resources in those regions?

4.2 Analyses, data type, and sources

The sets of analysis used in the paper to address the research questions are: per capita groundwater withdrawal in different states (m³/capita/annum); the intensity of groundwater use in different states of India (m³/sq.m of cultivated land); per capita arable land in different states (ha/capita); per capita effective renewable water availability per unit of arable land in selected basins of India (m³/capita per annum); and per capita agricultural water demand (m³/capita/annum) in these river basins (8 of them).

The secondary data used for the analysis included district-wise utilizable groundwater resources and groundwater draft (year 2019); the rate of siltation of major Indian reservoirs; the gross area irrigated by different sources in different states of India (year 2015–16); the minor irrigation census data of 2013–14 for selected Indian states; and the utilizable surface water resources of major Indian river basins (CWC, 2017). The secondary data were collected from a wide range of sources. They are: the Central Ground Water Board; the Central Water Commission; the Ministry of Water Resources; and the Ministry of Agriculture. Several published research papers, including those from the authors, and the most recent international literature on the related topics were also used. In addition, recent field visits were conducted in different parts of rural India including command areas of large surface irrigation projects such as the Mahisagar irrigation project and the Sardar Sarovar project in Gujarat.

4.3 Future of India's irrigation: canals or wells?

Surface irrigation systems provide more dependable sources of water than groundwater-based systems in most parts of India[1] (Kumar et al., 2012a). For

---

1. The authors here do not refer to reliability of water supplies, but dependability of the source of water, such as reservoirs, aquifer. It is understood that reliability of water supplies is better for groundwater based irrigation schemes.

flow irrigation, there should be a dependable source of water and a topography permitting the flow of water by gravity to the places of demand. Ideally, the design itself ensures sufficient yield from the catchment to supply water to the command areas for an estimated duty of the designated command, or, in other words, the design command is adjusted to match the flows available from the catchment. Hence, the design life of the scheme will be more or less realistic for reliable "dependable yield" estimates unless major changes occur in the catchment that changes the flow regimes and silt load (Kumar et al., 2012a).

But, in the case of groundwater, thousands of farmers dig wells drawing water from the same aquifer. Given the open access nature of the aquifer, all the agricultural pumpers of groundwater operate individually. Obviously, the "safe yield" of the aquifer is not reckoned with while designing the well. So, the productive life of a well depends on the characteristics of the aquifer, wells and total abstraction (Kumar et al., 2012a). Two third of India's geographical area is underlain by hard rock formations with poor groundwater potential (CGWB, 2019). Most of the peninsular and central India and some parts of western India are underlain by hard rock aquifers of basaltic and granitic origin.

The highly weathered formations of the hard rocks, which yield water, have small vertical extents of up to 30 m. When the regional groundwater level drops below this zone, farmers are forced to dig bore wells, tapping the zone with poor weathering. These bore wells have poor yields, unlike the deep tube wells in alluvial areas such as north Gujarat, alluvial Punjab, Uttar Pradesh, and Haryana. With an increase in well numbers, the yield of wells goes down due to well interference, which is widespread in these hard rock areas (Kumar, 2007). For instance, as analysis of the 5th Minor Irrigation Census data (2013–14) shows, as many as 40% of the 85,601 deep bore wells (that are in use) in AP were not able to utilize their potential due to poor discharge. The figure was 59.9% for Maharashtra, which has basalt formations in nearly 95% of its geographical area. The figure was nearly 19.1% for Rajasthan, which has both semi-consolidated and hard rock aquifers (GoI, 2017). Therefore, in spite of the explosion in well numbers, the well irrigated area has not increased here during the past decade.

Secondly, as shown by Fig. 2.3 of Chapter 2, the growth rate in well irrigation is almost decelerated in most parts of India since the 90s. Table 4.1 shows that most of the well irrigation in India is concentrated in the arid and semi-arid regions of northern, north-western, western, central, and peninsular India. It is quite low in the states of Odisha and Jharkhand in eastern India. Amongst this, intensive well irrigation in terms of per capita groundwater withdrawal per annum is highest in some of the northern and north-western States, viz., Punjab (1289 m^3/capita/annum), followed by Haryana (493 m^3 per annum) and Madhya Pradesh (260 m^3 per annum) (Fig. 4.1).

Intensive irrigation could sustain for many decades only in a few pockets, such as alluvial Punjab and Haryana, and Uttar Pradesh. This is because these regions are underlain by very good deep alluvial aquifers that are regionally extensive (GoI, 1999). These regions are already saturated in terms of irrigated areas. Further, expansion of the irrigated area is not possible in these areas.

TABLE 4.1 Gross irrigated area and well irrigated area for major Indian states.

Sr. no.	Name of the state	Gross irrigated area	Gross groundwater irrigated area	Percentage contribution of groundwater
1	Andhra Pradesh	3,634,910	1,592,357	43.81
2	Bihar	5,413,509	3,457,888	63.88
3	Chhattisgarh	1,838,067	819,910	44.61
4	Gujarat	7,001,500	4,560,300	65.13
5	Haryana	9,511,882	7,362,081	77.40
6	Jammu and Kashmir	501,274	27,088	5.40
7	Jharkhand	235,128	117,241	49.86
8	Karnataka	3,639,043	2,001,215	54.99
9	Madhya Pradesh	11,385,090	7,761,718	68.17
10	Odisha	1,151,171	182,156	15.80
11	Punjab	7,660,738	5,808,065	75.82
12	Rajasthan	10,603,502	7,232,470	68.21
13	Tamil Nadu	3,277,745	2,132,530	65.06
14	Telangana	3,172,401	2,315,936	73.00
15	Uttar Pradesh	20,881,918	17,306,180	82.88
16	Uttarakhand	542,866	390,841	72.00

From: http://data.icrisat.org/dld/src/crops.html and https://knoema.com/INMOSPIAIIG2019/area-under-irrigation-by-source-in-india.

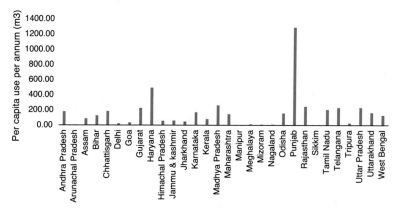

FIGURE 4.1   Intensity of groundwater withdrawal in per capita terms. (*From: based on CGWB, 2019 and Census of India, 2011 population data*).

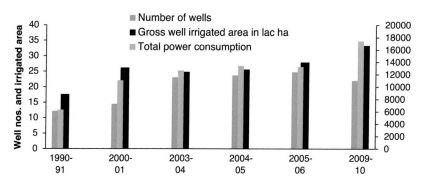

FIGURE 4.2   Well-groundwater irrigated area nexus.

Whereas in Rajasthan, Gujarat, Andhra Pradesh, and Tamil Nadu, problems of over-exploitation halt further growth in well irrigation. Most of the untapped groundwater is in the eastern Gangetic plain, devoid of sufficient arable land, and lies un-watered (Kumar & Singh, 2005; Kumar et al., 2008b; Shah & Kumar, 2008). Peninsular India and central India have a lot of un-irrigated land. Agriculture is prosperous in this part of the country, and demand for water is only going to grow. But, well irrigation is experiencing a "leveling off" and sometimes a decline due to "over-exploitation" and monsoon failure.

Analysis of time series data on well irrigation from Andhra Pradesh illustrates this point. The data presented in Fig. 4.2 (source: based on data provided in Busenna, 2016; GOAP, 2010; Jain, 2008; Swarna Sri & Narasimhan, 2011) shows that the gross well irrigated in the state peaked in 2000–01. But, by 2003–04, the irrigated area dropped by nearly 1.6 lac ha. Since then, there has been only a minor increase in gross irrigated area till 2005–06, with the highest figure recorded in 2005–06. By 2009–10, however, the gross irrigated area increased by around 5.5 lac ha. But what is interesting is the fact that this had no correlation with the growth in the number of wells. There were only 14.48 lac wells in the state in 2000–01. However, they together irrigated 26.18 lac ha of land in gross terms. The average area irrigated by a well was 1.8 ha. But, this declined to a record low of 1.07 ha in 2003–04, with the growth in the number of wells (by 8.62 lac), not getting translated into proportional growth in irrigated areas. This is a common phenomenon in hard rock areas occurring as a result of well interference. With an increase in the number of wells, the influence area of a well increases, and the available groundwater gets distributed among a larger number of wells (Kumar, 2007). The average area irrigated by a well recorded a minor improvement in 2005–06 (1.12 ha). This could be attributed to factors such as an increase in recharge from rainfall, a change in cropping patterns, and an increase in groundwater pumping in command areas, resulting from reduced surface water release from canals for irrigation (Kumar et al., 2011).

The figures show that by 2000–01, the state had an optimum number of wells, i.e., 14.28 lac in terms of providing maximum irrigation by a single well. The only apparent benefit achieved through an increase in the number of wells is distributional equity, with more farmers getting direct access to groundwater for irrigation. It is not to say that distributional equity is not an important concern in irrigation. In fact, most of the farmers who are late entries in well irrigation are small and marginal farmers. But, with such a high density of wells in hard rock areas, the economic efficiency of groundwater abstraction becomes extremely low. This is evident from two important facts: (1) the rate of well failures is becoming alarming; (2) the area irrigated by a well has reduced to 1.12 ha in 2005–06, down by nearly 40% from the highest figure of 1.8 lac hectare experienced in 2000–01, and then picked up in 2009–10 to 1.51 ha, owing to a reduction in the number of wells to 22 lac; (3) the power consumption for irrigating one hectare with groundwater has increased consistently, from 3569.0 kWh in 1990–91 to 4222.7 in 2000–01, and 5225.9 kWh in 2003–04 and then to 5117 kWh in 2009–10[2]. A lot of this increase can also be attributed to the replacement of diesel engines or energization of manually operated wells. But, one would expect such a process of electrification of wells to have taken place by the time the groundwater irrigated area touched almost the peak (i.e., in 2000–01). Therefore, it is safe to conclude that the increase in energy requirement for groundwater pumping after 2000–01 is because of the increase in well numbers resulting from the fast drawdown of water levels in wells due to the phenomenon of well interference.

Analysis carried out for the hard rock areas of the Narmada river basin in Madhya Pradesh showed similar trends. The average area irrigated by a single

---

2. The consumption went down slightly in 2005–06. This could be due to improvements in groundwater conditions, in the form of a rise in water levels, which can actually reduce the energy required for lifting a unit volume of groundwater.

well has declined over a 25-year period (see Chapter 2). Such a phenomenon is occurring due to well interference, a characteristic feature of hard rock areas, which starts occurring when all the groundwater that can be tapped is already tapped. In such situations, an increase in the number of wells does not result in an increase in the total irrigated area (Kumar, 2007). Hence, it is wrong to assume that well irrigation in India could sustain the same pace of growth in the coming years.

The spatial imbalance in resource availability and demand is aggravated by the uneven distribution of surface water resources spatially. Nearly 69% of India's surface water resources are in the Ganga-Brahmaputra-Meghna (GBM) basin (CWC, 2017). In the GBM basins, the water demand for agriculture is far less than the total renewable water resources[3], which is the sum of both renewable surface water and groundwater. As compared to this, in the five basins of south, western, and central India, viz., the Sabarmati, Narmada, Krishna, Cauvery, and Pennar, the water demand for agriculture alone exceeds the renewable water resources (Kumar et al., 2012a).

This imbalance can be effectively addressed only by large surface water projects and not by groundwater projects (Kumar et al., 2012a)[4]. Historically, water was taken from the rich upper catchments of river basins, which formed ideal locations for storages (Verghese, 1990). Surface irrigation can expand in the future with investments in large reservoirs and transfer systems that can take water from the abundant regions of the north and east to the parched but fertile lands in the south, though their economic viability and social costs and benefits will have to be ascertained. It goes without saying that there is an influential lobby in India and in neighboring countries such as Bangladesh that criticizes a water management solution that involves large-scale water transfer from north to south on scientific, technical, social, financial, ecological, economic and environmental grounds, though without much scientific data to support their argument (see Kumar et al., 2008c for details).

But such transfer solutions do not hold much water in the case of wells, as the engineering feasibility of transferring groundwater in bulk is itself a questionable proposition. The greatest example is the hyper-arid north-western Rajasthan. The six districts of this region, which have endogenous surface water and saline groundwater, are now irrigated by water from the Indira Gandhi Canal, which carries water from the Sutlej River in the Shivalik hills of Himachal Pradesh in north India. It irrigates a total of 2.035 m.ha of land.

---

3. The water demand for agriculture was estimated as the reference evapotranspiration (ET0) for the crop growing season, whereas the effective water resource availability was estimated by dividing the total renewable water resources per unit area of the cultivated land (see Kumar et al., 2012a for details).

4. This, however, does not negate the role of demand management in regions where demand exceeds supplies.

4.4 How far are surface irrigation systems inefficient?

Engineering efficiencies in large surface irrigation projects in India are much lower than those of well irrigation schemes (Koech & Langat, 2018). But such comparisons are used by some researchers and activists to build the argument that surface irrigation projects are performing poorly and that the government investment in surface irrigation should be diverted to better management of aquifers (see, IWMI, 2007; Shah, 2009, 2016). While it goes without saying that the management of canal irrigation leaves much to be desired, such arguments about the bad performance of surface irrigation systems are based on obsolete irrigation management concepts that treated the water diverted from reservoirs in excess of crop water requirements as "waste" (Howell, 2001; Perry, 2007; Perry & Kumar, 2022; Seckler, 1996). But these comparisons are not reflections of the economic efficiency of the entire system, as the wastewater gets re-used in the downstream part of the same system by well irrigators (Chakravorty & Umetsu, 2003).

As Seckler (1996) notes, the fundamental problem with this concept of water use efficiency based on supply is that it considers inefficient both evaporative loss of water and drainage. It is not well informed by the water use hydrology of surface irrigation systems. Most of the seepage and deep percolation from flow irrigation systems replenish groundwater and are available for reuse by well owners in the command (Allen et al., 1997; Seckler, 1996; Perry & Kumar, 2022). This recycling process not only makes many millions of wells productive but also saves the scarce energy required to pump groundwater by lowering pumping depths. This is one reason why well irrigation can sustain in many parts of Punjab and Haryana, the Mulla Command in Maharashtra, the Krishna river delta in Andhra Pradesh, the Mahi and Ukai-Kakrapar commands in south Gujarat (Kumar et al., 2010), and the SSP command in north, central and south Gujarat (Jagadeesan & Kumar, 2015).

A recent study of the field water balance of the Ludhiana district of Punjab developed on the basis of a water accounting framework using data on water level fluctuations during monsoon and during the year, as well as data on rainfall, irrigation water import, and groundwater draft, estimated the total irrigation return flow to the aquifer from gravity and well irrigation to the tune of 2570 MCM in a wet year and 1759 MCM in a dry year, while the rainfall recharge was 1000.5 MCM and 744 MCM, respectively during the wet year and dry year (Table 4.2).

Late Prof. B. D. Dhawan, one of the most renowned irrigation economists in India, looked at the economic returns from surface irrigation systems in his book, where he examined the merits of the claims and counter-claims about the benefits of big dams. He had highlighted the social benefits generated by large irrigation schemes through positive externalities such as improvements in well yields, a reduction in the incidence of well failures, and an increase in the overall sustainability of well irrigation by citing the example of the Mulla command in Maharashtra (see Dhawan, 1990).

TABLE 4.2 Parameters of water balance and water accounting of Ludhiana district, Punjab.

S. no.	Parameter	Values	
1	Type of hydrological year	Wet year	Dry year
2	Year considered	June 2011–May 2012	June 2016–May 2017
3	Annual rainfall (mm)	1218.40	506.23
4	Monsoon rainfall (mm)	1089.92	383.11
5	Surface water import (MCM)	717.34	717.34
6	Annual W.L.F (as per INDIAWRIS): m	0.58	−2.25
7	Irrigation consumptive use (MCM)	1478.96	2341.58
8	Industrial and domestic use (MCM)	106.05	163.30
9	Annual storage change as per India-WRIS (MCM)	267.06	−1043.53
10	Total rainfall recharge as per GEC method (MCM) of CGWB	646.85	NA
11	Net storage change during monsoon using W.L.F (MCM)	960.35	−352.51
12	Recharge during monsoon (MCM)	960.35	717.92
13	Non-monsoon rainfall recharge (MCM)	40.17	26.08
14	Total estimated rainfall recharge (MCM)	1000.52	744.00
15	Estimated storage change (MCM)	132.84	−1043.53
16	Estimated storage change as per CGWB recharge Est.	−220.82	NA
17	Total draft (MCM)	3438.35	3547.07
18	Total estimated irrigation return flow (MCM)	2570.67	1759.53

From: Author's own analysis.

The social benefits (positive externalities) these canals generate by protecting groundwater ecosystems are immense (Jagadeesan & Kumar, 2015; Rotiroti et al., 2019; Shah & Kumar, 2008), with reduced energy costs for pumping groundwater being one (Jagadeesan & Kumar, 2015; Vyas, 2001). The likely impact of this on the energy economy of the country will be evident from the fact that the total power subsidy to the agriculture sector in India stood at

87,900 crore rupees (US $11.72 billion) during BE 2015–16 (Chand, 2018). Most of this goes to subsidizing pump irrigation in the states having large areas under well irrigation, such as Punjab, Gujarat, Haryana, Madhya Pradesh, Maharashtra, Rajasthan, Tamil Nadu, Karnataka, and Andhra Pradesh. It is also seen that the subsidy went up in nine years from just Rs. 7335 crore in 1992–93 (source: Planning Commission, 2002). Hence, for sustaining groundwater use and augmenting groundwater recharge in the semi-arid and arid regions which now have a precarious balance between recharge and draft, surface irrigation will play an enormous role.

However, irrigation planners have, by far, nearly failed to capture these positive externalities in the cost-benefit calculations (Shah & Kumar, 2008). The data from erstwhile Andhra Pradesh shows that the command irrigated regions of the state have the lowest number of groundwater "over-exploited" mandals. The tail-end regions of the canals have a sufficient number of bore wells, which actually reap the benefit of return flows from canals and thus have good yields.

4.5 Is recharging groundwater using local runoff viable in the arid and semi-arid regions?

It is often suggested that flows from the small canals (Times of India, 2008) or small water harvesting/artificial recharge structures (GoI, 2007; Shah, 2009) should be used for recharging aquifers. However, the arid and semi-arid regions where aquifers are depleting (CGWB, 2019) have extremely limited surface water (Kumar et al., 2008a), and hence these arguments are fallacious. Fig. 4.3 shows the over-exploited districts in India (source: CGWB, 2019). They are in western and central Rajasthan; almost the entire Punjab; alluvial north Gujarat; and parts of Andhra Pradesh, Madhya Pradesh, Maharashtra, and Tamil Nadu. The surface water resources in the basins where these districts fall are extremely limited and are already tapped using large and medium reservoirs (source: CWC, 2017). Hence, the over-exploited regions coincide with the regions of surface water shortage.

Any new interventions to impound water would reduce the downstream flows, creating a situation of "Peter taking Paul's water." Such indiscriminate water harvesting is also leading to conflicts between upstream and downstream communities, as reported by Ray and Bijarnia (2006) for Alwar in Rajasthan and Kumar et al. (2008a) for Saurashtra in Gujarat. Water harvesting and recharge have poor physical feasibility and economic viability in semi-arid and arid regions, according to Kumar et al. (2008a), and have a negative impact on access equity in water.

The idea of dug well recharging was pushed for the hard rock areas on the false notion that it is a cheap (costing only Rs. 4000 per well), easy and safe method of groundwater banking, unlike what is being done in the United States and Australia (Shah et al., 2009). But this is far from the truth. Collecting runoff

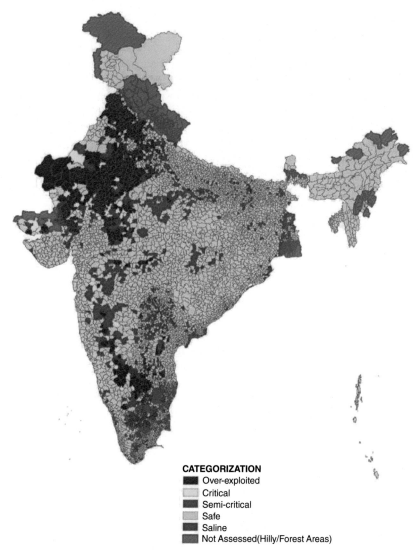

CATEGORIZATION
- Over-exploited
- Critical
- Semi-critical
- Safe
- Saline
- Not Assessed(Hilly/Forest Areas)

FIGURE 4.3   Over-exploited, critical and semi critical districts in India. (*From: CGWB, 2019*).

from the lowest points in the farm, channelizing it to the well location, and then filtering it before finally putting it in the well could be quite expensive, as land leveling and filter box construction would cost high depending on the farm size and soil type (Kumar et al., 2010). Over and above, spending such large sums in no way guarantees environmental safety as the runoff would contain fertilizer and pesticide residues from the field, which cannot be removed using filters (Kumar, 2000). Further, as Kumar et al. (2008a) argued, the central government's

Rs. 1800 crore scheme to recharge groundwater through four million open wells in hard rock districts of the country, if implemented, would render many small and large reservoirs unproductive. Hence, well irrigation in peninsular and western India cannot be sustained unless water is brought from surplus basins in the east and north for recharging the aquifers there.

While the policy makers in the country, who had planned a large-scale initiative for groundwater recharge in rural areas, seem to be caught unaware of these undesirable consequences, what happens on the ground is even more deplorable. In most cases, the farmers take the benefit of this scheme using their land records and misappropriate the government funds, while field verification of the recharge structures is almost nonexistent.

Bringing water from water-surplus basins to peninsular India would require large head works, huge lifts, long canals, intermediate storage systems, and intricate distribution networks. As we have argued, recharge schemes using local water are economically unviable. The reason is that while the cost per cubic meter of recharge is abnormally high, the returns from irrigated crop production are far less (in the range of Rs. $1-17/m^3$), as found in a study of irrigation water productivity of various crops in nine agro climatic sub-regions of the Narmada river basin in Central India (Kumar et al., 2008a). The need for vast precious land for spreading water for recharge would also make it socially unviable while further increasing economic costs.

Since the aquifers in hard rock areas of India have extremely poor storage capacities due to low specific yields, and because actual recharge would depend on the depth to water table before the onset of monsoon (Kumar et al., 2022), efficient recharge would require synchronized operation of recharge systems and irrigation wells so as to keep sufficient space for the infiltrating water to remain in the aquifer. This would call for advanced hydraulic design of the recharge structures and sophisticated system operation. Therefore, such an approach of using imported surface water for recharge would sound like "catching the crane using butter." The fact that practicing environmentally sound artificial groundwater recharge in these water-scarce hard rock areas is a very expensive and complicated affair requiring the application of advanced science, is yet to be appreciated by a section of the water community (Kumar et al., 2012a).

Hence, the best option would be for the farmers to use this expensive canal water for applying to the crops in that season of import (mainly the monsoon season) and use the recharge from natural return flows for growing crops in the next season (Kumar et al., 2012a). Opportunities for using water from "surplus basins" for recharging depleted aquifers exist at least in some areas. Examples are alluvial north Gujarat and north-central Rajasthan. Ranade and Kumar (2004) have proposed the use of surplus water from the Sardar Sarovar Narmada reservoir during years of high rainfall for recharging the alluvial aquifers of north Gujarat through the designated command area in that region (Ranade & Kumar, 2004). They proposed the use of the existing Narmada Main Canal and the rivers and ponds of north Gujarat for this, and their analysis showed that it is economically viable. It can protect groundwater ecology by reducing pumping; reduce revenue losses in the form of electricity subsidies; and increase the flows in rivers that face environmental water scarcity, apart from giving direct income returns from irrigation. But one cannot agree more on the point raised by Rath (2003) that only crops having very high water use efficiency will have to be promoted in the commands receiving such waters so as to generate sufficient returns from irrigated production. But, this will be possible only if the price of irrigation water is pitched at a level that begins to reflect the scarcity value of the resource.

4.6 Is the contribution of surface systems to India's irrigation declining?

In the year 2010–11, wells accounted for nearly 61.4% of India's net irrigated area (39.05 m.ha), with the rest coming from canals and tanks (24.5 m.ha) (source: Agricultural Census, GoI, 2012). But to make a choice between a surface scheme and groundwater scheme based on crude numbers of "irrigated area by source" is "hydrologically and economically absurd." Which model of irrigation is best suited for the area in the future can be judged by the nature of topography, hydrology and aquifer conditions. For instance, in rocky central and peninsular India, only imported surface water can sustain and expand well-irrigation. Indiscriminately embarking on well irrigation would only ruin the rural economy. Farmers in these regions desperately drill bore holes to tap water with high rates of failure (see Chapter 3), resulting in farmer suicides.

At least some scholars have begun to use "declining area under canal irrigation" to build a case for stopping investments in surface irrigation (Mukherjee & Facon, 2009; Shah, 2009, 2016; TOI, July 10 & 17, 2008). They seem to argue that the change in cumulative area irrigated by canals is a good indicator of the return on investment in surface irrigation systems. But this is a clear case of the misuse of statistics. Such arguments come from a poor understanding of how surface irrigation systems work. The fact is that while completion of a new irrigation scheme increases the irrigated area of a particular locality or region, this may not show up in the time series data on aggregate irrigated area at the national level. This phenomenon can be better understood if we look at the real factors that influence the irrigation performance of surface systems.

First: as a study in the Narmada river basin in central India shows, increased pumping of groundwater in upper catchments for agriculture can significantly reduce stream flows in basins where groundwater outflows contribute to surface flows (Kumar, 2010), thereby affecting the inflows into reservoirs. In fact, as data from the agricultural census shows (Chapter 2), the whole country experienced a quantum jump in well-irrigated area during the 1970s through the 1990s. In a small watershed called Maheshwaram in Andhra Pradesh with a drainage area of 64 sq.km (6400 ha), between 1975 and 2002, a total of 707 wells came up, all in the valley portions (source: National Geophysical Research Institute, 2002, as cited in Armstrong, 2004). Now, rights to groundwater are not well-defined in India, and landowners enjoy the rights to use the groundwater underlying his/her piece of land (Cullet, 2014). Since agricultural wells are mostly private, *de facto* groundwater is private property. Therefore, it is beyond the institutional capacity of state irrigation bureaucracies to control such a phenomenon occurring in the upper catchments of their reservoirs. Also, as is evident from the earlier discussions and from some studies, small water harvesting systems are adding to the reduction in inflows into reservoirs (Glendenning & Vervoort, 2011; Kumar et al., 2008a; Ray & Bijarnia, 2006).

Second: farmers in most surface irrigation commands install diesel pumps to lift water from the canals and irrigate the fields. Such instances are increasing with the pump explosion in rural India. Over a period of 40 years, the number of irrigation pump sets in India went up from a mere 0.7 million in 1970 to 28 million in 2010 (Singh, 2015). As per the latest data, nearly 30 million irrigation pumps have been installed in the country (MNRE, 2017; Jain & Agarwal, 2018). The better control over water delivery, which farmers can secure by doing this, is the reason for their preference for energy-intensive lifting over gravity flow. Another important reason for installing pump sets is to prevent the illegal water diversion that is rampant in canal irrigation. The pumping devices enable the illegal diversion of water for irrigating plots that are otherwise out of command due to topographical constraints. This was found to be rampant in many large irrigation commands[5].

In India's irrigation statistics, the area irrigated by farmers who have land on both sides of the canals and who lift water from them using pumps is generally not counted as "canal irrigated area." It is put under the heading "irrigation from other sources" in the macro level statistics. Reporting of this area as "canal lift" can put the WRD and revenue officials in a dilemma, as lifting water from canals in most Indian states is un-authorized irrigation and not permitted by the law (GoG, 2019; GoK, 1994; GoM, 2013). The national level irrigation census (MoSPI, 2017) shows irrigation statistics under four headings, viz., "area irrigated by wells," "area irrigated by canals," "area irrigated by tanks," and "area irrigated by other sources." It shows that the net area irrigated by "other sources" is around 7.5 m.ha, against 59.8 m.ha of net area irrigated by all sources. So, this is a considerable area that cannot be kept unaccounted for, and this includes the area irrigated through canal lift, river lifting, lifting from drains, and direct lifting from reservoirs. With the exponential increase in the number of irrigation pump

---

5. To name a few, there is the Dharoi irrigation command in north Gujarat; the Mahi irrigation command in south-central Gujarat; the Mulla-Mutha command in Maharashtra; and the Sardar Sarovar Narmada project in Gujarat. Most of the delivery canals of this gravity irrigation scheme are contour canals. This meant that only the lower side of the canals would have the command. But, extensive surveying carried out in the area showed that farmers whose land is located on the higher side lift the canal water using diesel pump sets and irrigate their land. This is a widespread phenomenon (Kumar et al., 2010).

sets in rural areas, the area irrigated through such methods might have increased substantially over time (Kumar et al., 2021). This can substantially reduce the reported "canal irrigated area."

Third: large reservoirs, primarily built for irrigation in this country, are being increasingly used for supplying water to big cities and small towns, as recent studies show. A recent analysis involving 301 cities/towns in India shows that with an increase in city population, the dependence on surface water resources for water supply increases, with the dependence becoming as high as 91% for larger cities (Mukherjee et al., 2010; Kumar, 2014). Many large cities depend almost entirely on surface water imported from large reservoirs. Some examples are Bangalore, Ahmedabad, Chennai, Rajkot, and Coimbatore, with contributions ranging from 91% to 100% (Seetharam, 2007). Many of them used to depend on local tanks, ponds, and bore wells in the past for meeting their water needs (Kumar, 2014).

Fourth: farmers in canal command areas, especially at the head reaches, tend to put more area under water-intensive crops, ignoring the cropping pattern considered in the design. This is also one of the reasons for shrinkage in the irrigated command area (Kumar, 2018).

Fifth: water from many large surface irrigation systems in many parts of India is used to feed tanks and ponds in the command area and also along the canal alignment when the farmers do not need water at the time of its release in canals. This water is subsequently lifted using pumps to irrigate crops when there is no water release from the canals. This gets counted as area irrigated by tanks/ponds and not as "canal irrigated area." These tanks/ponds also become ideal for raising fish and prawns, as found in the Godavari delta in Andhra Pradesh and Mahi command in Gujarat.

Last, but not least, reservoir siltation is adversely affecting the performance of reservoir-based surface irrigation schemes (Kumar et al., 2012b). Because of increased silt deposition, the live storage capacity of most of the reservoirs in India has considerably declined. The data from 238 major reservoirs (the surveys were conducted during the last four decades) from 15 states show an average 15% loss of storage capacity, with a minimum of 2% in the case of Chhattisgarh to a maximum of 30% in the case of Andhra Pradesh. Since the hydrographical surveys were carried out at different points in time for different reservoirs, the average annual rate of loss in capacity is important. The average annual percentage loss in gross storage ranged from a low of 0.06 for Madhuranthagam reservoir on the Killivaru River in Tamil Nadu to a high of 21.03 for a Teesta Low Dam Project-IV reservoir on the Teesta River in West Bengal (Central Water Commission, 2015).

Therefore, if nothing is done to restore the original storage capacity of the reservoirs, the area under surface irrigation would decline in the natural course. Therefore, it is quite obvious that while the cumulative investments in surface irrigation systems go up with time, there may not be a proportional rise in the surface irrigated area (Kumar, 2018). Hence, the performance of surface schemes vis-à-vis return on investment can be evaluated only by looking at the performance of individual schemes considering these factors. At least some of the above facts are compelling reasons for fresh thinking on the planning and implementation of irrigation in India. Instead of writing off surface systems, there should be planned investments in surface irrigation in the groundwater over-exploited regions of India wherever hydrological opportunities exist, based on sound economic evaluation that considers the positive as well as negative externalities in the interest of sustainable groundwater use.

While these researchers lament the "dismal" performance of canal irrigation schemes in India and stress for giving impetus to well irrigation (Mukherjee & Facon, 2009; Shah et al., 2009, p. 13; Shah, 2016), what is more noteworthy is the fact that the area under surface irrigation, which includes canal irrigation, tank irrigation and irrigation through canal and river lifting, has been steadily increasing during the past five and a half decades and peaked in 2006–07 (at 25.1 m.ha), in spite of the myriad of problems. Though there was a minor short-term decline observed during 1993–94 and 2002–03, this decline was due to many factors. Three of them are: lack of adequate investments for new schemes (source: Planning Commission, 2008), droughts and increasing diversion of water from reservoirs to urban areas. At the same time, the growth in well irrigation declined significantly after 2000, with a growth rate of 0.18 m.ha/year against 1.05 m.ha/year observed during 1987–88 and 1999–00, and 0.634 m.ha/year observed during 1967–68 and 1987–88. Sustaining well irrigation growth is a matter of concern, as 15% (839) of the blocks/talukas/mandals in the country are over-exploited, 4% are critically exploited, and 10% (550) are in the semi critical stage, and these regions contribute very significantly to India's well irrigation (Kumar et al., 2012b).

4.7 Ground water quality problems

An aspect that is least appreciated by policy makers is the differences in the quality of water from aquifers and canals. Canal water, which originates from forested catchments of rivers and glaciers, carries a whole range of micro and macro nutrients. These nutrients get deposited on agricultural land, and over the years they make the land more productive. A recent comparative study of canal irrigation and well irrigation, carried out in Nawah Shehar district of Punjab showed that the canal-irrigated paddy gave higher yield and water productivity than well-irrigated fields despite the lower reliability of supplies. The differential yield came from the better soil nutrient regime in canal-irrigated fields (Trivedi & Singh, 2008).

Another important point is that several of the vegetables, fruits and flowers currently grown in the hot and arid, water-scarce regions cannot be raised with the native groundwater there, as it contains a considerable amount of salt. They are grown using water from wells replenished by return flows from canal-irrigated fields and seepage from large reservoirs. The farmers growing these crops are able to generate very high returns from irrigation with this freshwater, in economic terms. Return flow from canals and seepage from reservoirs reduces the salinity of the water, whereas access to wells enhances control over water delivery. With rising market demand for high-value fruits, vegetables, and flowers, there is an increasing preference among farmers for water free from salts (Kumar, 2018). Recent field work carried out in Mahi irrigation command in Anand district of Gujarat showed farmers' preference for canal water for growing crops such as bananas, vegetables and paddy, attributed to the better chemical quality of canal water when compared to water from the tube wells[6].

Against this, groundwater resources in many semi-arid and arid regions have high levels of mineral contaminants (CGWB, 2010). More areas are getting affected by water quality problems over time (Kumar, 2007; Kumar & Shah, 2004). These minerals can increase the soil salinity, and some crops would not grow under such saline conditions, particularly when the soil is heavy and rainfall is low. Arsenic present in the groundwater of the plains of West Bengal can pose serious crop production and food safety risks, as rice plants irrigated with groundwater are known to absorb arsenic from soils, get into plant tissues, and also affect yields (Heikens, 2006). Widespread contamination of groundwater is already posing a big challenge for the rural drinking water supply in India, as nearly 80% of the rural water supply in the country is from ground water based sources such as open wells, hand pumps, bore wells, and tube wells (Kumar et al., 2022).

4.8 Conclusion and policy inferences

In recent years, a myopic view favoring only private well irrigation in preference to canal irrigation has emerged among some irrigation researchers in India. They try to weigh surface irrigation against groundwater irrigation (Kumar, 2018). This has led to several fallacies. Evidence available from both Indo-Gangetic plain and in peninsular India suggests that there is a strong nexus between surface irrigation development and the sustainability of well irrigation. It is not prudent to invest in well irrigation without investment in large surface reservoirs and conveyance systems in semi-arid and arid areas. The risks associated with such irrigation development policies are more prevalent in the hard rock areas, as illustrated by the evidence from Andhra Pradesh (in this chapter) and Madhya Pradesh (in Chapter 3). The spatial imbalance in water resource availability and water demand in India, which creates water-surplus regions and water-scarce regions, can be addressed only through surface water transfer projects that include gravity irrigation as a component.

Application of outdated irrigation management concepts leads to an under-valuation of the benefits of surface irrigation (Howell, 2001; Perry, 2007;

---

6. Source: primary data collected by Amit Patel, Research Assistant, IRAP.

Perry & Kumar, 2022; Seckler, 1996) and the important role that it plays in sustaining well irrigation and managing groundwater in semi-arid and arid regions. The positive externalities (social benefits) generated by surface irrigation, such as enhanced recharge of aquifers resulting from excessive return flows that sustain well irrigation and drinking water supplies, savings in cost of energy used for pumping groundwater, and improved food security resulting from a lowering of cereal prices, if considered in the economic evaluation of large gravity irrigation systems, would support the cause of well irrigation and groundwater management.

The surface irrigation systems often cater to large urban water demands, which generate great social values (Mukherjee et al., 2010), but in the process reduce their irrigation potential; farmers irrigate their land through canal lifting, which gets recorded as lift irrigation; and the capacity of reservoirs declines over the years due to the natural process of siltation. Also, increases in groundwater draft in upper catchments and intensive water harvesting reduce stream flows in rivers, affecting reservoir storage and irrigation potential. It is beyond the institutional capacity of state irrigation agencies to control such phenomena. It is obvious that for evaluating the performance of surface schemes vis-à-vis return on investment, the performance of individual schemes should be looked at, keeping in view these factors. It is high time for the proponents of well irrigation to understand that water, whether well water or canal water, has to come from the same hydrological system (Kumar, 2018).

Groundwater in many semi-arid and arid regions suffers from poor quality owing to high mineral content and toxicity. Such regions include large parts of the Indus basin area in Pakistan, covering a large proportion of Sindh province and a relatively smaller proportion of Punjab province (Lytton et al., 2021; Qureshi, 2020). They can pose new risks to crop production and food safety by affecting soils and plant tissues, and drinking water security in rural and urban areas. Further, in regions where groundwater has salinity, it is not possible to grow vegetables, fruits, and flowers with well water. In those areas, return flows from canal-irrigated fields and seepage from the reservoirs reduce the salt concentration of native groundwater, which enables farmers to grow such crops using well water.

To conclude, instead of comparing surface irrigation and groundwater irrigation vis-à-vis their physical performance using simplistic and outdated criteria, which is unnecessary, planned investments in gravity irrigation should be made as a way to address the problems facing the groundwater sector, including resource over-exploitation and water quality deterioration that threaten not only agricultural production and food security but also drinking water security, apart from doing it as a means to provide direct irrigation benefits.

Yet such a myopic view favoring well irrigation is quite dominant even in Pakistan (Hussain & Abbas, 2019), a country that has the world's largest gravity irrigation network. While crediting the large expansion in well irrigation over the past five decades to "more dependable groundwater" (Qureshi, 2018), the fact that the proportion of canal irrigation to that country's irrigation landscape is declining has led them to argue that future investments in irrigation should focus on groundwater (Hussain & Abbas, 2019; Qureshi, 2020). This is in spite of the wide recognition that a large proportion of the groundwater recharge in Pakistan is due to return flows from canal irrigated areas (Lytton et al., 2021; Qureshi et al., 2008; Qureshi, 2018, 2020). For instance, Lytton et al. (2021) estimate that out of the total groundwater recharge of 61 BCM in the Indus basin in Pakistan, the contribution from canal seepage is 27 BCM and that from irrigation return flow is 17 BCM. From these figures, it is extremely clear that judicious allocation of surface water over the canal commands, with less water allocated to the head reaches of the canal and more water to the tail reaches, will play a crucial role in the sustainable management of groundwater in such regions where the former face waterlogging problems and the latter groundwater over-draft and water quality deterioration problems.

References

Allen, R. G., Willardson, L. S., & Frederiksen, H. (1997). Water Use Definitions and Their Use for Assessing the Impacts of Water Conservation. In J. M. de Jager, L. P. Vermes, & R. Rageb (Eds.), *Proceedings ICID Workshop on Sustainable Irrigation in Areas of Water Scarcity and Drought* (pp. 72–82). England: Oxford. September 11-12.

Amarasinghe, U., Shah, T., & McCornick, P. (2008). Seeking calm waters: Exploring policy options for India's water future. *Natural Resource Forum, 32*(4), 305–315.

Armstrong, J. (2004). *International trends in water: A world and nation under stress* [Presentation]. Water Summit 2004: Hyderabad, India.

Bhalla, S. S., & Mookerjee, A. (2001). Big dam development: Facts, figures and pending issues. *International Journal of Water Resources Development, 17*(1), 89–98. https://doi.org/10.1080/713672558.

Bhattarai, M., & Narayanamoorthy, A. (2003). Impact of irrigation on rural poverty in India: an aggregate panel data analysis. *Water Policy, 5*(5–6), 443–458.

Busenna, P. (2016, May). Agricultural wells in rural India: A case study of Ungaranigundla Village, Kurnool district of Andhra Pradesh. *Agricultural Situation in India, 73*(2), 22–27.

Census of India. (2011). Census of India, Government of India, https://censusindia.gov.in/census.website/.

Central Ground Water Board. (2010). *Ground water quality in shallow aquifers of India*. Central Ground Water Board, Ministry of Water Resources.

Central Ground Water Board. (2019). *National compilation on dynamic ground water resources of India 2017*. Central Ground Water Board, Dept. of Water Resources, RD & GR, Ministry of Jal Shakti.

Central Water Commissions. (2015). *Compendium on silting of reservoirs in India*. Watershed and Reservoir Sedimentation Directorate, Central Water Commission.

Central Water Commission. (2017). *Reassessment of water availability in India using space inputs, basin planning and management organization*. Central Water Commission.

Chakravorty, U., & Umetsu, C. (2003). Basin-wide water management: A spatial model. *Journal of Environmental Economics and Management, 45*(1), 1–23. https://doi.org/10.1016/s0095-0696(02)00012-8.

Chand, R. (2018). *Innovative policy interventions for transformation of farm sector* [Presidential address]. 26th Annual Conference of Agricultural Economics Research Association (India).

Cullet, P. (2014). Groundwater law in India: Towards a framework ensuring equitable access and aquifer protection. *Journal of Environmental Law, 26*(1), 55–81. https://doi.org/10.1093/jel/eqt031.

Deb Roy, A., & Shah, T. (2003). Socio-ecology of groundwater irrigation in India. In R. Llamas, & E. Custodio (Eds.), *Intensive use of groundwater: Challenges and opportunities* (pp. 307–335). Swets and Zetlinger Publishing Co.

Dhawan, B. D. (1990). *Big dams: Claims, counterclaims.* Commonwealth Publishers.

Glendenning, C., & Vervoort, R. (2011). Hydrological impacts of rainwater harvesting (RWH) in a case study catchment: The Arvari River, Rajasthan, India: Part 2. Catchment-scale impacts. *Agricultural Water Management, 98*(4), 715–730.

Government of Andhra Pradesh. (2010). *Agricultural statistics at a glance: Andhra Pradesh 2009–10.* Directorate of Economics and Statistics, Govt. of Andhra Pradesh.

Government of Gujarat. (2019). *Gujarat Irrigation and Drainage (Amendment) Act 2019.* Govt. of Gujarat.

Government of Karnataka. (1994). *Karnataka Irrigation Act, 1965 — With Rules and Notifications.* Vijaya Publications.

Government of Maharashtra. (2013). *The Maharashtra Management of Irrigation Systems by Farmers Act, 2005. (As modified up to the 7th May 2013).* Law and Judiciary Department, Govt. of Maharashtra.

Government of India. (1999). *Integrated Water Resource Development A Plan for Action, Report of the National Commission on Integrated Water Resources Development:* I. Ministry of Water Resources.

Government of India (GoI). (2007). *Report of the expert group on groundwater management and ownership.* Planning Commission.

Government of India. (2012). *Report on Agricultural Census of India - 2010-11.* New Delhi: Government of India.

Government of India (GoI). (2017). *Report of 5th census of minor irrigation schemes.* Minor Irrigation Statistics Wing, Ministry of Water Resources, River Development and Ganga Rejuvenation, Government of India.

Heiken, A. (2006). *Arsenic contamination of irrigation water, soils and crops in Bangladesh: Risk implications for sustainable agriculture and food safety in South Asia* [RAP publication 2006/20]. Food and Agriculture Organization of the United Nations.

Howell, T. (2001). Enhancing water use efficiency in irrigated agriculture. *Agronomy Journal,* (93), 281–289.

Hussain, A., & Abbas, H. (2019, September 27). To save Pakistan, look under its rivers. *The Third Pole.* https://www.thethirdpole.net/en/climate/pakistans-riverine-aquifers-may-save-its-future/.

Hussain, I., & Hanjra, M. A. (2003). Does irrigation water matter for rural poverty alleviation? Evidence from South and South-East Asia. *Water Policy, 5*(5–6), 429–442. https://doi.org/10.2166/wp.2003.0027.

International Water Management Institute. (2007). *Water figures: Turning research into development.* Newsletter of the International Water Management Institute (IWMI).

Jagadeesan, S., & Kumar, D. M. (2015). *The Sardar Sarovar Project: Assessing economic and social impacts.* SAGE Publications Pvt. Ltd.

Jain, A. K. (2008, June 24). *Challenges facing groundwater sector in Andhra Pradesh* [Presentation by the Special Secretary to Government of AP].

Jain, A., & Agarwal, S. (2018). *Financing solar for irrigation in India: Risks, challenges and solution.* Council on Energy, Environment and Water (CEEW).

Koech, R., & Langat, P. (2018). Improving irrigation water use efficiency: A review of advances, challenges and opportunities in the Australian context. *Water, 10,* 1771. https://doi.org/10.3390/w10121771.

Kumar, M. D. (2000). Dug well recharging in Saurashtra: are the benefits and impacts over-stretched? *Journal of Indian Water Resources Society* July, Roorkee, India.

Kumar, M. D. (2007). *Groundwater management in India: physical, institutional and policy alternatives* p. 354. New Delhi: Sage Publications.

Kumar, M. D. (2010). *Managing Water in River Basins: Hydrology, Economics, and Institutions.* New Delhi: Oxford University Press.

Kumar, M. D. (2014). *Thirsty Cities: How Indian Cities Can Meet their Water Needs.* New Delhi: Oxford University Press.

Kumar, M. D. (2018). *Water policy, science and politics: An Indian perspective.* Elsevier Science.

Kumar, M. D., Bassi, N., & Kumar, S. (2022). *Drinking water security in rural India: Dynamics, influencing factors, and improvement strategy (Water resources development and management series)* (1st ed.). Springer.

Kumar, M. D., Ghosh, S., Singh, O. P., & Ravindranath, R. (2006, June). *Changing surface water-groundwater interactions in Narmada River Basin, India: A case for trans-boundary water resources management* [Paper presentation]. III International Symposium on Transboundary Water Management, UCLM, Ciudad Real, Spain.

Kumar, M. D., Malla, A. K., & Tripathy, S. (2008c). Economic value of water in agriculture: Comparative analysis of a water-scarce and a water-rich region in India. *Water International, 33*(2), 214–230.

Kumar, M. D., Narayanamoorthy, A., & Sivamohan, M. V. K. (2010). *Pampered views and parrot talks: in the cause of well irrigation in India, Occassional paper # 1.* Hyderabad India: Institute for Resource Analysis and Policy.

Kumar, M. D., Patel, A. R., Ravindranath, R., & Singh, O. P. (2008a). Chasing a mirage: Water harvesting and artificial recharge in naturally water-scarce regions. *Economic and Political Weekly, 43*(35), 61–71.

Kumar, M. D., Sahasranaman, M., Verma, M. S., Kumar, S., & Narayanamoorthy, A. (2021). Getting the irrigation statistics right. *International Journal of Water Resources Development, 38*(3), 536–543. https://doi.org/10.1080/07900627.2021.1921711.

Kumar, M. D., & Shah, T. (2004). Groundwater contamination and pollution in India: Emerging challenge. In *Hindu survey of the environment.* Kasturi & Sons.

Kumar, M. D., & Singh, O. P. (2005). Virtual water in the global food and water policy making: Is there a need for rethinking? *Water Resources Management, 19,* 759–789.

Kumar, M. D., & Singh, O. P. (2008). How serious are groundwater over-exploitation problems in India? A fresh investigation into an old issue. In M. D. Kumar (Ed.), *Managing water in the face of growing scarcity, inequity and declining returns: Exploring fresh approaches.* 7th Annual Partners' meet of IWMI-Tata Water Policy Research Program, ICRISAT, Patancheru, AP, 2–4 April 2008.

Kumar, M. D., Sivamohan, M. V. K., & Narayanamoorthy, A. (2008b). *Irrigation water management for food security in India: The forgotten realities* [Paper presentation]. International Seminar on Food Crisis and Environmental Degradation-Lessons for India. Greenpeace.

Kumar, M. D., Narayanamoorthy, A., & Singh, O. P. (2009). Groundwater irrigation versus surface irrigation, *44*(50), December 12.

Kumar, M. D., Sivamohan, M. V. K., Niranjan, V., & Bassi, N. (2011, February). *Groundwater management in Andhra Pradesh: Time to address real issues* [Occasional paper # 4]. Institute for Resource Analysis and Policy.

Kumar, M. D., Sivamohan, M. V. K., & Narayanamoorthy, A. (2012a). The food security challenge of the food-land-water nexus. *Food Security Journal, 4*(4), 539–556.

Kumar, M. D., Narayanamoorthy, A., & Sivamohan, M. V. K. (2012b). Key issues in Indian irrigation. In M. D. Kumar, M. V. K. Sivamohan, & N. Bassi (Eds.), *Water management, food security and sustainable agriculture in developing economies*. Earthscan/Routledge.

Kumar, M. D., Jagadeesan, S., & Sivamohan, M. V. K. (2014). Positive externalities of irrigation from the Sardar Sarovar Project for farm production and domestic water supply. *International Journal of Water Resources Development, 30*(1), 91–109. doi:10.1080/07900627.2014.880228.

Llamas, M. R. (2002). *Groundwater irrigation and poverty* [Keynote presentation]. International Regional Symposium Water for Survival, Central Board of Irrigation and Power, and Geographical Committee of International Water Resources Association, New Delhi.

Lytton, L., Ali, A., Garthwaite, B., Punthakey, J. F., & Saeed, B. (2021). *Groundwater in Pakistan's Indus Basin: Present and future prospects*. World Bank. https://openknowledge.worldbank. org/handle/10986/35065.

Ministry of Non Renewable Energy (MNRE). (2017). *Annual Report 2016–2017, Ch. 4: National Solar Mission*.

Ministry of Statistics and Programme Implementation. (2017). *Irrigation-statistical year book India 2017*. Ministry of Statistics and Programme Implementation, Govt. of India.

Morris, G. L., & Fan, J. (1998). *Reservoir sedimentation handbook: Design and management of dams, reservoirs and watershed for sustainable use*. McGraw Hill Book Co.

Mukherji, A. (2003). *Groundwater development and agrarian change in eastern India* [IWMI-Tata comment # 9]. Based on Vishwa Ballabh, Kameshwar Chaudhary, Sushil Pandey and Sudhakar Mishra, IWMI-Tata Water Policy Research Program.

Mukherji, A., & Facon, T. (2009). *Revitalizing Asia's irrigation: To sustainably meet tomorrow's food needs*. International Water Management Institute and Food and Agriculture Organization of the United Nations.

Mukherjee, S., Shah, Z., & Kumar, M. D. (2010). Sustaining Urban Water Supplies in India: Increasing Role of Large Reservoirs. *Water Resources Management, 24*(10), 2035–2055.

Perry, C. J. (2001). World commission on dams: Implications for food and irrigation. *Irrigation and Drainage, 50*, 101–107.

Perry, C. (2007). Efficient irrigation; inefficient communication; flawed recommendations. *Irrigation and Drainage, 56*(4), 367–378. https://doi.org/10.1002/ird.323.

Perry, C. J., & Kumar, M. D. (2022). Hydrology versus ideology: Flawed concepts underlying water sector reform proposal. *Economic and Political Weekly, 57*(6), 66–68.

Planning Commission. (2002, May). *Annual report on working of State Electricity Boards and Electricity Departments*. Planning Commission, Government of India.

Planning Commission. (2008). *Eleventh five year plan, 2007–2012: Agriculture, rural development, industry, services and physical infrastructure*. Oxford University Press.

Qureshi, A. S. (2018). Managing surface water for irrigation. In T. Oweis (Ed.), *Water management for sustainable agriculture*. Burleigh Dodds Science Publishing Limited.

Qureshi, A. S. (2020). Groundwater governance in Pakistan: From colossal development to neglected management. *Water, 12*(11), 3017. https://doi.org/10.3390/w12113017.

Qureshi, A. S., Gill, M. A., & Sarwar, A. (2008). Sustainable groundwater management in Pakistan: Challenges and opportunities. *Irrigation and Drainage, 57*, 1–10.

Ranade, R., & Kumar, M. D. (2004). Narmada water for groundwater recharge in North Gujarat: Conjunctive management in large irrigation projects. *Economic and Political Weekly, 39*(31), 3510–3513.

Rath, N. (2003). Linking rivers: some elementary arithmetic. *Economic and Political Weekly, 39*(29), 3033.

Ray, S., & Bijarnia, M. (2006). Upstream vs downstream: Groundwater management and rainwater harvesting. *Economic and Political Weekly, 41*(23), 2375–2383. http://www.jstor.org/stable/4418325.

Rotiroti, M., Bonomi, T., Sacchi, E., McArthur, J. M., Stefania, G. A., Zanotti, C., Taviani, S., Patelli, M., Nava, V., Soler, V., Fumagalli, L., & Leoni, B. (2019). The effects of irrigation on groundwater quality and quantity in a human-modified hydro-system: The Oglio River basin, Po Plain, northern Italy. *Science of the Total Environment, 672*, 342–356. https://doi.org/10.1016/j.scitotenv.2019.03.427.

Seckler, D. (1996). *The new era of water resources management: From dry to wet water saving* [Research report 1]. International Water Management Institute.

Seetharam, K. E. (2007). *2007 benchmarking and data book of water utilities in India.* Asian Development Bank.

Shah, M. (2016). *A 21ˢᵗcentury institutional architecture for water reforms in India: Final report submitted to the Ministry of Water Resources.* River Development & Ganga Rejuvenation, Government of India.

Shah, T. (2001). *Wells and welfare in Ganga Basin: Public policy and private initiative in Eastern Uttar Pradesh, India* [Research report 54]. International Water Management Institute.

Shah, T., Mukherji, A., Qureshi, A. S., & Wang, J. (2003). Sustaining Asia's Groundwater Boom: An Overview of Issues and Evidence. *Natural Resource Forum, 27*(2), 130–141.

Shah, T. (2009). *Taming the anarchy: Groundwater governance in South Asia.* Resources for Future and International Water Management Institute.

Shah, T., Kishore, A., & Pullabhotla, H. (2009). Will the impact of 2009 drought be different from 2002? *Economic and Political Weekly, 44*(37), 11–14.

Shah, Z., & Kumar, M. D. (2008). In the midst of the large dam controversy: Objectives, criteria for assessing large water storages in the developing world. *Water Resources Management, 22*(12), 1799–1824. https://doi.org/10.1007/s11269-008-9254-8.

Sikka, A. K. (2009). Water productivity of different agricultural systems. In M. D. Kumar, & U. Amarasinghe (Eds.), *Water sector perspective plan for India: Potential contributions from water productivity improvements* [Draft report]. International Water Management Institute.

Singh, G. (2015). Agricultural mechanisation development in India. *Indian Journal of Agricultural Economics, 70*(1), 64–82.

Swarna Sri, K., & Narasimham, S. V. L. (2011). Realistic estimate of agricultural power in Andhra Pradesh (India): A case study. *ARPN Journal of Engineering and Applied Sciences, 6*(8), 31–36.

Times of India. (2008, July 10 and 17). Wasting $50 billion dollar in major irrigation, *Swaminomics.*

Trivedi, K., & Singh, O. P. (2008). Impact of quality and reliability of irrigation on field-level water productivity of crops. In M. D. Kumar (Ed.), *"Managing water in the face of growing scarcity, inequity and declining returns: Exploring fresh approaches," Proceedings of the 7ᵗʰAnnual Partners' Meet of IWMI-Tata Water Policy Research Program ICRISAT Campus, Patancheru, Hyderabad, 2–4 April, 2008, Vol. 1.* International Water Management Institute (IWMI).

Verghese, B. G. (1990). *Waters of Hope: Himalaya-Ganga development and cooperation for a billion people* p. 462. New Delhi: Oxford & IBH Publishing Pvt. Ltd.

Vyas, J. (2001). Water and energy for development in Gujarat with special focus on the Sardar Sarovar project. *International Journal Water Resources Development, 17*(1), 37–54.

Chapter 5

Farm ponds and solar irrigation pumps: Economics and implications for groundwater management policy

5.1 Introduction

In response to growing public concerns about groundwater depletion and its likely fallout on the livelihoods of farmers, there were several attempts by the state and central government to find alternatives to reduce the pressure on groundwater for irrigation (see, World Bank, 2020). The numerous schemes launched by the governments of several water-scarce states of India on decentralized rainwater harvesting is an example. While many of these schemes are more than two decades old, there is no *ex-ante* or *ex post facto* evaluation of them.

There were also government initiatives to reduce the increasing revenue burden on the power utilities due to the heavy subsidies being offered to farmers for the electricity supplied for pumping groundwater, which involved offering alternative sources of power. The schemes launched by the government of India and the state governments on solar power irrigation are a good example (Kishore et al., 2014). But there are no sound economic analyses that support investments in solar irrigation pumps in the Indian context (Bassi, 2018). On the contrary, most policies and projects promoting solar-based groundwater pumping for irrigation through subsidies and other incentives in India and many other countries overlook the real financial and economic costs of this solution as well as the potential negative impacts on the environment caused by groundwater over-abstraction (Closas & Rap, 2017). Yet all the academic and popular literature discuss the clean energy benefits but by and large remain silent about the high capital investments required and, therefore, whether the systems are cost effective even after considering the clean energy benefits or not. The other was the "direct delivery of power subsidy to farmers" tried in Punjab, which we will discuss in Chapter 9.

There were two specific schemes launched in the recent past. They are the Maharashtra farm pond scheme and the solar irrigation cooperative in Gujarat.

Groundwater Economics and Policy in South Asia. DOI: https://doi.org/10.1016/B978-0-443-14011-2.00017-6
105

The first one is entirely a government scheme and was implemented on a large scale in Maharashtra by the state government there. The second one was conceptualized by an international research institution and later adopted by a state government agency. The first one was aimed at providing an alternate, decentralized source of water for irrigation. The second has the twin objectives of offering an alternate energy source for the farmers, who are already connected to the power grid, to pump groundwater while incentivizing them to sell the surplus electricity to the power utility. While there are six models available under the scheme with schemes designed for each, one of them is the SPaRC model implemented in Dhundi village in Anand district of Gujarat by the International Water Management Institute in collaboration with Madhya Gujarat Vidyut Company Ltd. (MGVCL).

Soon after their implementation, both schemes became controversial, with critics appearing in popular magazines and journals (Bassi, 2015; Closas & Rap, 2017; Kale, 2017; Sahasranaman et al., 2018, 2021). In this chapter, we will critically evaluate both the models for the potential impacts they could create on groundwater and their welfare effects. The first critique (farm ponds in Maharashtra) is based on an analysis of primary data collected from the field. The second critique is based on a synthesis of a series of articles published on the concepts underlying the model and the workings of the model by the author and others.

5.2 Maharashtra's farm ponds

Over the last one and a half decades, farmers in the drought-prone areas of Maharashtra were motivated to take up farm ponds which were initially promoted under the National Horticulture Mission, RKVY (Rashtriya Krishi Vigyan Yojna) and MGNREGA schemes of the Government of India. Under the National Horticulture Mission (NHM), a subsidy to the tune of 50% was provided for excavation as well as for the purchase of polythene sheets. The objective of the scheme was to capture the runoff. The subsidy varied with size. A farm pond of size 30 m × 30 m × 4.5 m was provided with a subsidy component of INR 3.55 lac. The state government has also launched the construction of farm ponds under the *Magel Tyala Shet Tale* scheme. The normal size of the pond permitted under the scheme is 30 m × 30 m × 3 m.

The size of the reservoirs that are built by the farmers in their plots ranged from small (10 m × 10 m) to large ones (40 m × 40 m)—with storage capacity ranging from less than 0.01 to 0.10 MCM. These surface reservoirs are built for providing irrigation to the horticultural crops—with pumps and pipes. It is mostly used for watering pomegranate fields and sugarcane. The region is known for large-scale production of pomegranate, which is exported to many countries as well as other parts of India, and for some of the oldest sugar cooperatives of the country. In fact, first-time growers of pomegranate in other parts of India (especially Gujarat and Rajasthan) visit this place to fetch the best saplings from the nurseries around and to learn about the cultivation of this fruit from the progressive farmers of this region.

There are several hundreds of these tiny reservoirs filled with clean water, spread over a not-so-large geographical region, which is generally desiccated as it is part of a rain shadow area of Ahmednagar district. Most of these reservoirs are well constructed, rectangular in shape, with large bunds, and all of them were lined with High Density Poly Ethelene (HDPE) film to prevent seepage of water. The height of the bunds and clean water essentially meant that the farmers were not harvesting runoff water from their fields and neighboring areas. Had it been the case, the water would have been unclean with heavy silt and clay particles. In fact, an article that appeared in the *Economic and Political Weekly* in 2017 talked about how the scheme of the government of Maharashtra to promote the construction of farm ponds for on-farm water security had been flawed. The article argued how the farmers across Maharashtra were using the funds available from the scheme essentially to build surface storage reservoirs to store pumped groundwater for use during the summer months, when the wells run dry, spelling doom for groundwater (Kale, 2017).

However, there is no quantification of the amount of water stored by them or their overall impact on farming enterprises.

5.2.1 Results from a field study

A field study was carried out in 2018 in Ahmednagar district to assess the impact of the farm ponds on farming and groundwater use. The study covered three blocks of the district, viz., Sangamner, Parner and Nagar. A total of 48 farmers who adopted farm ponds were interviewed. Of the total of 48 farm pond owners, 7 (14.6%) had large size farm ponds (size ranging from 0.05 to 0.1 MCM), 24 (50%) farmers had medium size farm ponds (size ranging from 0.01 to 0.05 MCM) and the remaining 17 (35.4%) farmers had small farm pond (size less than 0.01 MCM). Most of the farm ponds fall in the medium to big category. None of the farm ponds had an outlet for discharging surplus water (Table 5.1).

5.2.1.1 Impact of farm pond adoption on cropping pattern

The cropping pattern of the farm pond adopters had changed considerably in the post-farm pond construction scenario. Earlier, the focus was on growing fodder-yielding cereal crops so as to meet both the fodder and grain needs of the households. For a large number of households, Kharif cultivation was largely for household needs, and for most of the households, it was subsistence farming. Those with irrigation facilities devoted some area to cash crops such as a small patch of pomegranate orchards (if so), sugarcane, and onions. Summer crops were taken only by a few households. Table 5.2 clearly shows the overall increase in the area under irrigation. While the area for pearl millet decreased from 5.74% to 0.83%, the area under pomegranate increased from 3.6% to

TABLE 5.1 Capacity and cost of farm ponds.

Farm pond size	Average storage capacity of farm pond (m³)	Average effective storage capacity of farm pond (MCM)	Average cost per cubic meter of effective storage (Rs)	The average total volume of water pumped into the irrigation field (m³)
Small	0.0041	0.0311	11.51	27,129
Medium	0.0188	0.038	13.21	19,802
Large	0.0700	0.09	12.11	19,377

From: Analysis of primary field survey data.

13.10% after adoption. Similarly, the percentage area under brinjal, onion, and tomato also increased considerably. The area under green fodder during summers has increased from 0.5% to 4.64%.

5.2.1.2 Impact on crop yield and income

Table 5.3 shows that there is an efficient use of water, leading to not only an extension of the total area under cultivation of winter, summer, and perennial crops but also an improvement in the average crop yield from onion (Rabi) using drip-increased from 9374 to 12,678 kg per acre, which is an increase of 35.2%. Similarly, the yield of winter tomato increased from 8165 to 18,143 kg per acre, which again is 122%. The access to a supplementary source of water (water from ponds) enabled farmers to provide optimum irrigation to the crops, thereby obtaining higher yields. Due to the increase in yields, the net returns from crop production per unit cropped area have also gone up considerably for some crops (onion, tomato, soya bean, wheat, and pomegranate). Though the cost of cultivation had also gone up in the post-adoption scenario (in most cases), the increase in revenue had offset that. It is also to be seen that the increase in yield could not be attributed entirely to the farm pond since other activities such as mulching, drip, and change of seed variety, etc. must also have led to an increase in yield. In some cases, the net income dropped even after the adoption of drip irrigation (like in the case of drip-irrigated kharif onion). Factors such as the fluctuations in the farm gate price of the crop also might have influenced this, especially for crops like onion, whose price fluctuates widely.

5.2.1.3 Impact on water use and crop water productivity

We have seen that post the adoption of the farm ponds, the total area under crop and crop yields had increased for almost all crops, except some low-value ones. Hence, it is important to see how the use of water, the most crucial input, had changed and what has been its impact on crop water productivity

TABLE 5.2 Change in cropping pattern in Ahmednagar district.

Crop name	Total cropped area (Acre)		Total irrigated area (Acre)		% of the area under irrigation		% of total cropped area	
	Before farm pond	After farm Pond	Before farm pond	After farm pond	Before farm pond	After farm pond	Before farm pond	After farm pond
Gooseberry	2	0	2	0	0.39		0.50	
Banana	3	2	3	2	0.59	0.31	0.75	0.33
Bean	0	1	0	1		0.15		0.17
Brinjal (Kharif)	7	4	7	4	1.38	0.61	1.75	0.66
Brinjal (Winter)		1		1		0.15		0.17
Brinjal (Summer)		5		5		0.77		0.83
Brinjal (Perennial)	2	4	2	4	0.39	0.61	0.50	0.66
Cabbage (Kharif)	6	7	6	7	1.18	1.07	1.50	1.16
Cabbage (Winter)	4	5	4	5	0.79	0.77	1.00	0.83
Carrot		1		1		0.15		0.17
Chickpea (Kharif)	7	1	4	0	1.38	0.15	1.00	0.00
Chickpea (Winter)	29.5	18.0	29.5	18	5.82	2.76	7.36	2.98
Chili (Kharif)	3	3	3	3	0.59	0.46	0.75	0.50
Chili (Winter)	6	2	6	2	1.18	0.31	1.50	0.33
Chili (Summer)		4		4		0.61		0.66
Coriander		0.5		0.5		0.08		0.08
Cucumber (Kharif)		1		1		0.15		0.17

(continued on next page)

TABLE 5.2 Change in cropping pattern in Ahmednagar district—cont'd

Crop name	Total cropped area (Acre)		Total irrigated area (Acre)		% of the area under irrigation		% of total cropped area	
	Before farm pond	After Pond	Before farm pond	After farm pond	Before farm pond	After farm pond	Before farm pond	After farm pond
Cucumber (Winter)		1		1		0.15		0.17
Cucumber (Perennial)		1		1		0.15		0.17
Custard apple		5		5		0.77		0.83
Fenugreek		2		2		0.31		0.33
Flower (Kharif)		3		3		0.46		0.46
Flower (Winter)	1		1		0.20		0.26	
Green fodder (Kharif)	25	14	21	14	4.93	2.15	5.24	2.34
Green fodder (Winter)	4	12	4	12	0.79	1.84	1.00	1.99
Green fodder (Summer)	2	35	2	28	0.39	5.37	0.50	4.64
Green pea (Kharif)		2		2		0.31		0.33
Green pea (Winter)	3	3	3	3	0.59	0.46	0.75	0.50
Green pea (Summer)	2	3	2	3	0.39	0.46	0.50	0.50
Guava	2	2	2	2	0.39	0.31	0.50	0.33
Hulge Madag		5		5		0.77		0.83
Maize		6		6		0.92		0.99
Mango	2		2		0.39		0.50	
Green gram (Kharif)	10	14	15	9	3.75	2.15	3.74	1.49

(continued on next page)

Crop							
Green gram (Winter)	1	3	1		0.15		0.17
Muskmelon	3	3	3	0.59	0.46	0.75	0.50
Onion (Kharif)	95	93.5	95	18.44	14.59	23.32	15.73
Onion (Winter)	83.5	47.5	83.5	9.37	12.82	11.85	13.83
Onion (Summer)	14	0.5	14	0.10	2.15	0.12	2.32
Onion (Perennial)	8	1	8	0.20	1.23	0.25	1.32
Orange	4	2	4	0.39	0.61	0.50	0.66
Paddy	2		2		0.31		0.33
Pearl millet/Bajra (Kharif)	30.5	23	5	16.96	4.68	5.74	0.83
Pearl millet/Bajra (Rabi)	2	2		0.39		0.50	
Pigeon pea (Kharif)	1		1		0.15		0.17
Pigeon pea (Winter)	1		1		0.15		0.17
Pomegranate	85.35	18.35	83.35	3.62	13.10	4.58	14.13
Ridge gourd	2.5		2.5		0.38		0.41
Sorghum (Kharif)	15	4		2.96		1.00	
Sorghum (Winter)	20	7	11	4.73	3.07	1.75	1.82
Sorghum (Summer)	2		2		0.31		0.33
Soybean (Kharif)	14	10	17	2.76	2.61	2.49	2.82
Soybean (Winter)	6		6		0.92		0.99
Sugarcane (Winter)	2		2		0.31		0.33

(continued on next page)

TABLE 5.2 Change in cropping pattern in Ahmednagar district—cont'd

Crop name	Total cropped area (Acre)		Total irrigated area (Acre)		% of the area under irrigation		% of total cropped area	
	Before farm pond	After Pond	Before farm pond	After farm pond	Before farm pond	After farm pond	Before farm pond	After farm pond
Sugarcane (Winter)		3		3		0.46		0.50
Sugarcane (Summer)		2		2		0.31		0.33
Sugarcane (Perennial)	2.5	3.5	2.5	3.5	0.49	0.54	0.62	0.58
Tomato (Kharif)	19.5	27	19.5	27	3.85	4.15	4.86	4.47
Tomato (Winter)	12	15.5	12	15.5	2.37	2.38	2.99	2.57
Tomato (Summer)		2		2		0.31		0.33
Tomato (Perennial)	2	8	2	8	0.39	1.23	0.50	1.32
Watermelon		4		4		0.61		0.66
Wheat (Winter)	34.6	41	34.6	41	6.83	6.29	8.63	6.79

From: Analysis of primary field survey data.

TABLE 5.3 Net change in return with and without farm pond in Ahmednagar district.

Crops name	Average input cost (Rs)		Average crop Yield (kg/Acre)		Average gross return (Rs/Acre)		Average net return (Rs/Acre)	
	Before farm pond	After farm pond	Before farm pond	After farm pond	Before farm pond	After farm pond	Before farm pond	After farm pond
Pearl millet	18,180	17,207	1768	1885	22,291	25,848	4111	8640
Green gram	32,038	31,792	2892	4512	70,377	81,775	38,339	49,983
Onion (Kharif)	31,717	39,059	5703	7227.5	85,538	93,958	53,821	54,898
Onion (Kharif)–Drip	39,523	41,697	7915	8535	110,803	110,955	74,967	69,258
Soya bean (Kharif)	8593	12,930	568	1150	13,829	25,760	5236	12,830
Tomato (Kharif)	83,320	85,166	10,345	11,724	186,220	187,584	102,906	102,417
Tomato (Kharif)–Drip	87,389	87,838	11,612	12,122	192,755	193,952	105,366	106,114
Chickpea	11,460	13,983	822	1023	26,992	30,188	15,532	16,205
Onion (Rabi)	43,732	51,166	8285	8350	83,486	112,725	39,754	61,558
Onion (Rabi)–Drip	55,780	80,108	9374	12,678	121,865	170,362	66,085	90,254
Sorghum	17,067	17,850	1465	1580	23,735	27,270	6670	9419
Tomato (Rabi)	70,600	71,910	9072	16,163	199,580	193,954	128,980	122,043
Tomato (Rabi)–Drip	76,122	86,501	8165	18,143	200,034	235,868	123,912	149,367
Wheat	12,923	15,268	1777	1816	28,690	33,748	15,767	18,480
Pomegranate	105,133	106,213	10,589	11,422	459,303	510,653	354,170	404,440

From: Analysis of primary field survey data.

and aggregate water use at the farm level. Table 5.4 shows changes in the average amount of water applied to the crop, average crop productivity, and average economic water productivity both with farm ponds and without the same. For Kharif crops, the average water per unit area has largely remained the same except for onion and tomato grown with drip. Winter tomato grown with drip has shown a considerable reduction in water application per unit area, whereas, in the case of pomegranate with drip per unit area, the application of water has increased. Basically, with access to water in the pond throughout the year, the farmers are able to provide a sufficient amount of watering to the crop, which was not possible prior to adoption. Average economic water productivity before and after the farm ponds for major crops has remained largely unchanged.

If the change is assessed on a household level with the farm pond owners, the total area under winter crops has increased to the total cultivable area, whereas the acreage under summer crops has increased from almost zero to full acreage due to the presence of orchards (see Table 5.2). Drip irrigation has been taken up in a big way. Drip systems were installed for irrigating perennial crops as well as all other winter and summer crops. This includes tomato and brinjal (for which lately even mulching has been used), onion, sugarcane, watermelon, and muskmelon besides pomegranate cultivation on a large scale. Generally, the households have maintained about 20–40% of the total area under perennial crops, where both the gestation period and the risks are high, but at the same time, returns are also exponentially higher.

Post-farm pond the water productivity of crops both in physical (kg/m^3 of water) and economic terms (Rs/m^3 of water) had increased. However, the increase is significant only for some seasonal crops such as winter onion, winter onion with drip, winter tomato with drip, green gram, and pomegranate. This is not very high when compared to the cost of supplying the water through the pond. It is estimated that the cost per m^3 of water ranges from Rs. 11.5 to Rs. 13.2/m^3 of water. If we consider the average life of the system to be five years, the cost of water supplies from the farm pond works out to be around Rs. 4/m^3 (at a discount rate of 6%). Though not all that water used for crop production comes from the pond, the increase in water productivity in economic terms is not high enough to justify the huge investment.

5.2.2 The concerns about farm pond construction

The Marathwada region, which has witnessed the largest number of farm ponds being built, has a very fragile ecology. It is underlain by hard rocks (basalt) with extremely limited groundwater potential. The groundwater levels in the region witness sharp seasonal falls after the monsoon ends, as indicated by the time series data on (spatial) average water level fluctuations during the monsoon in three blocks (see Fig. 5.1), with the drying up of wells as a result of excessive pumping for irrigation. Fig. 5.1 shows that the water level fluctuations are very

TABLE 5.4 Water application and crop water productivity of different crops in Ahmednagar district.

Crop name	Average water applied to the crop (m³)		Average crop water productivity (kg/m³)		Average economic water productivity (Rs/m³)	
	Before farm pond	After farm pond	Before farm pond	After farm pond	Before farm pond	After farm pond
Pearl Millet/Bajra	2372	3176	0.75	0.59	1.73	2.72
Green gram	2667	2598	1.08	1.74	14.36	19.24
Onion (Kharif)	6137	5996	0.93	1.20	8.77	9.16
Onion (Kharif)–Drip	5842	5485	1.35	1.56	12.83	12.63
Tomato (Kharif)	8047	7667	1.28	1.53	12.79	13.36
Tomato (Kharif)–Drip	7637	7190	1.52	1.69	13.80	14.76
Chickpea	1892	1378	0.43	0.74	8.20	11.75
Onion (Rabi)	6393	6612	1.29	1.26	6.22	9.31
Onion (Rabi)–Drip	5487	5217	1.71	2.43	12.04	17.30
Sorghum	6016	6135	0.21	0.28	0.54	0.96
Tomato (Winter)	7548	7593	1.20	2.13	17.09	16.07
Tomato (Winter)–Drip	6460	5393	1.26	3.36	19.18	27.70
Wheat	5350	5461	0.33	0.33	2.95	3.38
Pomegranate–Drip	7285	7955	1.45	1.44	48.62	55.52

From: Analysis of primary field survey data.

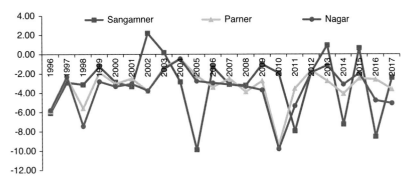

FIGURE 5.1   Groundwater fluctuation in three blocks of Ahmednagar (1996–2017).

high and can even drop during the monsoon in droughts (like in the year 2002, which was a drought year). Here, a negative fluctuation in water level indicates a water level rise during monsoon and vice versa.

The Ahmednagar district had earlier witnessed several initiatives from Civil Society Organizations for watershed management, with the pretext that such interventions would improve and augment groundwater recharge and raise the productivity of rain-fed crops. Electricity for pumping groundwater is almost free. On top of it, the government's launching of the scheme enabled the well-owning farmers to construct such reservoirs and pump out more water than what they would have otherwise been able to abstract.

The question is, how do the farmers get extra water to pump? This is where the unique geohydrology of the region becomes important. During the rainy season, given the limited storage capacity of the shallow basalt aquifers, if the monsoon is good, the wells overflow as there is no pumping due to a lack of demand for water in irrigating the Kharif crops. But the storage reservoirs have enabled them to pump water from the wells during this season and store it for future use. This water (also called the "rejected recharge" in hydrology) otherwise would have flown into the neighboring streams as base flow, contributing to the stream flows in the river channels. The region is dotted with many large and medium reservoirs on the tributaries of the Godavari River basin. With the construction of farm ponds, all this water gets converted into groundwater. Farmers also pump water from the wells during the winter and summer seasons (depending on availability) and store it in these ponds for irrigation.

Three aspects of this scheme are absolutely worrisome. *First*: the public funds are used for the benefit of private individuals who are largely wealthy. The subsidy is quite handsome, depending on the size of the pond. It is quite obvious that farmers with large holdings would go for large-sized farm ponds, getting larger subsidy support. Hence, there is an inherent problem in the scheme design itself. The subsidy amount ranges from Rs. 22,110 for a pond of size 15 m × 15 m × 3 m to Rs. 50,000 for a pond of size 30 m × 30 m × 3 m.

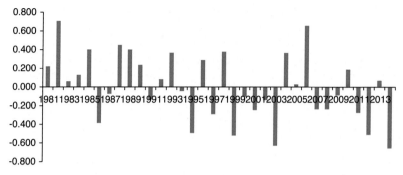

FIGURE 5.2 Rainfall deviation from mean annual rainfall (as a fraction of the mean) in Parner Block of Ahmednagar District, Maharashtra.

The very fact that all those who are building such farm ponds are rich farmers, means that the subsidy benefits get cornered by the rich and wealthy. The small and marginal farmers, who do not own wells (in fact, as we have seen in Chapter 2, only a small fraction of them own wells and pump sets), would never go for such farm ponds, knowing fully well that with high aridity and low rainfall, there is too little runoff that can be harvested from their fields. As Fig. 5.2 (which shows the annual rainfall for 34 years in Parner Block) suggests, droughts are very common in this region.

Second: the "common property" surface water, which otherwise would be contributing to the inflows in many large reservoirs in the Marathwada region, is now increasingly being appropriated by private individuals through their wells during the monsoon season. Moreover, the groundwater draft during the summer season also becomes large, with the adoption of perennial tree crops like pomegranate.

Third: these decentralized surface reservoirs evaporate a lot of water. A pond having a water-spread area of 1000 sq.m will evaporate nearly 2500 m^3 of water annually if we consider the potential evaporation to be of the order of 2500 mm per annum. This is sufficient to provide full irrigation (of nearly 500 mm) to a crop on half a hectare of land. When we consider the fact that several thousand such ponds are already built in various districts of Maharashtra, the devastating ecological damage of the scheme becomes a matter of anyone's guess.

Our field study shows that the water from these farm-level reservoirs had been put to highly productive use, as indicated by the increase in yield and income and slightly higher water productivity of crops. But it benefits a few large and medium well-owning farmers who grow high-value crops and who are already wealthy. A good, well-irrigated pomegranate farm produces fruits, with a net income return of Rs. 48–50/m^3 of water, against Rs. 4–5/m^3 for wheat. But the price that society has to pay for this is enormous. As per official statistics of the government of Maharashtra, a total of 51,500 such ponds have already been constructed in the state and the total expenditure would be anywhere in the range of 150–160 crore rupees (1.5–1.6 billion rupees). On top of it is the cost of resource depletion, in the form of millions of people in the villages being deprived of water for basic needs.

While the objective of the scheme is noble, i.e., to promote in situ water conservation for protective irrigation of seasonal crops, the feasibility of in situ water harvesting was not examined. What the scheme has done in reality is just the opposite. The availability of a large lined pond enables the farmers to mine groundwater and stock it for later use, and expand the area under perennial crops like pomegranate and sugarcane. This is a clear case of schemes being designed with no proper mechanism put in place for monitoring to ensure that the funds are not misused.

5.3   Solar irrigation pumps

5.3.1   The solar irrigation cooperatives of Gujarat

As the idea of rainwater harvesting structures for recharging aquifers is being put to dust, the new fashion is "solar irrigation pumps." Even while serious concerns are raised about their economic viability and environmental sustainability in water-scarce regions with limited groundwater (Closas & Rap, 2017), some water professionals from India think that growing crops would no longer be profitable and that the farmers in India, who own wells, should now produce "solar crops." According to them, the farmers should run their wells using solar energy and sell the surplus energy to the utilities, and a large capital subsidy should be made available from the government for the purchase of solar PV systems. To support this idea, a narrative is created that the power distribution for the agriculture sector in Gujarat is too poor and unreliable (see Durga et al., 2021; Shah et al., 2017), though in another article (see Shah & Verma, 2008) the same authors claimed that the 8-hour power supply in Gujarat for the agriculture sector is one of the best in the country.

The business model envisioned for the solar irrigation cooperative is as follows. The government gives a subsidy of Rs. 40,000 (US $500) (actually it is far higher now, and the government gives an 85% subsidy in some cases as the cost per kW comes to around Rs. 1 lac, i.e., US $1250!) for a solar PV system having a power generation capacity of 1 kW, which is sufficient to replace a connected load of 0.50 kW in the existing power distribution system. If the farmer has a 5-HP (3.75 kW) electric pump, it will be replaced by a 10-HP (7.5 kW) solar PV system, for which the government subsidy would be Rs. 4 lac (US $5400).

The inbuilt assumption is that the farmer would then stop buying electricity from the utility and instead would also start supplying the surplus electricity (after using it for running his/her own pump) to the Utility. The condition to make this model work is that since the farmer is providing yeoman service in protecting the environment, the Utility should purchase this power at a price 10–12 times higher than the price at which it now offers electricity to the same farmer (50 paise per kWh). The virtues attached, according to the proponents of the scheme, are numerous: reduction in carbon emissions from agriculture; reduced groundwater pumping; reduced energy subsidy; and a win-win situation for the farmer and the electricity utility (Durga et al., 2021; Shah et al., 2017).

Such claims are quite bizarre. With 16 million such pumps (as proposed by its proponents), the government will have to spend INR. 6.4 lac crore (20 times the amount spent on NREGS every year), i.e., US $85 billion, while the total cost of production and supply of electricity in agriculture is less than US $ 8 billion/year. There cannot be a better recipe for ruining the government finances and emptying the coffers. It is a different matter if the government does decide to spend this much amount of funds, a lion's share of it would end up with the rich (large and medium) farmers, who would go for large-capacity solar PV systems.

Further, the assumption that farmers would be able to run their motor as and when needed using a solar PV system is a flight of fancy. The availability of solar energy required to produce the electricity for running the pump can be hampered by cloud cover and poor sunlight, respectively. The minimum requirement for the farmer to make his energy production system self-sufficient is to have large-capacity batteries to store the electricity produced during hours of good sunshine. On the other hand, in the summer season, while the crop water demand increases sharply, the performance of the photovoltaic cells suffers due to high temperatures, with a sharp reduction in voltage (Kumar, 2019). The storage batteries are prohibitively expensive. In the absence of batteries, the farmer will have to depend on the power supply from the grid. The argument that the Utility has to buy power from the farmers at a rate far higher than the price at which they supply power is untenable. In fact, the farmers are quite aware of the impracticality of depending entirely on solar PV systems for running their irrigation pumps. Because of this reason, they are unwilling to surrender their grid connections wherever such systems exist.

Even if we assume that the utility agrees to such a power purchase model, it would require "net metering" (two-way metering). That being the case, there is no reason to believe that the farmer would not tamper with the meters (while using power from the grid and also selling power to the grid). As a matter of fact, in this case, there is a greater incentive to tamper with meters. That being the case, it is somewhat strange that such proposals have come from researchers who have been arguing against metering agricultural power connections in India for nearly two decades, citing the same reason (see Shah, 1993; Shah et al., 2004). On the contrary, if we have faith in the farmers, why cannot we just install meters for every electric pump and supply reliable and high-quality power to the farmers and charge for it? This way, we can find solutions to a few of the problems facing the groundwater and energy sectors.

Obviously, under both scenarios, monitoring would be required to make sure that there is no tampering with meters. Hence, it is far more sensible to start metering electricity consumption in the farm sector to improve electricity use efficiency and cut down power consumption in the agriculture sector, thereby improving the financial workings of power utilities. As a result, we can cut down carbon emissions. For this, we need the political will and not large public funds to spend on subsidies for unproven technologies like the "solar PV systems for small Indian farms." The route, which is suggested now, is expensive, indirect, and non-workable. On the contrary, it would lead to further depletion of groundwater, a huge burden on the government exchequer (for subsidizing solar pumps), and drive the power utilities to total bankruptcy.

But, a series of articles were published in the Economic and Political Weekly and some national dailies with the aim of promoting this model (see, Durga et al., 2021; Kishore et al., 2014; Shah et al., 2014, 2017; Verma et al., 2019). The entire agenda seems to be the privatization of electricity generation (leaving it in the hands of rich and influential farmers) and the junking of the power utilities. One such article published in Economic and Political Weekly (Shah et al., 2017) had the sole aim of establishing the economic viability of solar pump irrigation but was based on flawed analysis using many faulty assumptions.

The article begins with the statement: "Anand, the Gujarat town that gave India its dairy cooperative movement, has now spawned in Dhundi village the world's first solar cooperative that produces Solar Power as a Remunerative Crop" (Shah et al., 2017, p. 14). While the dairy cooperatives were founded at a time when there was an acute shortage of milk in the country and dairy farmers were being exploited by middlemen, there is no scarcity of electricity today, especially in states like Gujarat (Sahasranaman et al., 2018). The real issue is of farmers being provided electricity free or at nominal charge leading to inefficient energy use and depletion of ground water while increasing the financial burden on the exchequer (Chand, 2018). Hence, the two scenarios are drastically different, and there are no compelling reasons for the farmers in states like Gujarat to go for solar cooperatives when it is very easy to get power connections for agriculture from a well-established grid, except that the electricity produced is "clean." Yet no data are provided by the authors regarding the clean energy benefits of using solar power over fossil–fuel-based electricity.

Without checking for rationality, the authors claim that solar pumps else-where "continue to run whether the farmers need the power to irrigate or not since surplus solar energy goes waste anyway." It seems the authors mean to say that in the absence of a system of selling power to the grid, the farmers will keep on pumping out water whenever electricity is generated, even if it damages their pumps and floods their fields, destroying their crops. This is somewhat intriguing. Perhaps the authors assume that the irrigation pump is connected to the solar PV system without having any control device in place for switching on and off the power generation from the panel and power supply to the motor. Therefore, the intention seems to be to justify the feed-in-tariff (FiT) system introduced by the authors in their pilot project. Going by a report published in the Indian Express (Nair, 2016) and a discussion with the adopter farmers, the authors have obfuscated the fact that even with a working FiT, the farmer members of their cooperative decided to pump out water using the surplus electricity produced and sell it to their neighbors instead of selling power to the Utility, for the simple reason that the profit was higher. Thus, it belies the authors' own claim that farmers simply pump out water if electricity is available.

5.3.2 Misleading numbers

Shah et al. (2017) discussed six models for solar irrigation that are considered by non-governmental organization (NGOs) and state governments, including the Dhundi village model, for examining their pros and cons, and hurriedly concluded that the Dhundi model is the only model with no flaws and a successful pilot intervention. While some data on the economics of solar power are provided based on the 18-month experience in Dhundi, no such data were provided for the other models. They say that with the model being farmer-centric and the member farmers having a complete stake in the project, all that is needed for the project to become bankable and scalable is a "slightly higher capital cost subsidy of around INR 45,000/kWp to solar cooperatives and a guaranteed Feed in Tariff or FiT of Rs. 5/kWh". Interestingly, the subsidy obtained by the farmers to make the cooperative work is much higher. Using the data presented in the article, i.e., a capital cost subsidy of Rs. 46 lac for installing a system with a capacity of 56.4 kWp, the subsidy works out to be Rs. 82,000/kWp. The actual figure considered by the authors is Rs. 45,000/kWp (though it was wrongly put as Rs. 45/kWp), at such a low subsidy, there will be no takers for the solar PV systems.

The authors advise that India's policy framework for promoting solar pumps should aim to incentivize farmers to conserve energy and groundwater to control over-exploitation. But the proposal raises too many doubts about the economics of solar power and how it impacts groundwater and energy use. For instance, if all farmers were to use energy efficiently and feed the excess solar electricity generated into the grid, what should be the price for electricity that the farmer should receive? But the authors do not provide any numbers for the purchase price for electricity being produced by the farmers. As we will see in Section 5.3.4, even a price of Rs. 7.1/kWh, which includes the INR 4.60 offered by the power utility and Rs. 2.50 offered by IWMI, is insufficient to create an incentive among farmers to limit their pumping (Sahasranaman et al., 2018).

Further, what will be the cost to the exchequer by way of output subsidies? Can the grid take all the power that the farmers produce? The grid needs electricity at night as well, while solar power is available only during the daytime. This means batteries are required for storing the electricity generated by solar PV systems. Does the subsidy include the cost of the battery? Will non-farmers pay Rs. 7 plus for this power? If not, the government will have to subsidize this expensive power for other consumers.

Interestingly, in Dhundi, solar pumps replaced diesel pumps, and all the economics was worked out by the authors accordingly. But they go on to suggest the replacement of electric pump sets that are connected to the grid with solar pumps. Even if the government goes by the authors' prescription of promoting 15 million grid-connected solar pumps (under their SPaRC model), it would cost the exchequer nine lac crore rupees (app. US $110 billion) in the form of capital subsidies, whereas the benefits are far too low and also highly uncertain. Over and above this, the adverse impact such a step will have on our groundwater economy will be enormous (Sahasranaman et al., 2018). The reason is that even the millions of farmers who are currently buying water from well owners on an hourly basis and who use water efficiently and choose cropping systems to obtain higher water productivity will have little incentive to do so and instead switch over to some water intensive crops to maximize their returns per unit of land. This is what the evidence suggests (Kumar et al., 2010).

5.3.3   The logic behind subsidizing solar irrigation pumps

The proponents of solar irrigation, while pushing for an irrational capital subsidy, seem to cash in on the overall political climate in the country that favors clean energy over fossil fuel-based power rather than basing their calculations on environmental economics that consider social costs and benefits. The International Water Management Institute (IWMI) and CCAFS (the CGIAR program on Climate Change, Agriculture and Food Security), which piloted the Dhundi cooperative, offered farmers a green energy bonus of Rs. 1.25/kWh and a water conservation bonus of another Rs. 1.25/kWh, to take the total FiT to Rs. 7.13/kWh for the power sold. The total capital investment in installing the solar pumps, micro-grid, cabling, switches, transformers, and meters in Dhundi was INR 50.65 lac, with SPICE members contributing Rs. 4.65 lac and the CCAFS/IWMI contributing Rs 46 lac. At one place in the article, the authors say that a contribution of 25,000/kWp by the farmer is nearly 40% of the total investment, which means that the subsidy is 37,500/kWp. At another place, it says, "with a capital subsidy of INR 45,000/kWp, a generation factor of 4.5 kWh/kWp/day and a WARR of 6/kWh (which a FiT of 5/kWh would easily generate)," Dhundi-pattern SPICE would be bankable, with the economic IRR exceeding 21% over a 20-year project cycle. "The benefits to DISCOMs and water buyers would be extra" (Shah et al., 2017, p. 16). The fact remained that, for a total investment of 50.65 lac rupees, the cost per kWp works out to be Rs. 96,000.

Had the Dhundi SPICE members taken grid power connections, the authors say, even with the farmers using only two-thirds of their entitlements for irrigation, the Madhya Gujarat Vij Company Limited (MGVCL) would have had to bear a subsidy burden of over Rs. 4 lac per year and additionally would have been required to invest Rs. 12 lac to connect these tube wells to the grid, at an amortized annual cost of Rs. 1.2 lac. Thus, according to the authors, Dhundi SPICE has saved MGVCL all these costs. Also, the article suggests that Dhundi SPICE will leave MGVCL better off by Rs. 8 lac/year for 25 years under the purchase agreement of the farmers. As we will see in the subsequent paragraphs, the calculations are based on false assumptions.

A simple calculation shows that the capital cost for solar power (Rs. 50 lac at a discount rate of 8%) alone works out to be Rs. 4.4 lac per year for an anticipated life span of 25 years. If the calculations are done considering that solar panels have a useful life of only 12 years, which is a realistic estimate for Indian conditions with the market being flooded with poor quality solar panels and the fast rate of degradation of the panels owing to the hot and dusty conditions, the capital cost subsidy goes up to Rs. 6.2 lac per year. On the other hand, the estimates of cost-saving due to the non-supply of grid power to the farmers who are now using solar energy are based on the assumption of farmer entitlement to electricity.

In order to show that the reduction in subsidy burden for the utility due to the adoption of solar power by farmers is very large (Rs. 4 lac per year), the authors assumed that the farmers are entitled to 1,62,000 units of electricity annually from the grid. However, there is no such thing as "entitlement to electricity from the grid." The amount of electricity used by the farmers from the grid depends on their actual energy demand and ideally, that should form the basis for estimating the cost saving through subsidy reduction. The electricity used by the farmers for irrigation is only 45,350 kWh for a period of 18 months, as per the figures provided by the authors, which works out to about 30,000 kWh annually. Here again, this 30,000 kWh includes not just the electricity the farmers used for irrigating their fields but also that used to pump out water for sale to other farmers, from which they made a profit of about Rs 5 lac in 18 months, or an average of Rs 3 lac per year. In any case, the actual cost saving, therefore, is Rs. 1.2 lac only (for 30,000 kWh of electricity, with an average cost of power generation and supply of Rs. 4.5 per unit and a current price of Rs. 0.5 per unit for the farm sector). The direct benefit of power generation using a solar PV system is Rs. 270,000 per annum (60,000 × 4.5). The additional benefit of producing nearly 60,000 kWh of clean energy is Rs. 29,000 per annum.

In order to inflate the benefits of having solar power, a figure of Rs. 12 lac or Rs. 1.2 lac per annum is used by the authors as the capital investment needed to connect the tube wells to the grid. The actual connection charge in Gujarat (where there is a very dense network of tube wells and bore wells already connected to the power grid) is only Rs. 10,000 per farmer for getting a connection for a 7.5 HP pump. Hence, the cost saving due to this is Rs. 60,000 (for all six farmers put together) or approximately Rs. 5000 per year.

The average cost incurred by the Distribution Companies (DISCOMs) to provide a new electric tube-well connection with the transformer, switchgear, cables, and other fittings is about Rs. 1.75 lac. If a farmer presently using a diesel pump installs a solar pump, this cost can be avoided, so long as the farmer uses the electricity for personal use only. However, the moment farmer decides to connect to the grid to make use of the FiT system, these systems will have to be installed. So, there will be no savings for the DISCOMs in this scenario. Again, there will be no savings for the DISCOMs in the case of conversion of existing grid connected electric pumps to solar, as the transformer, switchgear, etc. are already installed incurring the above-mentioned costs.

The green energy bonus of Rs. 1.25/kWh and a water conservation bonus of another Rs. 1.25/kWh work out to a subsidy of Rs. 2.5 per unit of power produced. This is the minimum amount that the utility will have to shell out if it has to replicate the model in other areas. The power company gets the electricity from farmers at Rs. 7.13 per unit, whereas the average cost of producing electricity through conventional sources is only Rs. 4.5/kWh. This means every unit of electricity purchased from the farmers costs the Utility an extra Rs. 2.16, even after considering the clean energy benefit of Rs. 0.47/kWh (i.e., 7.13 – (4.5 + 0.47)). So, the net loss to power utility per year due to the purchase of solar power from the solar cooperative is Rs. 64,700 (30,000 × 2.16). Hence, the net income is Rs. 339,300 per annum.

Even under the most optimistic scenario of the solar PV system working efficiently for 25 years, the total of Rs. 3.39 lac of annual benefit against an annualized cost of Rs. 4.4 lac is simply unviable. This calculation does not include the one-time investment for creating the microgrid necessary for evacuating the power generated by the farmers of the solar cooperative, which costs two million rupees (Rs. 20 lac). Thus, the subsidy for the equipment that is used to generate power that helps farmers sell water to other farmers and feed the grid is money taken from the taxpayers and gifted to the "solar farmers."

5.3.4 Unreal groundwater and energy savings

Regarding groundwater use, Shah et al. (2017) claimed, "Although solar pumps became operational in January 2016, the power purchase agreement came into force only in May. Hence, before May 2016, they used all the energy generated for pumping groundwater." Were they producing more food/cash crops before May? If not, were they pumping out more water than required? According to the authors, at Rs 250/bigha, solar farmers earn 10/kWh by selling water to other farmers, 40% more than the FiT of 7.13/kWh that they earn by selling power to MGVCL. Without the power buy-back guarantee, the authors say, the adverse consequences would be much greater, resulting in as much groundwater depletion as the use of free grid power has caused throughout Western India. If so, why would they sell power to the grid if selling water is more lucrative? They will get paid immediately too. In fact, the farmers in Dhundi were selling water even after the power purchase agreement came into force (Nair, 2016).

If the whole purpose of FiT is to create an incentive for saving electricity, it is quite intriguing that the authors shy away from suggesting a far more straightforward option of introducing energy metering and a pro-rata tariff for agricultural power supply in western India. Clearly, the authors have not understood how the incentive structures work. Even with a power purchasing agreement, with a highly attractive tariff, the farmers will have little incentive to save groundwater, as the amount they can earn by irrigating some of the water-intensive crops per unit of electricity would be far greater. Even for the low water-efficient wheat, the net return per kWh of electricity used for irrigation works out to be around Rs. 50. For many cash crops like cotton, the net return per kWh of electricity will be far higher. This means the farmers would have used up much of the electricity and groundwater to irrigate their own and their neighbors' fields before they find the economic sense to start selling power to the grid (Sahasranaman et al., 2018). The condition for the sale of surplus electricity to the grid will exist only in areas with poor groundwater environments and where the entire resource gets depleted and irrigation wells dry up by the beginning of summer when solar power generation generally peaks.

To support the claim of groundwater and energy saving, an analysis of water use efficiency and water productivity in crop production between solar pump irrigators and neighboring farmers who run tube wells energized by the grid should have been provided. If the authors' claim of water and energy saving is to be valid, then the WUE and water productivity in crop production should be much higher for solar-pump irrigators. However, the required analyses were not attempted. Instead of coming up with a solution for the problem of groundwater exploitation, the authors come up with the suggestion of offering solar PV systems almost free, which would aggravate the current problem, and then persuade the government to solve it by offering an unviable price, which would only cause distortion in the energy market (Sahasranaman et al., 2018).

5.3.5 Summary

One of the main arguments for decentralized solar irrigation with FiT system seems to be that metering electricity supplied to farmers from the grid to ensure efficient energy use is a herculean task to accomplish, given the exceptionally high transaction cost and farmers' resistance to metering electricity (see Shah et al., 2017). If the excuse for getting rid of the meters for agro wells a few decades ago was the "transaction cost," the power utilities can use the same excuse to shy away from "net metering" (Sahasranaman et al., 2018). Also, the solar PV systems are very expensive, and the widely publicized benefits of producing "clean energy" are too meagre in economic terms to offset the huge capital costs (Bassi, 2018). It might be viable at the "utility level" (with economies of scale) and under certain conditions, particularly with sophisticated operation and maintenance, but not at the level of micro farms. The manufacturing of solar PV systems itself leaves a large carbon footprint. No matter what the transaction cost is, we will ultimately have to formalize our energy and water economies for better productivity, growth and sustainable resource use.

Our initial assessment suggests that the solar irrigation cooperative model piloted in Dhundi and later implemented in many other villages of south Gujarat is unviable. Most of the figures that are shown as benefits of using solar power are nothing but financial support from the agency that has promoted the experiment and government support that is not based on any scientific rationale. Such models are simply unsustainable for India's power sector, which is already bleeding. A system that is economically unviable at present, even after considering the positive externalities on the environment (in the form of clean energy) (Bassi, 2015, 2018), cannot be made viable or bankable through subsidies and cash doles. If the government is willing to pump in billions of dollars as input and output subsidies annually, any crop that receives such subsidy support would become remunerative for the farmers but will ruin the economy. The issue is what societal benefits or positive welfare effects are made of such actions. Instead of committing such economic blunders, we need to wait until solar PV systems become cheap through technological breakthroughs and market conditions become highly favorable. Even then, the threat of uncontrolled groundwater abstraction, especially in areas with limited groundwater potential, will be real (Closas & Rap, 2017). As it stands now, to keep such experiments afloat, a large amount of money will have to be pumped in. Replication of such experiments can only perpetuate pervasive subsidies, which, once rolled out on a large scale, will be difficult to revoke (Sahasranaman et al., 2018).

5.4 Concluding remarks

The farm pond model and the solar irrigation cooperative model, discussed in this chapter, raise two serious concerns. The first is about the design of schemes with huge built-in capital subsidies without any clarity on the real welfare effects, i.e., whether they are positive or negative. The question here is: on what ground can the huge subsidy given to the farmers (for constructing farm ponds and producing solar power) be justified when there is no water saving or power saving? Capturing water from a catchment (even if we assume that the farmers are harnessing runoff from their fields) in a water-scarce region with limited rains and surface water does not produce any public good *per se*.

On the other hand, producing expensive solar power (even after considering the clean-energy benefit) in a region that is a power-surplus also does not result in any public good. Both subsidize the rich. The second problem is the impact these schemes produce that is totally unintended and undesirable. Our analysis shows that in both cases, the scheme leads to increased use of groundwater, and in the case of solar irrigation pumps, it also leads to increased energy use by the farmers with larger government subsidies. Analysis carried out by Kumar (2019) on the techno-economic feasibility of solar-powered irrigation pumps in South Asia, considering four countries viz., India, Pakistan, Bangladesh, and Nepal, shows that it would work reasonably well only in the Terai region of Nepal and the eastern Gangetic plains of India because of the following favorable conditions: (1) the demand for irrigation water throughout the non-monsoon periods and in certain cases also during the monsoon, with standing crops in all three seasons

(Kharif, winter, and summer); (2) the availability of solar energy throughout the months of the year during which irrigation water demand for crops mainly exists; (3) the sufficient availability of groundwater throughout the year at reasonably shallow depths; (4) rural electrification is still very poor and obtaining diesel is difficult due to poor road connectivity; and (5) very small land holdings reduce the economic viability of irrigation using diesel engines.

References

Bassi, N. (2015). Irrigation and energy nexus. *Economic & Political Weekly, 50*(10), 63.

Bassi, N. (2018). Solarizing groundwater irrigation in India: A growing debate. *International Journal of Water Resources Development, 34*(1), 132–145. https://doi.org/10.1080/07900627.2017.1329137.

Chand, R. (2018, November 15). *Innovative policy interventions for transformation of farm sector* [Presidential address]. In *26th Annual Conference of Agricultural Economics Research Association (India)*.

Closas, A., & Rap, E. (2017). Solar-based groundwater pumping for irrigation: Sustainability, policies, and limitations. *Energy Policy, 104*, 33–37. https://doi.org/10.1016/j.enpol.2017.01.035.

Durga, N., Shah, T., Verma, S., & Manjunatha, A. V. (2021). Karnataka's "Surya Raitha" experiment: Lessons for PM–KUSUM. Economic and Political Weekly. https://cgspace.cgiar.org/handle/10568/116411.

Kale, E. (2017). Problematic uses and practices of farm ponds in Maharashtra. *Economic and Political Weekly, 52*(3), 20–22.

Kishore, A., Tewari, N. P., & Shah, T. (2014). Solar irrigation pumps: Farmers' experience and state policy in Rajasthan. *Economic and Political Weekly, 49*(10), 55–62.

Kumar, M. D. (2019, September 10-11). Solar powered irrigation systems (SPIS) in South Asia: When and where do they become best bet technologies? [Presentation]. In *Regional Dialogue on Strengthening Institutional Capabilities for SPIS in South Asia*. FAO and ICIMOD.

Kumar, M. D., Singh, O. P., & Sivamohan, M. V. K. (2010). Have diesel price hikes actually led to farmer distress in India? *Water International, 35*(3), 270–284.

Nair, A. (2016, August 14). Gujarat: Solar co-operative at Dhundi village sells water instead of electricity. *The Indian Express, Ahmedabad edition*. https://indianexpress.com/article/india/india-news-india/gujarat-solar-co-operative-at-dhundi-village-sells-water-instead-of-electricity-2974172/.

Sahasranaman, M., Kumar, M. D., Bassi, N., Singh, M. S., & Ganguly, A. (2018). Solar irrigation cooperatives: Creating a Frankenstein monster for India's groundwater. *Economic and Political Weekly, 53*(21), 65–68.

Sahasranaman, M., Kumar, M. D., Singh, Verma, M. S., Perry, C. J., Bassi, N., & Sivamohan, M. V. K. (2021). Groundwater-Energy Nexus: What Will Solar Pumps Achieve? *Economic and Political Weekly, 56*(11), 13 March.

Shah, T. (1993). Groundwater markets and irrigation development: political economy and practical policy, Oxford University Press, pp. 267, New Delhi.

Shah, T., Scott, C. A., Kishore, A., & Sharma, A. (2004). Energy irrigation nexus in South Asia: Improving groundwater conservation and power sector viability (Research Report 70). Colombo, Sri Lanka: International Water Management Institute.

Shah, T., & Verma, S. (2008). Co-management of electricity and groundwater: An assessment of Gujarat's Jyotirgram Scheme. *Economic and Political Weekly, 43*(7), 59–66.

Shah, T., Durga, N., Rai, G. P., Verma, S., & Rathod, R. (2017). Promoting solar power as a remunerative crop. *Economic and Political Weekly, 52*(45), 14–19.

Shah, T., Verma, S., & Durga, N. (2014). Karnataka's smart, new solar pump policy for irrigation. *Economic and Political Weekly, 49*(48), 10–14.

Verma, S. (2019). Solar irrigation pumps and India's energy-irrigation nexus. *Economic and Political Weekly, 54*(2), 62–65.

World Bank. (2020, February 17). *World Bank signs agreement to improve groundwater management in select states of India.* World Bank. https://www.worldbank.org/en/news/press-release/2020/02/17/improving-groundwater-management-india.

Chapter 6

Large reservoirs for improving groundwater sustainability in over-exploited regions

6.1 The background

Groundwater depletion has been a serious problem confronting the western Indian state of Gujarat for nearly four decades (see Bhanja et al., 2017; Chindarkar & Grafton, 2019; Kumar, 2007; Moench, 1994). During the past three decades, several researchers have analyzed these problems and come up with options and theoretical propositions to solve these problems. They include: a legal framework that enables local communities to control groundwater draft (Moench, 1994); instituting water rights (Moench, 1998), including tradable rights in groundwater (Kumar, 2000, 2007); pro rata pricing of electricity in the farm sector and energy rationing (Kumar et al., 2011, 2013); decentralized water harvesting structure (Shah et al., 2009); power supply rationing in the farm sector (Shah et al., 2004); inter-basin water transfers and recharge of depleted alluvial aquifers (Ranade and Kumar, 2004); and indirect recharge of groundwater through the supply of irrigation water through gravity (Kumar et al., 2014; Jagadeesan & Kumar, 2015). Parallelly, the state government has been responding to these problems mainly through technical solutions, with major schemes and programs designed to implement them.

The major schemes launched by the government of Gujarat are the Sardar Patel Participatory Water Conservation Programme launched in 1999; and the *Sujalam Sufalam Yojna*. Some researchers have analyzed these schemes and eulogized them as powerful instruments for replenishing depleted aquifers and triggering agricultural growth. Shah et al. (2009) reported that a participatory water conservation program involving the construction of several thousands of check dams and intelligent rationing of power supply to the farm sector under *Jyotigram Yojna* led to "miracle growth" in agriculture and improved groundwater conditions (Shah et al., 2009). Patel et al. (2020) tried to empirically show that the groundwater build-up in the Saurashtra region after 2002 was entirely due to decentralized water harvesting there. Both Shah and Verma (2008),

Groundwater Economics and Policy in South Asia. DOI: https://doi.org/10.1016/B978-0-443-14011-2.00003-6
129

and Bhanja et al. (2017) reported that the *Jyotigram Yojna*, which involved providing a 24×7 power supply to the rural domestic sector and separating the agricultural feeder line from that of the rural domestic supply, led to reduced electricity use in agriculture and a consequent reduction in groundwater draft and build-up of groundwater storage during 2002–14. Bhanja et al. (2017) further reported: "This groundwater storage rejuvenation may possibly be attributed to the implementation of ingenious groundwater management strategies in both Indian public and private sectors" (Bhanja et al., 2017, p. 5).

The conclusions drawn in these papers are of potential relevance to groundwater management policy as they dealt with the issue of how certain policy instruments and programs can be used to control groundwater over-exploitation in semi-arid regions. Stated briefly, these authors generally argued that in recent years, the long-term decline in Gujarat's aquifer stocks has been reversed and that according to Shah et al. (2009), also led to "miracle growth" in agriculture. The same strategy was therefore advocated for other semi-arid regions of the country. However, there has been no critical scientific evaluation of the schemes for their actual impacts on either groundwater or irrigated crop production dependent on groundwater wells.

Worldwide, the roles of large gravity irrigation systems for water banking (Rotiroti et al., 2019; Watt, 2008), enhancing the quantity and quality of groundwater, and raising water levels (Lytton et al., 2021; Rotiroti et al., 2019), improving the sustainability of well irrigation (Perry, 2007; Qureshi et al., 2010), and reducing the energy footprint of groundwater pumping (Jagadeesan & Kumar, 2015; Vyas, 2001) have long been well recognized. The water use hydrology of the Indus Basin Irrigation System, as briefly discussed in Chapter 2 and Chapter 4, amply demonstrates the importance of gravity irrigation in sustaining well irrigation in semi-arid and arid regions. Though Gujarat is known for several large gravity irrigation systems (IRMA/UNICEF, 2001), its role in improving groundwater balance in over-exploited regions has not been as much a subject of academic interest in the field of groundwater management as water harvesting and artificial recharge. This is in spite of the large-scale water import made possible through a large network of canals from the Sardar Sarovar project to the water-scarce regions, which also experienced groundwater over-draft and depletion.

On the basis of an analysis of empirical data on regional rainfall, groundwater level fluctuations during monsoon (prior to and post the policy interventions and programs), and surface water availability for irrigation, we show in this chapter that the groundwater build-up in the state during 2002–14 is falsely attributed to the above-mentioned programs as the analysis of the changes happening on the groundwater and agricultural fronts in terms of the "causes and effects" lacked a nuanced understanding of the hydrological regime, geohydrological setting, surfacewater–groundwater interactions and larger socio-economic changes.

6.2 Impacts of government policies and programs: tall claims

The water and agriculture sectors of Gujarat have been in focus for nearly a decade for the "rare feat" the state has achieved in decentralized water management and miracle growth in the agricultural sector (Kumar & Perry, 2019; Kumar et al., 2010; Mehta, 2013). Papers published since 2009 in this regard are quite unique in their tone and tenor, such that the claims made are very big. For instance, Bhanja et al. (2017) concluded that "…. where huge groundwater consumption is widely known to be leading to severe dwindling of groundwater resource in recent times, previously unreported, discernible GWS (read as Ground Water Storage) replenishment can also be observed in certain Indian regions. Specifically, in parts of western (HMZ B) and southern (HMZ E) India, GWS_{obs} decreased at the rate of $-5.81 \pm 0.38\,km^3$/year (in 1996–2001) and $-0.92 \pm 0.12\,km^3$/year (in 1996–2002), and reversed to replenish at the rate of $2.04 \pm 0.20\,km^3$/year (in 2002–14) and $0.76 \pm 0.08\,km^3$/year (in 2003–14), respectively".

A major inference of their analyses pertaining to Gujarat is that there had been a significant groundwater build-up (storage renewal trend of up to 4 m) in the state during the period 2002–14 and that the reduction in agricultural power supply from 16 billion units to 10 billion units between 2001 and 2006 might have resulted in a reduction in groundwater withdrawal for irrigation (as electricity-driven wells are responsible for over 90% of the groundwater withdrawal). Their findings pertaining to Gujarat are in line with the claim made earlier by other researchers in the context of Gujarat (see Shah & Verma, 2008).

The argument about the positive effect of agricultural power supply regulation on restricting groundwater abstraction by farmers was claimed to be validated by several statistical tests using sophisticated mathematical models to establish the potential effect of a reduction in agricultural electricity usage on aquifer storage change. What the authors have done is use a variety of statistical tests and a global hydrological model to establish a simple and straightforward relationship between two phenomena, which in reality, is very tenuous due to a variety of complex factors influencing groundwater storage changes and energy demand in agriculture, on which the authors seem to have no grasp.

Whereas Shah and Verma (2008) argued on the basis of very skewed primary data on hours of pumping and irrigated area from a tiny sample of farmers scattered across the state that, after the introduction of *Jyotigram Yojna*, the groundwater pumping for irrigation drastically reduced, and built a case that the restricted power supply to agriculture and the lack of opportunity to steal electricity led to farmers reducing their pumping, which in their opinion affected groundwater markets adversely. However, the analysis by Shah and Verma (2008) did not take cognizance of the larger changes happening in the rural farm sector in the state of Gujarat with hundreds of thousands of small and marginal farmers getting free power connections for well drilling, which resulted in a drastic reduction in demand for irrigation services from these former water buyers and a consequent reduction in the pumping by water-selling well owners (Kumar & Perry, 2019).

Shah et al. (2009), which examined the agricultural growth in the state during 2001–02 and 2007–08, on the other hand, reported that the state achieved a miracle growth of 9.6% in the primary sector during the said period and further argued that decentralized water harvesting sponsored by the state government was mainly responsible for this since 2001. According to their study, these structures enhanced the groundwater available through wells for irrigation expansion, though the study did not quantify the incremental recharge from such structures. This claim was contested by Kumar et al. (2010), which analyzed the historical trend in outputs of various crops and agricultural GDP in the state for nearly 60 years. The study showed that the estimated "miracle growth rate" was due to the selection of the wrong base year (i.e., 2000–01) for estimating agricultural growth, which happened to be a second consecutive year of drought. During this year (2000–01), the agricultural output in value terms dropped below that of 1998–99. In fact, during 1999 and 2000, Gujarat witnessed one of the most severe droughts of the century. This had a remarkable impact on the performance of agriculture, with the value of outputs declining from Rs. 23,098 crore in 1998–99 to Rs. 14,092 crore in 2000–01 (source: based on Dholakia & Datta, 2010).

6.3   The complex hydrological regime of Gujarat

Hydrologically, Gujarat is unique in many ways. It has the highest orographic gradient in the world, with the mean annual rainfall varying from 350 mm in Kachchh to around 2000 mm in certain parts of south Gujarat (Valsad and parts of Dangs district). The inter-annual variability in rainfall is also very high, with the coefficient of variation in annual rainfall generally showing an increasing trend with a decrease in mean annual rainfall (Kumar, 2002, 2004). The climate varies from hot and arid in Kachchh to hot and sub-humid in south Gujarat. As Fig. 6.1 shows, the geohydrology of the state varies from hard rock aquifers (basalt and limestone aquifers) to deep alluvial plains rich in groundwater and crystalline formations with extremely limited groundwater (Central Ground Water, 2021). Therefore, any analysis of the spatial and temporal changes in water resource stocks and agricultural outputs should take cognizance of these factors.

Yet, none of the studies mentioned above had tried to factor in these important variables while analyzing the effect of either policy changes or programs on the region's hydrology. Bhanja et al. (2017), for instance, put the entire state in one hydro-meteorological zone (HMZ B, source Fig. 6.1B of Bhanja et al. (2017)). Now that the authors have used such an approach, the (spatial) average of normal rainfall in the state is around 700 mm. But in the driest year (like 1999), it went down to 390 mm, and in a wet year (2007), it went up to 1180 mm (Dave et al., 2017). But, Bhanja and others provided quarterly rainfall figures, and according to them, the temporal fluctuation in the spatial average rainfall is

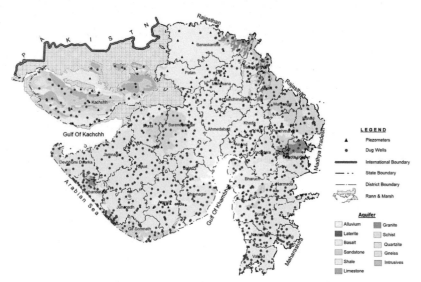

FIGURE 6.1   Geohydrology of Gujarat. (*From: Central Ground Water, 2021*).

between 150 mm and 200 mm. Had they used correct figures of rainfall, ideally, this range should have been 98–295 mm, and therefore their modeling studies would have produced entirely different results.

The presentation of rainfall on a quarterly basis raises another serious concern. In semi-arid tropics like Gujarat, where 95% of the rainfall occurs during the monsoon months and where, for at least two quarterlies, the total quantum of rainfall is nearly zero, rainfall data should be presented in annual terms. The authors surreptitiously presented quarterly rainfall figures (see Fig. 6.3A of Bhanja et al., 2017) to show that the variation of rainfall in the state was not significant enough to cause any effect on groundwater storage change. Such a sweeping statement, which is part of a weird attempt to attribute all that is happening with groundwater to policy change, over-simplifies an otherwise complex phenomenon of infiltration and recharge.

Gujarat is known for extreme climatic variability and high drought proneness (IRMA/UNICEF, 2001; Kumar, 2002, 2004). Dramatic fluctuations in agricultural outputs between years are characteristic of this region (Jagadeesan & Kumar, 2015; Mehta, 2013). Intriguingly, the works of Shah and Verma and Bhanja and others meant to examine the impact of schemes involving the comparison of water and agriculture situations at two different time periods, however, ignored the significant effect that temporal variation in precipitation can induce on both water availability and agricultural outputs.

A systematic comparative analysis of the rainfall data for two distinct climatic regions of Gujarat (the semi-arid region and the arid region) between two time-periods, i.e., from 1996 to 2002 and from 2003 to 2014 (Dave et al., 2017),

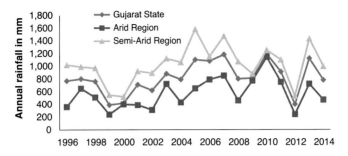

FIGURE 6.2   Temporal variations in rainfall of different regions of Gujarat.

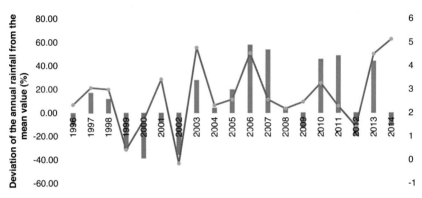

FIGURE 6.3   Deviation of spatial average annual rainfall from the mean value for the state and average groundwater level fluctuations.

shows that in all the regions, the mean value of the spatial average rainfall was much higher during the second period (Fig. 6.2 based on IMD (Indian Meteorological Dept. data for two physiographic units). The differences are very significant. For the semi-arid region, the difference in mean value was 299 mm (45% of the overall mean for the entire time period), and for the arid region, it was 257 mm, which is 30% of the overall mean for the entire time period (Kumar & Perry, 2019). Given the non-linear relationship between rainfall and runoff in the arid and semi-arid parts of Gujarat (Kumar, 2002), such differences are large enough to cause major differences in recharge and runoff. The rainfall difference corresponds to a total additional rainfall volume of 50 billion cubic meters. If we assume a mere 10% of this rainfall goes into recharge, the amount would be 5000 MCM (5 BCM) of water (Kumar & Perry, 2019).

A careful analysis of the data on groundwater level fluctuations during the monsoon season now available from India-WRIS shows the extent of the influence of rainfall on the overall change in groundwater storage during the monsoon. Fig. 6.3 shows the departure of the spatial average rainfall from the normal values for the state and the spatial average water level fluctuations during the monsoon. It clearly shows a positive correlation. In years when the monsoon rainfall was higher than the normal values, water level fluctuation was positive and very high. On the other hand, in most of the years when the rainfall was less than the normal value (–ve departure), barring 2014 and 1996, the water level fluctuation during monsoon was extremely low, indicating a very poor net change in storage. The correlation coefficient was 0.31 for rainfall and water level fluctuation. The weaker relationship is probably due to the regional variation in the rainfall and sharp variations in the aquifer characteristics, especially specific yield, both having direct implications for water level fluctuation (Central Ground Water, 2019; Kumar et al., 2022)[1]. This is in spite of the fact that the southern part of the state receives very high rainfall with low inter-annual variability, and is water-rich with a large amount of surface water and the presence of gravity irrigation schemes (Kumar, 2002).

6.4 Weak framework for analyzing groundwater storage change

In the recent past, many studies have attempted to analyze the groundwater rejuvenation and agricultural growth using empirical data (see Bhanja et al., 2017; Gulati & Shah, 2009; Patel et al., 2020; Shah et al., 2009). But they suffered from serious flaws either on the methodological front or on the analytical front or both.

Bhanja et al. (2017) are reported to have "…used seasonal *in situ* groundwater storage anomalies (GWSA_{obs}) and monthly groundwater storage anomalies (GWSA_{sat}, 2003–14), calculated using the Gravity Recovery and Climate Experiment (GRACE) satellite mission and land surface model-simulated soil moisture and surface water equivalents, to quantify the groundwater storage anomaly trends in the Indian region" (Bhanja et al., 2017, p. 3). Geology has a huge influence on groundwater level fluctuations in response to hydrological stresses such as recharge and abstraction. While analyzing the groundwater behavior, the authors missed the critical point about the huge heterogeneity in aquifer characteristics across the state of Gujarat. Saurashtra mainly has basalt (90%); the mainlands of north and central Gujarat mainly have deep alluvium; south Gujarat has alluvium on the western side and crystalline rocks on the eastern parts; and Kachchh has a mix of different geological formations, including coastal alluvium and basalt and sedimentary formations. As we have seen in Chapter 3, the groundwater level change in a hard rock area is not comparable with that in alluvial formations due to sharp differences in the specific yield of aquifers that significantly alter the water level fluctuation for the same quantum of net recharge, with higher water level fluctuation in hard rock

---

1. As pointed out by Kumar et al. (2022), while an increase in rainfall to a great extent can increase the recharge during monsoon depending on the storage space in the aquifer, a higher specific yield will result in a lower water level rise for the same quantum of recharge.

areas (Kumar et al., 2022). Hence, aggregated analysis of groundwater level data for the region as a whole, as done in the study by Bhanja et al. (2017), would offer highly misleading results. However, the paper does not even delve into this issue.

Shah et al. (2009), which examined the rise in pre-monsoon groundwater levels during 2001 and 2007 in the Saurashtra region, among others, simply did not consider the difference in rainfall between monsoons of 2000 and 2006, which would have seriously affected the total utilizable recharge and the groundwater withdrawal for agriculture during those years, and thus the pre monsoon depth to groundwater levels in the subsequent years (i.e., 2001 and 2007).

Patel et al. (2020) tried to build on the earlier argument by Shah et al. (2009) on the effect of the community-based recharge movement in Saurashtra. A competing hypothesis held that improvements in groundwater levels in Saurashtra were more due to a succession of good rainfall years during 2001–14, aided by the transfer of surface water from a big dam on the Narmada River, rather than the distributed recharge movement (Kumar & Perry, 2019). They developed and implemented a 2-way test of these competing hypotheses: First, they compared groundwater level fluctuations in Saurashtra during a recent period of high rainfall years with a similar period prior to the recharge movement; second, for both these high rainfall periods, we also compare groundwater recharge patterns in two other comparable aquifer and terrain regions, viz., Vidarbha and Marathawada in Maharashtra, which did not experience recharge movement on the same scale as Saurashtra did. The biggest flaw in their analysis however was the failure to capture the combined effect of rainfall magnitude and depth to groundwater table before monsoon on recharge during monsoon and thus water level fluctuations.

The authors compared the average water level fluctuations during two time periods, i.e., 1975–84 and 2000–09. They estimated recharge as a fraction of rainfall values to show that Saurashtra experienced improved groundwater recharge from a unit of rainfall, nearly two-fold between Period 1 and Period 2 (from 0.063 to 0.114). The author's contention was that the increases in RF in Marathawada (24%) and Vidarbha (13%) were much less than that of Saurashtra (80%). They concluded that: "…the runoff must have been controlled and diverted into groundwater recharge much better in Saurashtra than in Marathawada or in Vidarbha".

However, there are two major flaws in their analysis. First of all, it is based on the inherent assumption that aquifer recharge increases with rainfall (if the geology and all other conditions enabling recharge remain the same). This is incorrect. Generally, in hard rock areas, the ratio of recharge and rainfall (RF) will be low when rainfall is high (Kumar et al., 2022). Beyond a point, any increase in rainfall will not result in a rise in water level as once the aquifer gets saturated, the additional rainfall will be available as runoff if the pre-monsoon depth to water level is the same in both situations. So, comparing the recharge coefficient of Saurashtra with those of Marathwada and Vidarbha, which represent two different hydrological regimes, makes no sense.

The above phenomenon also explains how the recharge in Saurashtra had increased during the second period, while it had not happened for the other two regions. The pre-monsoon depth to water level has a great influence on recharge during the monsoon (Kumar et al., 2022). That said, Patel et al. (2020) showed that the difference in average pre monsoon depth to water level in Saurashtra between two time periods was around 3 m, whereas the difference was negligible for Marathwada and nil for Vidarbha.

6.5   Has there been a change in water management policy in Gujarat?

The policy debate on groundwater management in Gujarat is as old as the policy debate concerning groundwater at the national level (see Moench, 1994). The reason is that Gujarat had one of the basket cases of groundwater depletion, with the district of Mehsana becoming an infamous case of groundwater over-exploitation and aquifer mining, and research and action to arrest the problems starting way back in 1976 (with a project from united nations development program (UNDP) on artificial recharge of groundwater). Later on, many districts of the state (such as the erstwhile district of Junagadh, and the districts of Bhavnagar and Ahmedabad) received the attention of researchers and government agencies, with the problem manifesting itself in one form or the other in those areas (seawater intrusion, seasonal drops in water levels, etc.). Yet, in spite of these problems, there were no efforts to control over-development through policies and regulations. While many states passed legislation to check groundwater over-draft (such as Karnataka, Kerala, Madhya Pradesh, Goa, Delhi, and Maharashtra), no such efforts were forthcoming from the government of Gujarat.

However, in the recent past, many authors reported that there had been a change in farm power supply policy in the state of Gujarat around the year 2003 to address the problem of growing energy subsidies in agriculture (Bhanja et al., 2017; Shah et al., 2004; Shah & Verma, 2008). Bhanja et al. (2017) called it an "ingenious groundwater management strategy." But the fact of the matter is that no such policy was ever introduced in Gujarat. Instead, it was a change in policy with regard to the supply of electricity to the domestic sector in rural areas. The *Jyotigram* scheme was introduced to separate the feeder line for domestic purposes from that of agriculture as a way to ensure high-quality 24 × 7 three-phase power supply to the domestic sector. The work of feeder line separation was taken up due to the fear that if a common feeder line is used, it would lead to farmers pumping groundwater for extended time periods using the three-phase electricity (instead of the 8 hours for which they were supplied power till 2003–04). The *Jyotigram* scheme was partially introduced in 2004 (Kumar & Perry, 2019).

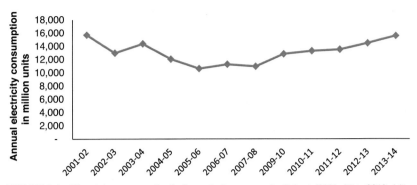

FIGURE 6.4   Electricity consumption in the agriculture sector in Gujarat (2001–02 to 2013–14).

Though some authors have claimed that this was a strategy adopted by the government of Gujarat to ration power supply to agriculture (Shah & Verma, 2008,), it is factually incorrect[2]. Even before the *Jyotigram* scheme, farmers used to get only eight hours of three-phase power for running their irrigation wells. Though they used to adopt an indigenous method (called *Thetta*) to convert single-phase electricity from the feeder (after the 8-hour long 3-phase power supply goes off) into three-phase electricity prior to *Jyotigram*, this was to a great extent stopped after the feeder line separation. However, in areas like north and central Gujarat with deep water table conditions, farmers resorted to using higher capacity (submersible) pump sets for pumping more water during the 8 hours without reporting to the power utility, as the "illegal tapping" option was not available. This is evident from the data presented in Fig. 6.4, which shows that after 2006 (the year in which Jyotigram was implemented state-wide), electricity consumption in agriculture has been going up steadily. Thereafter, the power distribution companies have become more vigilant, and the theft had to be curtailed by frequent raids. A total of two lac agricultural power connections were raided in 2012–13 alone, and the value of the electricity stolen was 200 million rupees (DNA, 2013). In sum, contrary to the claim by Shah and Verma (2008), there was no effective step to reduce electricity supply.

6.6   Had decentralized water harvesting made any impact on groundwater?

Attempts are made now and then to attribute the positive change in groundwater storage (Bhanja et al., 2017; Shah et al., 2009) and agrarian dynamism observed during 2001–02 to 2007–08 (Shah et al., 2009) to the hundreds of thousands of reservoirs created by the small water harvesting structures built in the state.

---

2. This point, made by M. Dinesh Kumar, was quoted in Malik (2009), which analyzed the effectiveness of various instruments for groundwater regulation in India.

However, there are no official statistics on the number of check dams built in Gujarat since the launching of the scheme (the Sardar Patel Water Conservation Programme) in the state in 1999, while the professionals who work on water management in the state agree that many such structures are built in Saurashtra and to an extent in the eastern hilly areas. So far as other regions (especially north and central Gujarat and south Gujarat) are concerned, the topography does not permit the construction of such structures, as the area is largely plain with no well-defined drainage network.

These check dams do store some water, but the volume of storage is limited to 1000–2000 m^3 per structure. Their "storage-volume/wetland area" ratio is very low, submerging large areas but contributing too little in terms of improving water balance when compared to a large reservoir (Perry, 2001). Almost all these structures were built in the hard rock regions of Saurashtra and eastern Gujarat, which have good drainage density. The alluvial areas of water-scarce north and central Gujarat have very poor drainage density and the topography does not permit the construction of such structures (Kumar & Perry, 2019).

The problem with such structures built in hard rock areas is that when the rainfall in their local catchment is large enough to generate sufficient runoff, the shallow aquifers having low effective porosity also get replenished, and the wells start overflowing due to aquifer saturation. As shown by Kumar et al. (2022), the storage space in the aquifer (governed by the depth to water table before the onset of monsoon and the specific yield of the aquifer) is a major determinant influencing the recharge from rainfall infiltration. This means that the check dams cannot induce any additional recharge during the monsoon season in wet years in the hard rock areas (Kumar et al., 2006). The water remains on the surface, and a large portion of it gets evaporated during the same season (around 750–900 mm during the season) given the high aridity in the region, reducing their overall effectiveness as artificial recharge structures. As shown by Kumar (2002), the monsoon season accounts for nearly 33–39% of the total annual reference evapotranspiration in the state, which is a strong indicator of the evaporation that can occur from water bodies (see Fig. 6.5, which shows the ET_0 values for four locations in Gujarat). In low rainfall years, when the recharge becomes very essential, there is no water generated in the catchment that can be used for artificial recharge. Studies carried out by international water management institute (IWMI) in the Saurashtra region clearly show that due to the construction of too many structures, the inflows into several medium irrigation schemes are adversely affected (Kumar et al., 2008). Hence, the real contribution of these structures to improving groundwater balance is very doubtful (Kumar & Perry, 2019).

Also, these structures have a very short life of 3–4 years, and many of them get washed away by monsoon (flash) foods. Many a time, the structures get rebuilt, and such statistics of the cumulative numbers of structures built in an area over the years are probably used to demonstrate widespread impact of these structures at the regional level. The actual number of structures on the

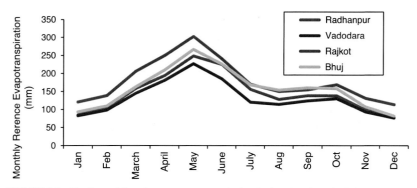

FIGURE 6.5 Total monthly reference evapotranspiration in four locations in Gujarat. (*From: Kumar, 2002*).

ground could be a mere 10–15% of the figures provided by proponents of check dams.

6.7 Factors that caused groundwater rejuvenation

Understanding water level changes in aquifers involves complex considerations of various inflows into and outflows from aquifers (Custodio, 2000; Kumar et al., 2022). If we want to understand the reasons for (positive) groundwater storage change in Gujarat during 2002–14, we need to recognize the fact that groundwater storage change is the net effect of net recharge and abstraction and not merely the effect of a change in abstraction or change in recharge. There is natural recharge occurring during the monsoon due to rainfall, and this is quantified by the Central Ground Water Board using the water level fluctuation approach[3]. Bhanja and others, who used the GRACE data to estimate storage change, believe that positive change in storage is due to a reduction in abstraction as they rule out any difference in annual recharge caused by variation in annual rainfall. But this is far from the truth, as is evident from the spatial average of the pre and post monsoon water level fluctuation for the state during 1996–2001. As we have seen earlier, during the drought years (1999 and 2000), the water level fluctuation was very low, which must be due to very limited recharge, followed by pumping during the monsoon season to irrigate crops (wherever aquifers had stock of water).

During 1996–2001, while there were two consecutive years of severe drought (i.e., 1999 and 2000), during 2003–14, most years were extremely good in terms of average annual rainfall, with the exception of 2012, which was a "severe

---

3. There are some problems with the methodology, especially when it is used for hilly regions. However, for plains, which constitute a major share of Gujarat's geographical areas, this methodology should give some reliable results (Kumar and Singh, 2008, and discussions in Chapter 3).

drought" year. The years, 2005, 2006, 2007, and 2013 were exceptionally wet years, with a positive departure of the rainfall from the normal values by 20%, 58%, 54%, and 44%, respectively (see Fig. 6.2). Hence, the average annual natural recharge (during 2002–14) should have been much higher than that of 1996–2001. It is not clear whether the authors used rainfall data in the global hydrological model for simulating policy change. Even if they do, since the rainfall values are distributed equally over the four quarters of the year, the model is unlikely to show its effect on change in groundwater storage.

Apart from the direct positive effect it has on infiltration and groundwater recharge, rainfall has an indirect negative effect on abstraction. Groundwater abstraction significantly reduces in high rainfall years. This is because the irrigation water demand of crops per unit cropped area drastically reduces in high rainfall years, with high precipitation during monsoon resulting in better soil moisture storage for monsoon crops and higher residual moisture for winter crops. Hence, the net change in groundwater balance due to changes in rainfall will be disproportionately higher than the effect of rainfall changes on natural recharge. In fact, in drought years, the groundwater storage change (pre-monsoon–post-monsoon depth to water level in wells) in some of the low rainfall regions of Gujarat and many parts of India is negative, as recharge becomes small and farmers start pumping water during the monsoon season itself due to moisture stress. This trend reverses during wet years.

A major cause of groundwater storage change in Gujarat is the introduction of canal water from the Sardar Sarovar Project (SSP). The area under irrigation in the designated command area has been steadily increasing since 2001–2002, when water started flowing through the canals. As per the estimates (for January 2017) from a study done by BISAC in Gujarat, the SSP irrigates a total of 8.88 lac ha in the command area and 0.45 lac ha in the fringes (so the total is 9.28 lac ha), which is a little over 50% of the design command area (Kumar & Perry, 2019).

A map of Gujarat showing the canal network of SSP is given in Fig. 6.6. As it clearly shows, the water from the Sardar Sarovar reservoir goes to north Gujarat, central Gujarat, eastern Gujarat, and some parts of Saurashtra and Kachchh. Most of the area, which receives Narmada water for gravity irrigation, is underlain by alluvium, providing the ideal condition for recharge from canal seepage and irrigation return flows (Jagadeesan & Kumar, 2015). As per the data provided by Sardar Sarovar Narmada Nigam Ltd., a total volume of 7932 MCM of water was diverted to different regions of Gujarat for irrigation and other uses during 2014–15.

In addition to irrigating the land by gravity, canal water from Sardar Sarovar reservoir is also diverted into the rivers of north Gujarat through Narmada Main Canal, and the alluvial river beds are also inducing a good amount of recharge. There is a significant amount of recharge induced by the gravity-irrigated areas of SSP, directly benefiting the wells, along with a reduction in dependence on well water for irrigation for many farmers. Primary data collected

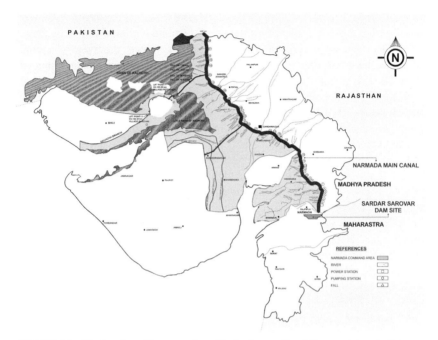

FIGURE 6.6 The location of the reservoir and Sardar Sarovar Canal Command Area in Gujarat.

from thousands of farmers in the Sardar Sarovar canal command area in six districts of Gujarat (Ahmedabad, Mehsana, Bharuch, Panchmahals, Vadodara, and Narmada) showed improvement in the areas irrigated by the wells and a reduction in the incidence of well failures in the command area after Narmada water introduction in 2002 (Jagadeesan & Kumar, 2015; Kumar et al., 2014), a clear validation of the claim by researchers about the role of irrigation return flows in sustaining well irrigation (Kumar et al., 2021; Perry, 2007).

The reduction in well failure was found to be very remarkable in the Vadodara district, falling from 9.5 per year to almost nil. In Ahmedabad and Narmada districts, the extent of the reduction was lower. In Panchmahals, there has been an increase in the incidence of well failures from almost 0 to 4.0. On the other hand, the command area of wells has increased post Narmada canal water in four locations. While it remained more or less the same in one location, i.e., Bharuch, it reduced in one location, i.e., Panchmahals (Table 6.1). One reason for the negative trend in Panchmahals vis-à-vis the incidence of well failures and well command area could be the substantial increase in the number of wells tapping the hard rock aquifers of the district, which perhaps is disproportionately higher than the additional recharge available from the small canal irrigated area, resulting in well interference. It can also be seen from the table that wherever the incidence of well failures reduced, the command area of wells increased, showing the natural tendency.

TABLE 6.1 Changes in incidence of well failures and irrigation potential of wells in Narmada Command, post canal irrigation.

Name of location	No. of incidence of well failure in the village per year		Average command area of a well (ha)	
	Pre-Narmada	Post-Narmada	Pre-Narmada	Post-Narmada
Ahmedabad	3.00	1.2	36.6	37.8
Bharuch	4.75	5.2	6.1	6.05
Mehsana	0.00	1.2	14.7	15.1
Narmada	2.40	1.5	2.39	3.14
Panchmahals	0.00	4.00	5.10	3.90
Vadodara	9.50	0.00	4.40	4.50

From: Kumar et al. (2014).

Indications are that the SSP helps reduce the demand for water from wells significantly by providing cheap surface water to hundreds of thousands of farmers while augmenting the recharge. Hence, as argued by Kumar and Perry (2019), one should expect a change in groundwater storage in Gujarat as a net result of:

- higher rainfall during the period 2002–14, as compared to 1996–2001, resulting in greater natural recharge and reduced groundwater demand for irrigation per unit cultivated land;
- induced recharge from canal irrigated fields in over 10,000 sq.km area of the state and the canal network; and
- the reduced groundwater pumping owing to a significant part of the irrigation and domestic water demands being met by surface water release from the SSP.

These changes had affected electricity usage for agriculture in more than one way. First: they reduced the demand for groundwater (owing to higher rainfall and supply of irrigation water from the SSP) and probably resulted in reduced electricity demand for agricultural groundwater pumping if other determining factors remained the same. Second: the rise in water level in many areas where the water table had been dropping consistently in the past due to SSP-induced recharge and rainfall infiltration meant that the electricity demand for unit volume of groundwater pumped had reduced. Yet the changes (reduction in electricity demand) are not expected to be very dramatic because the overall demand for irrigation water in the state has increased over time (especially after 2007) owing to farmers bringing more area under irrigation, preventing any large-scale cut in groundwater draft. So, on the one hand, the groundwater abstraction had increased over time; on the other hand, "total recharge" from rainfall and gravity irrigation had also increased. Interestingly, the map presented by Bhanja et al. (2017) to depict the impact of electricity supply regulation on groundwater storage change shows that groundwater build-up is mainly confined to parts of south, central and north Gujarat (the main command of SSP) (Fig. 6.1E of Bhanja et al., 2017), the areas that have been receiving water from the SSP for many years.

6.8   What can we learn from Gujarat's experience of managing groundwater?

Gujarat is an infamous case for groundwater over-exploitation and mining for irrigated agriculture. With no effective state regulation of groundwater withdrawal, for nearly two decades from the early 1990s, this phenomenon of groundwater over-draft was perpetuated by connected load-based pricing of electricity for agricultural use. However, during the past one and a half decades, there have been notable improvements in groundwater conditions in the state (Bhanja et al., 2017; Kumar & Perry, 2019). This changing groundwater dynamics in Gujarat had become a hot topic for debate among researchers and policy makers (see Chindarkar & Grafton, 2019). Chindarkar and Grafton (2019), who analyzed the effect of JGY on groundwater storage change, contended that JGY has been implemented without adequate consideration of (1) a publication bias, whereby researchers have a greater likelihood of having their results published if they are statistically significant and show a positive outcome; and (2) a "barrier" effect, which implies that while communicating evidence across science and policy divides, the evidence may not be accepted, even when true, and this limits policy advice and options.

The analysis presented in this chapter shows that the changing groundwater dynamic in Gujarat is the result of many complex factors operating in conjunction. Some of them are: i) the effect of change in rainfall on irrigation water requirements of crops, which reduced the pressure on groundwater for irrigation; ii) the effect of rainfall on natural recharge to groundwater; the additional recharge to groundwater from irrigation return flows in newly introduced gravity irrigation from SSP; iii) the effect of change in groundwater levels on energy required to pump groundwater; and, iv) the effect of geology in determining the impact of recharge on groundwater levels. None of the past studies that attempted to explain groundwater storage build-up or agricultural growth in Gujarat had considered the dynamic variables that take into account these factors.

Having said that, the real water-electricity-agriculture dynamics in Gujarat are that initially there was some reduction in demand (during 2002–2006) and then a decline in the growth rate of demand for electricity for groundwater pumping due to a variety of reasons. They include: (1) overall better rainfall conditions during the past one and a half decades, as compared to the previous decade; (2) the introduction of surface water from SSP for irrigation through gravity reducing the pressure on groundwater resources for meeting irrigation demands;

and (3) improved groundwater conditions (water level rise), reducing the energy requirement for pumping a unit volume of groundwater (Kumar & Perry, 2019). While doing analyses of energy and water use involving considerably long time periods, we also need to keep in mind the fact that the gross demand for water will also be on the rise in regions like Gujarat on a long-term basis, and energy use will increase if water is available in the aquifer if there is no restriction on energy supply induced by the feeder separation. This is exactly what is found. While on the one hand, recharge of groundwater has increased, there is a sufficient amount of energy supply available to pump that water out.

Analysis of Gujarat's experience of groundwater rejuvenation clearly shows the importance of large water transfer projects and the performance of the monsoon rains in changing groundwater balance in arid and semi-arid tropical regions that experience high variation in annual precipitation but are also often importers of large volumes of water for irrigation. It also points to the fact that one needs to exercise caution so as to capture the effects of variables (such as rainfall, aquifer properties and changes in overall water balance due to the introduction of surface water for irrigation) while analyzing the impact of programs and policies on the condition of a resource as complex as groundwater.

The experience of Gujarat also shows how various options can be explored for improving the groundwater condition in other semi-arid and arid regions of South Asia in the future. If such regions have large gravity irrigation systems, then water allocation from such systems to the command areas can be well-programmed with due consideration for the spatial differences in groundwater conditions across the command area vis-à-vis water levels and water quality. The basic strategy should be to allocate more water to the tail reaches of the canals (and less water to the head reaches), which would help not only reduce the dependence of the tail reach farmers on groundwater but also increase the recharge of the groundwater through return flows in those areas. It can then become a powerful tool for managing groundwater. Extending the canal networks of such large irrigation systems to areas of groundwater mining to be used as replacement water for the water pumped from underground for irrigation is another possibility.

References

Bhanja, S. N., Mukherjee, A., Rodell, M., Wada, Y., Chattopadhyay, S., Velicogna, I., Pangaluru, K., & Famiglietti, J. S. (2017). Groundwater rejuvenation in parts of India influenced by water-policy change implementation. *Scientific Reports, 7*(1), 7453. https://doi.org/10.1038/s41598-017-07058-2.

Central Ground Water Board. (2019). National Compilation on Dynamic Ground Water Resources of India *2017*. *Central Ground Water Board, Dept. of Water Resources*. Faridabad, India: RD & GR, Ministry of Jal Shakti.

Central Ground Water Board. (2021). *Ground water year book 2019-20 Gujarat state*. Regional Office Data Centre, Central Ground Water Board West Central Region.

Chindarkar, N., & Grafton, R. Q. (2019). India's depleting groundwater: When science meets policy. *Asia & the Pacific Policy Studies, 6*(1), 108–124. https://doi.org/10.1002/app5.269.

Custodio, E. (2000). *The Complex Concept of Over-exploited Aquifer, Secunda Edicion*. Madrid: Uso Intensivo de Las Agua Subterráneas.

Dave, H., James, M. E., & Ray, K. (2017). Trends in intense rainfall events over Gujarat State (India) in the warming environment using gridded and conventional data. *International Journal of Applied Environmental Sciences, 12*(5), 977–998.

Deccan News Agency. (2013, 21 November). Power theft of 201 crore caught last year. *Deccan News Agency*.

Dholakia, R., & Datta, S. (Eds.). (2010). *High growth trajectory and structural changes in Gujarat agriculture*. Macmillan Publishers India Ltd.

Gulati, A., Shah, Tushaar, & Shreedhar, G. (2009). *Agriculture performance in Gujarat since 2000: can it be a divadandi (lighthouse) for other states? Anand, Gujarat, India: International Water Management Institute (IWMI)* p. 24. New Delhi, India: International Food Policy Research Institute (IFPRI).

Institute of Rural Management Anand/UNICEF. (2001). *White paper on water in Gujarat*. Final report submitted to the Dept. of Narmada, Water Resources and Water Supply, Govt. of Gujarat, Gandhinagar.

Jagadeesan, S., & Kumar, M. D. (2015). *The Sardar Sarovar Project: Assessing economic and social impacts*. Sage Publications.

Kumar, M. D. (2000). Institutional framework for managing groundwater: A case study of community organisations in Gujarat, India. *Water Policy, 2*(6), 423–432.

Kumar, M. D. (2002). *Reconciling water use and environment: Water resource management in Gujarat, resource, problems, issues, strategies and framework for action*. Report prepared for State Environmental Action Project supported by the World Bank, Gujarat Ecology Commission.

Kumar, M. D. (2004). Roof water harvesting for domestic water security: Who gains and who loses? *Water International, 29*(1), 43–53.

Kumar, M. D., Ghosh, S., Patel, A., Singh, O. P., & Ravindranath, R. (2006). Rainwater harvesting in India: Some critical issues for basin planning and research. *Land Use and Water Resources Research, 6*, 1–17. https://doi.org/10.22004/ag.econ.47964.

Kumar, M. D. (2007). *Groundwater management in India: Physical, institutional and policy alternatives*. Sage Publications.

Kumar, M. D., & Singh, O. P. (2008). How serious are groundwater over-exploitation problems in India? A fresh investigation into an old issue. In M. D. Kumar (Ed.), *Managing water in the face of growing scarcity, inequity and declining returns: Exploring fresh approaches. 7th Annual Partners' meet of IWMI-Tata Water Policy Research Program. Patancheru, AP: ICRISAT 2–4 April 2008*.

Kumar, M. D., Patel, A., Ravindranath, R., & Singh, O. P. (2008). Chasing a mirage: Water harvesting and artificial recharge in naturally water-scarce regions. *Economic and Political Weekly, 43*(35), 61–71.

Kumar, M. D., Narayanamoorthy, A., Singh, O. P., Sivamohan, M. V. K., Sharma, M. K., & Bassi, N. (2010). *Gujarat's agricultural growth story: exploding some myths, Occasional Paper # 2*. Hyderabad: Institute for Resource Analysis and Policy.

Kumar, M. D., Scott, C. A., & Singh, O. (2011). Inducing the shift from flat-rate or free agricultural power to metered supply: Implications for groundwater depletion and power sector viability in India. *Journal of Hydrology, 409*(1–2), 382–394.

Kumar, M. D., Scott, C. A., & Singh, O. (2013). Can India raise agricultural productivity while reducing groundwater and energy use? *International Journal of Water Resources Development, 29*(4), 557–573.

Kumar, M. D., Jagadeesan, S., & Sivamohan, M. V. K. (2014). Positive externalities of irrigation from the Sardar Sarovar Project for farm production and domestic water supply. *International Journal of Water Resources Development, 30*(1), 91–109. https://doi.org/10.1080/07900627.2014.880228.

Kumar, M. D., & Perry, C. J. (2019). What can explain groundwater rejuvenation in Gujarat in recent years? *International Journal of Water Resources Development, 35*(5), 891–906. https://doi.org/10.1080/07900627.2018.1501350.

Kumar, M. D., Bassi, N., & Verma, M. S. (2021). Direct delivery of electricity subsidy to farmers in Punjab: will it help conserve groundwater? *International Journal of Water Resources Development, 38*(2), 306–321. https://doi.org/10.1080/07900627.2021.1899900.

Kumar, M. D., Kumar, S., & Bassi, N. (2022). Factors influencing groundwater behaviour and performance of groundwater-based water supply schemes in rural India. *International Journal of Water Resources Development,* 1–21. https://doi.org/10.1080/07900627.2021.2021866.

Lytton, L., Ali, A., Garthwaite, B., Punthakey, J. F., & Saeed, B. (2021). *Groundwater in Pakistan's Indus Basin: Present and future prospects.* World Bank. https://openknowledge.worldbank.org/handle/10986/35065.

Malik, R. P. S. (2009). Energy regulations as a demand management option: Potentials, problems and prospects. In M. R. Saleth (Ed.), *Promoting irrigation demand management in India: Potentials, problems and prospects. Strategic analysis of National River Linking Project Series 3.* International Water Management Institute.

Mehta, N. (2013, September 10-12). An investigation into growth, instability and role of weather in Gujarat agriculture, 1981-2011. In *21st AERA Annual Conference to be held at Sher-e-Kashmir University of Agricultural Sciences & Technology of Kashmir, Srinagar (J&K).*

Moench, M. (1994). Approaches to groundwater management: To control or enable? *Economic and Political Weekly, 29*(39), A135–A146.

Moench, M. (1998). Allocating the common heritage: Debates over water rights and governance structures in India. *Economic and Political Weekly, 33*(26), A46–A53.

Patel, P. M., Saha, D., & Shah, T. (2020). Sustainability of groundwater through community-driven distributed recharge: An analysis of arguments for water scarce regions of semi-arid India. *Journal of Hydrology: Regional Studies, 29*, 100680. https://doi.org/10.1016/j.ejrh.2020.100680.

Perry, C. (2007). Efficient irrigation; inefficient communication; flawed recommendations. *Irrigation and Drainage, 56*(4), 367–378. https://doi.org/10.1002/ird.323.

Perry, C. J. (2001). World Commission on Dams: Implications for food and irrigation. *Irrigation and Drainage, 50*, 101–107.

Qureshi, A. S., McCornick, P. G., Sarwar, A., & Sharma, B. R. (2010). Challenges and prospects for sustainable groundwater management in the Indus Basin, Pakistan. *Water Resource Management, 24*, 1551–1569.

Ranade, R., & Kumar, M. D. (2004). Narmada water for groundwater recharge in North Gujarat: Conjunctive management in large irrigation projects. *Economic and Political Weekly, 39*(31), 3510–3513.

Rotiroti, M., Bonomi, T., Sacchi, E., McArthur, J. M., Stefania, G. A., Zanotti, C., Taviani, S., Patelli, M., Nava, V., Soler, V., Fumagalli, L., & Leoni, B. (2019). The effects of irrigation on groundwater quality and quantity in a human-modified hydro-system: The Oglio River basin, Po Plain, northern Italy. *Science of the Total Environment, 672*, 342–356. https://doi.org/10.1016/j.scitotenv.2019.03.427.

Shah, T., Scott, C., Kishore, A., & Sharma, A. (2004). Energy-irrigation nexus in South Asia: Improving groundwater conservation and power. In sector *(viability (Vol. 70))*. Colombo, Sri Lanka: International Water Management Institute.

Shah, T., & Verma, S. (2008). Co-management of electricity and groundwater: An assessment of Gujarat's Jyotigram scheme. *Economic and Political Weekly, 43*, 59–66.

Shah, T., Gulati, A., Padhiary, H., Sreedhar, G., & Jain, R. C. (2009). Secret of Gujarat's agrarian miracle after 2000. *Economic and Political Weekly, 44*(52), 45–55.

Vyas, J. N. (2001). Water and energy for development in Gujarat with special focus on the Sardar Sarovar Project. *International Journal of Water Resources Development, 17*(1), 37–54. https://doi.org/10.1080/07900620120025042.

Watt, J. (2008). *The effect of irrigation on surface-ground water interactions: Quantifying time dependent spatial dynamics in irrigation systems.* School of Environmental Sciences, Faculty of Sciences, Charles Sturt University.

Chapter 7

Addressing groundwater management issues in semi-arid hard rock regions

7.1 Introduction

The erstwhile state of Andhra Pradesh (currently Telangana and Andhra Pradesh) is one of the most important agrarian states in India. Agriculture accounts for nearly 25% of the state's gross domestic product (GDP). Nearly 70% of the population is dependent on agriculture for livelihoods (Amarasinghe et al., 2007). The undivided Andhra Pradesh was also the "rice bowl" of the country. But the region lags behind other peninsular states such as Karnataka and Tamil Nadu in terms of human development and economic growth. The state's total cropped area is a percent of the cropped area of the country.

The two states of Telangana and Andhra Pradesh belonging to undivided Andhra Pradesh put together have one of the largest irrigated areas. With a gross irrigated area of 6.28 m.ha, the two states account for nearly 7.3% of the total irrigation in the country. Groundwater is the major source of irrigation in the state, with nearly 49% of the net irrigation from wells and tube wells (source: Amarasinghe et al., 2007; Reddy, 2007). The rest of the irrigation is from sources such as canals, tanks, and other sources. The state is mostly underlain by hard rock aquifers, with very poor storage and yield potential. Most parts of the state do not provide a favorable environment for intensive use of groundwater resources. Yet, limited access to water from surface irrigation systems such as reservoir and canal-based systems and tanks makes farmers have to resort to well irrigation through open wells.

Energization of wells was made possible by rural electrification and subsidized and free power to farmers. This triggered exponential growth in well irrigation during the 80s through the mid-90s. Along with this, the negative effects of intensive groundwater use had also started surfacing with drops in groundwater levels, the drying up of open wells, and seasonal and permanent failures of shallow and deep bore wells. These observed ill effects have forced development activists, scholars, and policymakers alike to raise alarm bells. Suggestions made over the past 10–15 years to tackle groundwater depletion

Groundwater Economics and Policy in South Asia. DOI: https://doi.org/10.1016/B978-0-443-14011-2.00012-7
149

problems and to sustain groundwater irrigation for socio-economic development largely focused on community regulation of groundwater use at the village level; electricity metering and energy pricing; a shift in cropping pattern with replacement of irrigated paddy by dry land crops; watershed management and artificial groundwater recharge; and large-scale adoption of micro irrigation systems and increased utilization of groundwater in canal command areas.

But, a close look at the growing literature on the topic of water management in India shows that these management solutions are not really based on scientific consideration of the real factors that determine the technical feasibility and socio-economic viability of these solutions, but instead are *run-of-the-mill* ideas based on popular perceptions. In this chapter, a critical assessment of the status of groundwater development in the undivided state of Andhra Pradesh is made at the outset from a multi-disciplinary perspective. The key factors that would greatly influence the success and failure of various water management interventions, that are in development parlance, are described; and new options for management based on the macro realities of the hydrology, geo-hydrology, and the socio-economic and policy environments are explored.

7.2 Groundwater development in the erstwhile Andhra Pradesh: how far are the estimates reliable?

Arguments in the past were made that the assessment of groundwater over-exploitation should involve complex hydrology, hydrodynamics, and geological, economic, social, and ethical considerations (Custodio, 2000; Kumar et al., 2001). Kumar et al. (2001), using the illustrative case study of the Sabarmati river basin in Gujarat, showed how using such complex considerations would provide an altogether different picture of groundwater exploitation from what the official estimates generally provide. Case studies of many Indian states show that the problems of over-exploitation in India are far more serious than what official estimates project them to be, as according to them, they fail to incorporate the complex physical and socio-economic factors that actually determine the degree of over-exploitation of aquifers (Kumar & Singh, 2008).

The only assessments of groundwater development in the erstwhile Andhra Pradesh are the official estimates of the central groundwater board and state minor irrigation departments. These estimates are based on a simplistic consideration of only the recharge and abstraction, done at the Mandal level. The recharge is the net of recharge from rainfall, percolation from natural and artificial water bodies, and irrigation return flows from both well irrigation and surface irrigation. In fact, the return flow coefficient for paddy is assumed to be 0.60 (APGWD, 1977, as cited in Marechal et al., 2003). This means that of the total renewable groundwater recharge estimated for the erstwhile state, a significant chunk would come from irrigated paddy fields. The abstraction estimates are based on norms for the average annual groundwater pumping by wells and bore wells.

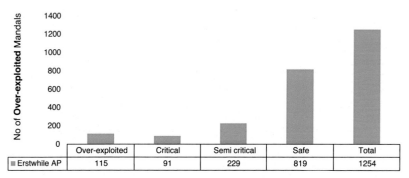

FIGURE 7.1   Stage of groundwater development in erstwhile Andhra Pradesh. (*From: CGWB, 2019*).

As per the latest CGWB estimates, 9% of the mandals in the erstwhile state of AP are over-exploited. Around 7.5% are in the critical stage, with the average annual abstraction in the range of 90–100% of the average annual recharge. Nearly 18% of the blocks are in the semi-critical category. Hence, as per the estimates, nearly 35% of the assessment blocks are facing over-development problems. Fig. 7.1 shows the stage of groundwater development in Andhra Pradesh at the block level. Out of the 115 over-exploited mandals, 70 are in the present Telangana state, and only 45 are in the present Andhra Pradesh. Thirty of the 45 over-exploited blocks from the current AP are in the Rayalaseema region, which has a very small area under surface irrigation. Twenty-one of the mandals are in the Anantapur district alone. There are very few mandals from coastal Andhra Pradesh, known for large canal networks, which are falling in the over-exploited category (source: based on CGWB, 2019).

The immediate reaction to this problem of increasing groundwater imbalance and "over-development" is to reduce groundwater abstraction and augment recharge through water harvesting and artificial recharge. But no systematic understanding is developed about the real causes of the groundwater imbalance. As is evident from some of the recent research, part of the problem is the failure to estimate the actual abstraction and recharge, watershed, or *mandal*-wise. This reduces the ability to foresee the various components of groundwater balance in the hydrological units under consideration and fix the real "malice." As Kumar and Singh (2008) point out, in many watersheds of hard rock regions of central and peninsular India, the outflows from groundwater to the surface streams are very significant. This component, for the non-monsoon season, is not considered in the groundwater estimation (Kumar & Singh, 2008). The extent to which this can influence groundwater recharge and abstraction estimates depends on the scale at which assessments are carried out. The smaller the unit, the greater will be the error it can induce, as outflows from one watershed can become inflows for another watershed. This phenomenon perhaps explains why, in spite of large-scale problems of well failures reported from rural areas of the state, they are not reflected in the official estimates of groundwater balance.

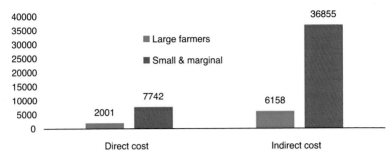

FIGURE 7.2 Equity impacts of groundwater depletion: direct and indirect costs (Rs/acre).

This lack of ability to estimate some of these important components of groundwater balance in some regions leads to under-estimation of groundwater over-exploitation problems as the outflows from the system always get under-estimated (Kumar & Singh, 2008). This was illustrated with the example of the Narmada river basin in MP, where estimates of the stage of groundwater development continue to show a positive balance in groundwater, while in many districts well failure and a reduction in well yields are rampant. As noted by them, a significant percentage of open wells (17.3%) in AP had failed by 2000–01 (Kumar & Singh, 2008), though the level of groundwater development in the state was only 45% even as per 2005 estimates (GoI, 2005). The magnitude of the problems of over-exploitation is also quite clear from the fact that the net area irrigated by open wells has been drastically and consistently declining in the state from 1.03 m.ha to 0.616 m.ha from 1998–99 to 2005–06, the year for which data are available (source: GOAP, 2007 as shown in Amarasinghe et al., 2007). With the open wells fast drying up, the farmers have resorted to drilling deep bore wells, with a consequent increase in their contribution to the well irrigated area in the state.

Groundwater depletion has serious negative impacts on access equity in groundwater (Kumar, 2007; Narayanamoorthy, 2014; Sarkar, 2011). The small and marginal farmers bear a much higher cost of resource depletion both in direct and indirect terms. This is because they have to incur as much cost for drilling wells to access groundwater as the large farmers, while the area brought under irrigation is much smaller. A study carried out in three villages in erstwhile AP showed that the cost per acre of irrigation was much higher for small and marginal farmers as compared to large farmers, and the differences widened with an increase in the magnitude of water scarcity (see Fig. 7.2, based on Reddy, 2005), Further, both the direct and indirect cost of groundwater degradation (due to well failure, and decline in crop yield, respectively) was estimated to be very high in the water-scarce village, with a much higher cost being borne by the small and marginal farmers as compared to the large farmers (source: based on Reddy, 2005, Table 12, p. 552).

7.3 Can well irrigation grow in the future?

The aggregate statistics on groundwater development show that there are large areas in the state, particularly canal command areas which have abundant groundwater resources that are under-utilized (Jain, 2008). It is viewed that well irrigation in Andhra Pradesh is poised to expand further, with negative outcomes for tank-irrigated areas. Such predictions were based on the data on irrigated areas from groundwater. However, such views shun real examination of the historical growth of groundwater irrigation in accordance with the growth in well numbers.

The data presented in Chapter 4 shows that after the gross well irrigated area in the state peaked in 2000–01, the irrigated area dropped by nearly 1.6 lac ha in 2003–04. Since then, there has only been a minor growth in the gross irrigated area, with the highest figure recorded in 2005–06. But what is interesting is the fact that this had no correlation with the growth in the number of wells, unlike what many scholars had argued in the past. There were only 14.48 lac wells in the state in 2000–01. But, they together irrigated 26.18 lac ha of land in gross terms. The average area irrigated by a well was 1.8 ha. However, this declined to a record low of 1.07 ha in 2003–04, with the growth in the number of wells to the tune of 8.62 lac, not translating into growth in irrigated areas. This is a common phenomenon in hard rock areas that occurs as a result of well interference. With an increase in the number of wells, the influence area of a well increases, and the available groundwater gets distributed among a larger number of wells (Kumar, 2007). The average area irrigated by a well recorded some improvement (1.51 ha) in 2009–10. This could be attributed to factors such as an increase in recharge from rainfall, a change in cropping pattern, and an increase in groundwater pumping in command areas, resulting from reduced surface water release from canals for irrigation.

By 2000–01, the state had the optimum number of wells, i.e., 14.28 lac to provide maximum irrigation by an individual well (i.e., 1.8 ha). The only apparent benefit that is achieved through the increase in the number of wells witnessed, later on, is distributional equity, with more farmers getting direct access to groundwater for irrigation. With such a high density of wells in a hard rock area, the economic efficiency of groundwater abstraction becomes extremely low.

The changing share of groundwater in AP's irrigation landscape has been noted by researchers (see Amarasinghe et al., 2007). The time series data they presented (Fig. 7.3) on net groundwater irrigated areas in the state showed it peaking in 2000 and declining thereafter. They argued that in recent years (since 2002), the well irrigation growth has become very sensitive to rainfall. Thereby, we can safely assume that groundwater irrigation had almost reached a saturation point by the year 2000–01, which basically means that the groundwater irrigation potential created by farmers that year through wells and pump sets was sufficient

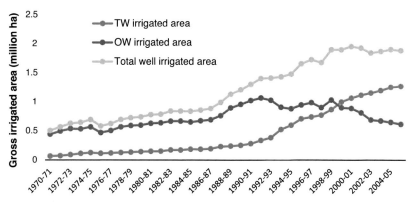

FIGURE 7.3   Net area irrigated by open wells and tube wells in Andhra Pradesh (1970–05).

to harness all the renewable groundwater resources, which occur in a good rainfall year. The state making any significant gain in its well irrigation in the long run, except through a shift in cropping pattern can be ruled out beyond any doubt. Further, a colossal drain in investment has occurred in the process of farmers' rat race of competitive drilling of wells for accessing groundwater. We estimate this wastage to the tune of 1648 crore rupees. This estimate is based on a realistic assumption that a farmer invests Rs. 16,000 for a bore well drilled in a hard rock area (source: based on Reddy, 2005). The total number of wells sunk between 2000–01 and 2003–04 is 10.31 lac by which there is hardly any additional area irrigated.

7.4   Negative externalities of well irrigation on tank irrigation

Researchers show tank irrigation has declined in South India. The reasons cited are: the onslaught of well irrigation in the region and the consequent lesser interest shown by farmers in tanks, owing to the inherent advantage of well irrigation over tank irrigation; the deterioration of tank catchments, including increased encroachment for cultivation; the silting up of tanks and supply channels; and the decline of traditional local institutions for tank management. But, this viewpoint has failed to look at the hydraulic interactions between aquifers and the tanks, and the physical externalities that changes in one can cause in the other. If one considers the period from 1970 till 2005–06, it can be seen that, like other south Indian states, tank-irrigated area in AP had also declined considerably. The net tank irrigated area declined from 1.11 m.ha in 1970–71 to 0.477 m.ha in 2004–05.

But what is striking is the phenomenon that growth in well irrigation is actually causing negative externalities on tank irrigation, as indicated by the strong inverse correlation that exists between the net tank irrigation area and the net well irrigated area at the state level ($R2 = 0.49$). This phenomenon can be

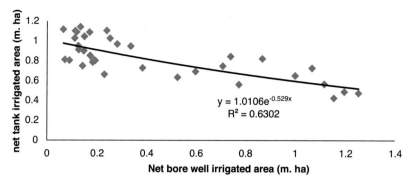

FIGURE 7.4 Tank irrigated area versus bore well irrigated area.

explained in the following way: In many areas, particularly those located in the upper catchments of rivers, a significant portion of the tank inflows come from shallow aquifers. Increased groundwater draft from aquifers, with the resultant lowering of the water table, can reduce the tank inflows. It can also activate the percolation of tank water into the underlying aquifers, particularly with extensive de-silting of tanks undertaken by the minor irrigation department of the AP government. This argument is strengthened by the inverse correlation between net bore well irrigated area and net tank irrigated area. The relationship was even stronger here ($R2 = 0.63$) (Fig. 7.4). This is quite understandable. Farmers generally resort to bore well irrigation in the hard rock regions when water levels in the upper aquifer drop drastically, which means a condition has already been reached where groundwater use has started influencing tank hydrology. Similar relationships were found for Chittoor ($R2 = 0.38$, and 0.53), Medak ($R2 = 0.55$, and 0.66), and Anantapur ($R2 = 0.29$ and 0.44) districts, for which district-level data were analyzed.

7.5 Effectiveness of water harvesting and artificial recharge in mitigating over-exploitation

There are many river basins in the undivided Andhra Pradesh. The most important of them are Godavari, Krishna and Pennar. In addition, there are many small east-flowing rivers between Krishna and Pennar, and Mahanadi and Godavari, and also south of Pennar (see Fig. 7.5). These east-flowing rivers carry large volumes of water. Of the total geographical area of the erstwhile AP, nearly 27% is in Krishna, and 17.5% is in Pennar. Another 27% of the area is in the Godavari basin. The rest of the area is in many of the east-flowing rivers from Mahanadi to Godavari, Godavari to Krishna, and Krishna to Pennar. Analysis of basin-wise runoff generation and runoff utilization shows that surface water utilization is highly skewed in the state (Fig. 7.6). Godavari is a water-abundant river basin with an annual surface runoff (dependable) of 41.9 BCM, of which only 40%

FIGURE 7.5 River basins of erstwhile (undivided) Andhra Pradesh.

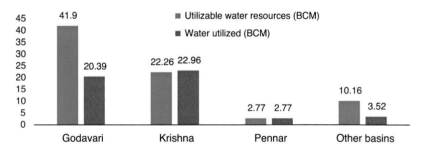

FIGURE 7.6 Annual surface water resources in the river basins of AP (utilizable and utilized).

of the total surface water is utilized. Therefore, the basin is still "open." So are the many small river basins between Mahanadi and Pennar. Here, the degree of utilization is only 35%. As against these, the two river basins, viz., Krishna and Pennar, are surface water-scarce and have already experienced a high degree of water utilization with a large number of reservoir and diversion schemes. The reservoir schemes include both small reservoirs like tanks and ponds and large modern reservoirs.

In erstwhile AP, the problems of groundwater depletion are mostly encountered in the regions that fall in the basins of Krishna and Pennar. This, however, does not mean that the whole area inside these two basins is experiencing problems. In fact, coastal areas, including alluvial areas of the Krishna delta, which receives water from the Nagarjuna Sagar scheme, Sriram Sagar, and the delta irrigation systems, are showing positive balance in groundwater, with the recharge exceeding the annual abstraction. The problems of depletion are concentrated in the areas that are not benefited by surface irrigation (Kumar et al., 2012). These areas are already facing a problem of acute shortage of surface water. Though the weighted average annual rainfall is 800 mm, it varies from a highest of 1200 mm in the high altitude tribal zone to 800–1000 mm in the coastal zone (Krishna, Godavari delta area) to a lowest of 500–600 mm in interior Rayalaseema region (source: based on Biggs et al., 2007). Hence, the regions that desperately require augmentation of groundwater are falling into "closed basins," having no surplus runoff for recharge. Using the runoff from local catchments for recharge would mean causing negative effects for the d/s storage and diversions schemes in these basins. Already intensive water harvesting activities have had a significant negative impact on inflows into tanks in the Deccan plateau area, which is used for irrigation and domestic purposes.

A counter-argument is that groundwater recharge as against storage in large surface reservoirs ensures greater returns from every drop of water used up and equity in access to water from the basin. But the aquifer underlying most of Andhra Pradesh does not have good storage space because of the consolidated nature of the crystalline and basalt rocks. Groundwater occurs only in fractures, fissures, and weathered zones. The constraints imposed by hard rock geology in recharge efforts through percolation tanks are: high depth to the water table below and around the recharge structure due to the occurrence of recharge mount and shallow bed rocks, which prevent the percolation of water (Muralidharan, 1990, as cited in Muralidharan & Athawale, 1998); and low infiltration capacity of the thin soils overlaying the hard rock formations. Due to low specific yield (0.01–0.03), a sharp rise in water levels is observed in aquifers during monsoon, leaving little space for infiltration from structures. While harnessing water for recharge is extremely important during normal and wet years, the natural recharge in hard rock formations is high during such years as it is a function of seasonal rainfall (based on regression equations shown in Fig. 7 of Athawale, 2003), further reducing the scope for artificial recharge.

Soils pose additional challenges. Results obtained from short-term infiltration tests carried out in dug wells in Andhra Pradesh in two different soil conditions show that the infiltration rate becomes negligible (<0.60 mm/hour) within 10 minutes of starting the test in the case of silty clay. If the infiltration rate approaches zero quickly, it will negatively affect the recharge efficiency of percolation ponds. As thin soil cover has low infiltration (Muralidharan & Athawale, 1998), the extent of the problem would be larger in hard rock areas (ideal for percolation ponds) with thin soil cover. Dickenson and Bachman (1994), based on several infiltration studies, shows that the rate of infiltration declines to a minimum value within 4–5 days of ponding. This will also have adverse effects on the performance of structures built in areas experiencing flash floods and high evaporation rates, for which solutions include wetting or drying of pond beds through regulation of inflows. On the other hand, storage of water on the surface in small reservoirs would result in huge evaporation losses.

7.6 Potential impact of crop shift in reducing groundwater withdrawal

Paddy is perhaps one of the most misconceived crops. Development activists and often agricultural scientists view paddy as a water guzzling crop and an enemy of the ecology, especially when it is grown in a semi-arid or arid environment (Shah et al., 2021). But, this is because of the poor understanding of the water use hydrology of paddy wetlands. There is little appreciation of the fact that the consumptive water use by this crop is only a fraction of the total water applied, particularly when it is grown under partially submerged conditions (Ahmad et al., 2004; Singh, 2005). In practice, the applied water in excess of the depleted water (also known as a consumptive fraction (CF), which is the sum of ET, non-beneficial evaporation from soil strata, and the non-recoverable deep percolation) would find its way to the groundwater system (Allen et al., 1997). There is no doubt that depletion of the water applied in the field could be more than the crop consumptive use, i.e., ET, and this extra depletion would also be determined by the climate, soil, and the depth of the dewatered zone below the crop root zone. The depletion of water can occur due to non-recoverable deep percolation and non-beneficial evaporation from the soil after the harvest of the crop (Kumar, 2009; Kumar et al., 2008a; Kumar & van Dam, 2009).

But, it is incorrect to assume that all the water applied in the field would eventually be depleted (Ahmad et al., 2004; Singh, 2005). In areas with a shallow groundwater table and permeable soils, the return flow component of the applied water could be quite significant (Watt, 2008). This is the water that actually improves the groundwater condition in canal-irrigated areas in the form of recharge to groundwater. It also explains the positive groundwater balance in the deltaic regions of coastal Andhra. A significant chunk of the renewable groundwater in alluvial Punjab is from irrigation return flows from both canal and well irrigation.

Obviously, in addition to the factors mentioned above, the return flow fraction of the irrigation water applied would also be determined by the dosage of irrigation (Kumar et al., 2009) and moisture conditions in the soil profile (Ahmad et al., 2004; Singh, 2005; Watt, 2008). In flow irrigation, the depth of individual watering is normally higher than that of well irrigation. The reason is the lesser control farmers have over irrigation water (Kumar et al., 2009). This facilitates a higher hydraulic gradient and hydraulic conductivity, increasing the velocity of the downward movement of water (Watt, 2008). Since generally, the groundwater table is high in canal commands, the return flow gets converted into recharge to groundwater, instead of remaining in the unsaturated soil zone contributing to moisture pressure in that zone. Even in well irrigation, dosages are quite high for paddy, due to the need for putting the field under partial submergence. Further in monsoon, the moisture levels in the soil profile would be generally good, which improves with applied water, increasing both the hydraulic gradient for movement of water, and soil hydraulic conductivity (Kumar & Bassi, 2011).

In order to understand to what extent paddy contributes to groundwater over-exploitation in Andhra Pradesh, it is important to know the characteristics of the regions where the crop is grown extensively with well water and in which season the crop is mainly irrigated. Here, we assume that the paddy field irrigated with canal water would eventually provide the return flows to groundwater, which would be available for reuse, and therefore would help augment groundwater. Similarly, paddy irrigated with groundwater during Kharif season would also provide substantial return flows, with the availability of rains. Experiences in paddy-irrigated areas in the state show that the area under upland winter paddy, which is grown with well-irrigation during the second crop season, is much less in comparison to irrigated paddy in the coastal districts (source: based on Murthy & Raju, 2009). Therefore, the argument that irrigated paddy depletes groundwater in erstwhile AP is open to question.

7.7 Maneuvering energy–groundwater nexus for co-management of electricity and groundwater

In the past few years, there has been growing recognition of the fact that subsidized and often free power supplied to agriculture is encouraging wasteful use of groundwater and electricity, apart from causing huge revenue losses to the state exchequer (see, for instance, Chand, 2018; Murthy & Raju, 2009; Saleth, 1997). This is evident from the data provided by Murthy and Raju (2009) for three districts of Andhra Pradesh. Their comparison of the energy requirement for groundwater pumping for irrigation and the actual energy consumed (Table 7.1) reveals a very high wastage of electricity used up in agricultural pumping. Part of the reason, is the zero marginal cost of electricity, while the other reason is the night power supply.

The total electricity subsidy for the agriculture sector in erstwhile Andhra Pradesh was Rs. 4176 crore annually in 2001–02 (GoI, 2002). The amount of subsidy might have only gone up with the total electricity consumption in agriculture going up to 17,441 million units in 2009–10 from 11,055 million units in 2000–01 (see Chapter 4). While the power supplied to agriculture being unreliable and of the poorest quality (with supply being provided during night

TABLE 7.1 Estimated energy requirements and actual energy consumed.

	Vijayanagaram	Visakhapatnam	East Godavari
Estimated energy requirement	6.83	11.03	27.74
Energy consumed	19.12	19.08	95.24
Energy saving potential	12.59	7.05	68.10
Percentage energy saving potential	65.85	38.90	71.06

From: Murthy and Raju (2009).

hours with voltage fluctuations and power cuts), farmers endure frustration and huge economic costs in the form of loss of production[1] (Monari, 2002), and thus virtually do not enjoy any subsidy. While this is true, what is concerning is the marginal cost of electricity under a flat rate system of pricing. It is still zero under the flat rate tariff. Here, both the government and farmers lose, while both the water and energy economies suffer.

Ideally, energy supply and power pricing in the farm sector can significantly influence the way groundwater is used for irrigation (Aarnoudse et al., 2017; Kumar, 2005; Kumar & Amarasinghe, 2009; Pfeiffer & Lin, 2014; Saleth, 1997; Zekri, 2008). The suggestions to regulate groundwater draft included improving the efficiency of pump sets (Rama Mohan & Sreekumar, 2009), scientific rationing of power supply and metering, and pro rata pricing of electricity (Shah & Verma, 2008; Shah et al., 2004). Among these three important suggestions, the second one, i.e., scientific rationing of power supply, has attracted attention in policy circles in the state. This has been propped up by the proponents of this idea as the second best solution for co-management of both the electricity and groundwater economy in South Asia, after metering and pro-rata pricing, which according to them is unlikely to work under South Asian conditions (see Shah et al., 2004). This is mainly because of the much-publicized "success" of *Jyotigram Yojna* in Gujarat, wherein the separation of feeder lines for agriculture and domestic sectors and the limited 8-hours of high-quality power was supposed to have resulted in the control of electricity theft and the consequent rational use of electricity by farmers for irrigated crops and therefore efficient use of water for crop production (Shah et al., 2009). Shah and Verma (2008) argued on the basis of a survey of farmers in different areas of Gujarat that this experiment has been highly successful in reducing electricity use and therefore groundwater pumping in the state. But, as we have seen in Chapter 6, the evidence available is a "far cry" from what their claim entails.

The fact is that the basalt and crystalline rock aquifers in the state are too poor to yield water for more than 2–3 hours a day (Marechal et al., 2003). Currently, power supply to agriculture is more a limited 5–6 hours a day. This is more than what farmers require to abstract the available water in their wells in most parts of Andhra Pradesh. The only exception would be coastal Andhra Pradesh in the deltaic region of basins such as the Krishna and Godavari, where groundwater is abundant. This region accounts for only a small percentage of the state's geographical area (see Table 7.2 for details). But, here again, groundwater is under-utilized and the demand for groundwater would be high only in years of drought when canal water release drops. Therefore, restrictions on power supply are not going to change the way farmers use groundwater for agriculture.

On the other hand, there is evidence from other parts of India on the role of pro rata pricing of electricity combined with a reliable power supply that

---

1. Since they are not able to irrigate the crop when it is most required, it has impact on the yield.

TABLE 7.2 Aquifers in the erstwhile state of Andhra Pradesh.

Sr. no	Type of aquifer	Approximate percentage area covered
1	Crystalline aquifer (consolidated)	50.0
2	Basalt formations (consolidated)	20.0
3	Deltaic alluvium (unconsolidated)	20.0
4	Hilly aquifers	10.0

From: Based on the aquifer profile of India, CGWB.

ensures that quality irrigation can bring about efficiency, sustainability, and equity in groundwater use (see Kumar, 2005; Kumar et al., 2011). Kumar et al. (2011) showed that when confronted with a positive marginal cost of using groundwater (as found in the case of well-owning farmers who pay pro-rata prices for electricity and farmers who use wells run on diesel engines), farmers use water more efficiently in physical terms, grow crops with higher water productivity in economic terms (Rs/m^3), improve their overall farming system, and secure higher net returns from a unit volume of water used at the farm level, as compared to farmers who use wells run with electric pump sets and pay for electricity on the basis of connected load. The greater control over irrigation water helps diesel engine owners apply water as and when required, thereby achieving better on-farm water management. Also, these farmers were also found to be using less amount of water per unit irrigated area, and the reduction was disproportionately higher than the reduction in net return from a unit area of land (Rs/ha), making it possible to sustain the net return from irrigated farming with much less use of groundwater. Research had also shown that the heavy subsidies well-owning farmers enjoy under the flat rate system of pricing electricity and the free electricity for groundwater pumping do not get transferred to the water-buying farmers (Kumar et al., 2013).

There are studies showing similar results from other parts of the world. Pfeiffer and Lin (2014) studied the effects of energy prices on a farmer's irrigation water pumping decision for the Kansas High Plains aquifer in the United States. Their results show that energy prices have an effect on both intensive and extensive margins. Increasing energy prices would affect crop selection decisions, crop acreage allocation decisions, and the demand for water by farmers. In particular, an increase in energy prices would decrease water use along both the intensive and extensive margins. The estimated total marginal effect, which sums the effects at the intensive and extensive margins, is that an increase in the energy price of $1 per million BTU would decrease groundwater abstraction by an individual farmer by 5.9 acre-feet per year, which is approximately 3.6% of the average amount pumped in a year by a farmer. The estimated elasticity of water extraction with respect to energy price is –0.26 (Pfeiffer & Lin, 2014).

7.8 Potential impact of drip irrigation on groundwater conservation

Expectations were raised regarding the potential of drip irrigation in Andhra Pradesh with the task force on micro irrigation set up under the chairmanship of Chandrababu Naidu, former Chief Minister of Andhra Pradesh. For instance, Narayanamoorthy (2009) assesses that the potential of a drip irrigation system in AP would be 1.68 m.ha. But, there has been no scientific assessment of the real potential for drip irrigation in the state, which involved data on cropping systems that exist in different regions of the state, and due consideration to the conditions under which drip system becomes the best bet technology. Researchers have argued that drip irrigation systems would become the best bet technologies when they are adopted in areas with semi-arid and arid climates with deep water table conditions, for row crops, and under well or lift irrigation. Further, it was argued that the water-saving benefit of drip irrigation will be significant in semi-arid and arid climates when used for row crops and in areas with deep unsaturated zones. The technical feasibility and economic viability of drip irrigation would be higher for well-irrigated crops, using independent pressurizing devices; and their economic viability would also be better for distantly spaced crops, for which the capital cost of the system would be less. The potential area under crops that are conducive to water-saving MI technologies, wherein the intended benefits could be derived, in the state was estimated to be only 5.57 lac ha (Kumar, 2009; Kumar et al., 2008a). Their estimate considered only the well irrigated areas in AP.

Unfortunately, none of these issues were given adequate attention while assessing the potential of drip irrigation systems in the state. A cursory look at the irrigation landscape of the state would show that nearly 3.84 million ha out of the total 6.28 m.ha of the irrigated land is under paddy, i.e., 61% (see Table 7.3 for details). One has to simply exclude this area when thinking about drip irrigation systems. While some experiments are on with drips for upland paddy and have been found to result in enhanced yield, they are still not found to be economically viable. The crops that are amenable to micro irrigation systems and that are grown extensively in Andhra Pradesh are groundnut, chilly, cotton, sugarcane, cotton, and tobacco. They together accounted for a total area of 1,684,900 ha

TABLE 7.3 Paddy irrigated area (2007–08).

Particulars	Autumn	Winter	Summer	Total
Cropped Area	2,577,609	1,406,334	0	
Irrigated Area (Ha)	2,443,633	1,406,334	0	3,849,967

TABLE 7.4 Area under crops amenable to micro irrigation systems.

Sr. no	Name of crop	Area irrigated (ha)
1	Groundnut	2,94,000.00
2	Chilly	1,69,000.00
3	Sugarcane	3,18,000.00
4	Cotton	2,55,000.00
5	Tobacco	47,000.00
6	Tomato	74,400.00
7	Banana	69,000.00
8	Mangoes	4,11,000.00
9	Onion	47,500.00
10	Total area under crops amenable to drips	16,84,900.00

From: Based on irrigated cropped area in AP in 2004–05, as cited in Reddy (2007) (for groundnut, chilly, sugarcane, cotton, and tobacco) and MOA and FW (2018) (for tomato, banana, mangoes, and onion).

(see Table 7.4). Among these crops, groundnut actually gives better results under micro sprinklers and not drips. Nevertheless, it is also considered in view of the water-saving benefit it can give when irrigated with the use of MI systems.

It is quite likely that some of these crops are also grown in canal command areas, making it difficult for the adoption of MI systems because of the need for intermediate storage systems (Kumar et al., 2008a). That would further reduce the potential of MI systems in the state. While MI systems are adopted extensively by farmers in some canal commands in Rajasthan[2] through the use of intermediate storage systems, such storage systems are viable by virtue of the large landholdings of the farmers in that region. Further, sprinklers make it possible to irrigate the otherwise difficult terrain with sloppy land and sandy soils. Hence, the actual potential of drip systems is quite likely to be close to the estimates provided by Kumar et al. (2008a). Finance is going to be another major impediment to MI adoption. The actual record of MI adoption in the state is a true reflection of these structural constraints facing the adoption of drip systems. Under the much-publicized Andhra Pradesh Micro Irrigation Project of the government of AP, a total area of 1.66 lac ha was covered under MI systems over a time period of nearly 30 months. The total subsidy benefit to farmers was to the tune of 209.96 crore rupees, which works out to be Rs. 18,070 per ha (Kumar & van Dam, 2009). The result is that most of the farmers have installed drip systems for fruit crops, for which returns, as noted by Dhawan (2000),

---

2. For instance, micro sprinklers are used in the IGNP command in Bikaner district with the help of *Diggies*, which are large surface storage tanks.

would anyway be high even without drips, and sprinklers, which are low cost but not technically efficient for small farms (Kumar et al., 2008a). Crops such as sugarcane, chilly, and groundnut would require much higher capital investment for installing drip systems as compared to fruit crops like sweet lime.

By 2016–17, an area of 8.51 lac ha was under drip irrigation and 2.56 lac ha under sprinkler irrigation in the present-day AP, while 1.08 lac ha was under drip irrigation and 0.24 lac ha under sprinkler irrigation in the present-day Telangana, as per the data provided by the Ministry of Agriculture, Government of India (source: https://pmksy.gov.in/microirrigation/AtGlance.aspx). Hence, the total area under micro irrigation from the region is 12.38 lac ha. The total area under drip irrigation, which can be considered as water-saving irrigation technology, is only 9.59 lac ha. This covers more than 58% of the potential area if we consider the nine crops listed in Table 7.4. However, we have not considered orchards, especially guava, sapota, custard apple, sweet lime, and papaya, and several vegetables and fruits (such as brinjal, potato, and watermelon), under the area amenable to micro-irrigation. The region has a significant area under these crops (MOA & FW, 2018), and drip irrigation is suitable for these crops. This means there is a lot of scope for expansion of the area under drip irrigation as farmers increasingly shift to high-value vegetables and fruit crops in the water-scarce regions. A lot of the investment is obviously coming from the Rayalaseema region, which is infamous for rampant well failures due to groundwater over-draft. The much-needed future investment is least likely to come from farmers given the current mode of pricing of electricity and water followed in the farm sector, which does not create any incentive for farmers to save water (Kumar, 2009; Kumar et al., 2008a).

7.9 Feasibility of community management of groundwater

Much has been written about the management of groundwater by local communities in India in general (Kulkarni & Vijay Shankar, 2009; Kulkarni et al., 2015; Mukherjee, 2021; Shah et al., 1998), and Andhra Pradesh in particular (Jain, 2008). Some of the initiatives, such as the APFMGS (Andhra Pradesh Farmer Management of Groundwater Systems) project (Shah, 2008) and the APWALTA (Andhra Pradesh Water, Land and Trees Act) project, have been widely acclaimed as innovative approaches for local management of groundwater by the rural communities, never before witnessed in any part of the developed world. The project was based on the premise that communities would observe self-regulation under an over-arching legal framework at the state level if supported by the village Panchayat and NGOs if they had proper knowledge about the resource. The thrust was on generating information about groundwater resources on a seasonal and annual basis, based on rainfall and monitoring of water levels in wells. Farmers' groundwater school was the cornerstone of this project. APFMGS envisaged that if farmers had the capacity to monitor groundwater resources in their localities, they would have great motivation to use them judiciously. Similar views are shared by researchers working in other parts of India (Jadeja et al., 2015, for Gujarat; Kulkarni et al., 2015; and Mukherjee, 2021, for India as a whole), who believe that knowledge of the aquifer is critical to organizing the communities around the resource.

So far as APWALTA is concerned, there are large-scale examples of communities undertaking self-regulation of groundwater, though there are some isolated cases of communities enforcing a ban on the drilling of deep bore wells in the village with the support of the Panchayats. But, unfortunately, there is no evidence of such measures addressing access equity concerns in groundwater, both intragenerational and inter-generational. As regards APFMGS, it did a commendable job of training farmers in hundreds of villages and handling instruments for groundwater monitoring. These farmers could also periodically monitor wells and generate data on groundwater levels in those villages. But, such steps have not resulted in measures for reducing groundwater draft, such as a shift in cropping pattern (from water-intensive crops to low water-consuming crops) or adoption of water-efficient irrigation technologies.

In spite of the almost three decades of work in participatory groundwater management initiated by NGOs in different parts of India (covering Gujarat, Andhra Pradesh, Karnataka, Madhya Pradesh, Maharashtra, and Rajasthan), there has not been a single case of communities managing the groundwater resources of their locality sustainably in a geographical area that is large enough to be significant as a successful initiative. What is missing in every such initiative is the ability of the community organization promoted by the external agency to define the abstraction limits of individuals and community groups volumetrically and enforce them. The growing reality that these local community-led actions cannot make much headway in ensuring sustainable groundwater use in their local areas in the absence of higher-order institutions that are legitimate and can enforce the rights of individuals and communities is not yet appreciated.

7.10 Future of water and agriculture in Andhra Pradesh and Telangana

There is very limited scientific literature available on the future of water and agriculture in AP and the challenges posed by the growing water scarcity. In spite of the fact that the total cultivable land in the state, in per capita terms, stands one of the highest among all the major states of India[3], and only 41% of this is irrigated currently, very little attention has been paid by international scholars working on water and food to the role of the state in addressing the future food security challenges facing the nation. With a major share of the cropped area under paddy, the contribution of the state to the central pool is remarkable. With 17.8 million tons of raw rice produced (in 2001), its contribution to

---

3. 0.19 ha per capita against a national average of 0.116 ha per capita (source: authors' own estimates).

TABLE 7.5 Shift in irrigated cropping pattern in Andhra Pradesh over five decades.

	1956–57	1960–61	1970–71	1980–81	1990–91	2000–01	2007–08	2008–09
Rice	81.23	79.26	78.64	77.82	71.32	68.31	61.26	63.05
Cotton	0.06	0.09	0.33	0.51	1.527	3.25	3.44	3.78
Chilly	1.56	1.44	2.23	1.59	2.59	2.62	2.83	2.51
Sugarcane	2.07	2.51	2.79	3.94	4.19	5.87	6.30	4.72

From: Andhra Pradesh Water Sector Reforms, Directorate of Economics and Statistics (2013).

TABLE 7.6 Stage of groundwater development over time in erstwhile Andhra Pradesh.

1980s	1990s	2002	2004	2007
28%	29%	43%	45%	41%

From: Jain (2008).

the national granary is 12.8%, against only 7% of the country's population (Kumar et al., 2008b). Therefore, depletion of groundwater can pose a major challenge not only to the national food security but also to the food security of millions of rural masses in the state.

We have already shown that paddy consumes only a fraction of the water that is applied, and the rest is available for re-use. A close look at the time series data on the problem of groundwater over-exploitation and area under paddy shows absolutely no correlation between the two. In fact, the percentage area under paddy had reduced significantly, while groundwater abstraction has increased remarkably from the 1980s through 2000.

The area under irrigated paddy went down from 81.23% of the gross irrigated area in 1956–57 to 63.05% in 2008–09 (Table 7.5). In aggregate terms, the area under irrigated paddy went up marginally from 3.83 m.ha in 1990–91 to 4.25 m. ha in 2008–09, a 10% growth over two decades, while the total irrigated area in the state increased from 5.37 m.ha in 1990–91 to 6.74 m.ha in 2008–09. Even here, the irrigated paddy area in 2007–08 was almost the same as that of 1990–91, while the state experienced a quantum jump in irrigation during these two decades (by nearly 9.15 lac hectares). Now it is unlikely that the area under canal-irrigated paddy has reduced over time in the state, as it is generally observed that farmers receiving canal water take up wet crops. If it is so, then it is unlikely that the area under well irrigated paddy went up over time. On the other hand, as shown in Table 7.6, the degree of groundwater utilization went up from 28% to 45% (Jain, 2008), which indicates a lack of connection between paddy irrigation and groundwater depletion. Still, academic and policy debates on sustaining Andhra Pradesh's groundwater irrigation economy have to a great extent centered on reducing the area under well-irrigated paddy (see Reddy & Kumari, 2007).

But, on the other hand, the area under highly water-intensive irrigated crops such as chilly and sugarcane had dramatically increased. In the case of sugarcane, it went up from 4.2% in 1990–91 to a high of 6.3% in 2007–08. The aggregate area added was 1.71 lac ha. The percentage area under cotton increased from 1.52 to 3.78, and in aggregate terms, the added area was 1.73 lac ha. Notably, there has been a phenomenal increase in area under irrigated maize in the state from 0.82 lac ha to 4.21 lac ha during the past two decades (from 1990–91 onwards), which accounted for 24% of the growth in irrigated areas. The crop requires nearly 540 mm of water when grown during the winter season in the parts of AP where it is grown today. If we assume that all this maize is grown with well irrigation, the added pressure on groundwater resources from this crop alone would be 1804 MCM per annum. This is based on the crop water requirement estimated for this crop for three districts of AP, viz., Hyderabad, Kurnool, and Nellore, estimated on the basis of FAO CROPWAT[4] for an additional area of 3.38 lac ha.

There are reasons to believe that the high incidence of well failures is forcing farmers to take up crops that give a very high return per unit area of land, which eventually consumes large amounts of water. Since the marginal cost of groundwater abstraction is zero, they are not concerned with the return per unit volume of water. The quest to recover the huge investments being made for well drilling is likely to lead the farmers to risky crops such as chilly and cotton. This can put farmers in a debt trap and sometimes resultant suicides, as has been found in some of the districts of Andhra Pradesh, like Anantapur. Naturally, the growing of paddy would be confined to some of the water-rich regions, such as coastal Andhra Pradesh, where water from canals replenishes the groundwater resources on a more reliable basis.

In spite of being a cereal surplus state and with significant exports, erstwhile Andhra Pradesh continues to be one of the most food insecure states in the country. The report on the state of food security in rural India prepared by the World Food Program of the United Nations and MS Swaminathan Research Foundation shows the average food insecurity index for AP to be in the range of 0.621–0.752 (MSSRF/WFP, 2001). This contradiction is occurring because of the lack of purchasing power of many people in both rural and urban areas. Keeping in view the larger concerns of national food security in general and the food security of poor people in the state in particular, it is important that the area under irrigated paddy be at least maintained at the current level, if not expanded. This can keep the production at satisfactory levels and keep the prices stable and affordable for

---

4. The estimated values of crop water requirement for winter maize are 503 mm for Nellore, 552 mm for Kurnool and 546 mm for Hyderabad.

the millions of poor people in the state. Paddy cultivation would also ensure the absorption of a large amount of the rural labor force in agriculture, generating income and thereby increasing the purchasing power. Given the problems of low success rates in drilling wells and the high incidence of well failure and reducing well yields, well irrigators will have the least incentive to take up paddy, which gives comparatively lower returns as compared to cash crops such as chilly, groundnut, cotton, tobacco, and sugarcane. Hence, what is needed is greater water security, which can come from a check on groundwater depletion.

7.11 Addressing the growing groundwater crisis in the region

It is quite evident that groundwater irrigation in the state has already reached a plateau and cannot expand further unless major improvements in irrigation efficiency occur. Efficiency improvements in groundwater irrigation in the state are a far cry given the fact that the state is indulging in more populist measures like free power for agriculture and heavy investments in popular welfare programs such as MREGA. Not only do these programs have a significant negative impact on energy (in the case of free power to agriculture) (Kumar, 2005, 2009) and the labor economy (in the case of MREGA) (Bassi & Kumar, 2010), but they can also cause huge damages to the groundwater economy in the long run. The farmers' increasing preference for growing water-intensive cash crops such as sugarcane and chilly is quite alarming.

Drip systems do not offer great potential in the state's context with a large area under irrigated paddy, as the maximum area that can be brought under drips is still around 7% of the gross cropped area and 15% of the gross irrigated area. Area under well-irrigated crops, which offer greater technical feasibility for MI adoption, would be even lower. Efforts aimed at artificial recharge of groundwater through watershed development, small water harvesting schemes and dug well recharge schemes are not likely to bring about any improvements in groundwater balance in the areas where it is required, due to the poor surface water potential in the river basins of these areas and the poor storage potential of the hard rock aquifers. Replacement of paddy by low water-consuming crops is not likely to have any significant impact on the groundwater balance, given the fact that most of the groundwater-irrigated paddy is grown in the Kharif season, during which a significant portion of the water applied to paddy fields gets back into the groundwater system as return flow, contributing to groundwater balance.

There are crops such as chilly, sugarcane, cotton, and tobacco, and winter paddy that are irrigated by groundwater and that deplete large amounts of groundwater. But, currently, most of the efficient micro irrigation systems are installed for orchard crops. It needs to be extended for other row crops such as sugarcane, chilly, cotton, and tobacco, which are also high-valued.

Pricing of electricity appears to be the only policy instrument that the government is left with to control the wasteful use of groundwater. The twin states (Telangana and Andhra Pradesh) should embark on metering and pro rata pricing of electricity in the farm sector, keeping in view of the fact that it would improve not only the energy economy but also the groundwater economy of the state. It is time to do away with the policy of free electricity to farmers, followed by both the states. With the introduction of pro rata pricing of electricity and improved quality and reliability of power supply, farmers would be encouraged to use water more efficiently in the field, thereby saving electricity and money. Improved quality of power would ensure that the farmers are able to irrigate the crop as and when required, protecting the crop from water stress and consequent yield losses. Also, such a pricing regime is going to create strong incentives among farmers to adopt micro irrigation systems for the highly water-intensive orchard crops and adopt on-farm water management practices such as zero tillage for maize. The water saved at the field level would translate into income savings.

The very fact that farmers' investments to the tune of thousands of crores are going to waste as a result of indiscriminate well drilling and high rates of well failures calls for major institutional interventions to avert an impending disaster. The only way competitive well drilling by farmers can be regulated is through the creation of groundwater rights, wherein the rights of individual farmers over the groundwater in aquifers underlying the land of hundreds of thousands of farmers have to be defined in volumetric terms. This has to be made tradable. This can be water use rights rather than absolute ownership rights over groundwater. The volumetric water rights can be fixed on the basis of the renewable groundwater resources that are utilizable, using equity and sustainability considerations (Hérivaux et al., 2020; Kumar, 2007; Zekri, 2008). Under proper market conditions, volumetric water rights, if made tradable, will also ensure the efficient use of water[5] (Kumar, 2005, 2007; Kumar & Singh, 2001; Rosegrant & Binswanger, 1994). This would encourage non-well-owning farmers to enter into water (physical) transfer agreements with well owners, thereby deferring the investments for new wells. It even provides opportunities for farmers to come together and decide on the number of wells to be operated at a time so as to mitigate the problems of excessive draw-downs in wells, which occur as a result of a large number of withdrawals happening simultaneously from a small area. Under such arrangements, the irrigation potential of energized wells could not only be enhanced but also fully utilized.

Both the fixed and variable costs of irrigation are likely to be low under this arrangement, as fewer wells will irrigate a larger area and the pumping cost will be lower owing to reduced draw-downs in wells. This would encourage the farmers to stick to the irrigation service providers rather than investing in their own wells and wasting money. The presence of a large number of potential

---

5. If markets are efficient, the price at which water is traded would reflect the opportunity cost of using water (Frederick, 1992; Howe et al., 1986). This would motivate farmers to either transfer water to grow crops that are economically efficient or use the water more efficiently for the existing crops, and sell the saved water at prices decided by the market to earn an income (Kumar, 2005, 2007; Kumar & Singh, 2001).

service providers might also ensure that the irrigation service (and not water) charges are highly competitive. But establishing and enforcing groundwater rights would be an enormous task, wherein one has to deal with the science of assessing the groundwater resources in different regions with reliability and also creating institutions for allocating and enforcement of water use rights.

Simultaneously, the twin states can also look at inter-basin water transfer options more seriously. The proposed 174 km long Polavaram-Vijayawada link canal, which is to take off from the proposed Polavaram reservoir on the Godavari River and supply water to Krishna River upstream of Vijayawada, would expected to have a total cultural command area of 1.4 lac ha under irrigation, in addition to releasing 2265 MCM of water into the Krishna River. Godavari being a water-surplus river basin (CWC, 2017), this project offers tremendous potential for not only reducing the demand-supply imbalance of water in Andhra Pradesh but also the environmental water scarcity in the lower stretches of the Krishna River. Studies by Sharma et al. (2008) on the groundwater externality of this water transfer project showed that the introduction of canal water would improve the groundwater regime, though it may not be adequate to meet the water requirements of the region. Releasing water through the link canal during Kharif for irrigation to provide supplementary irrigation to paddy would not only enhance the economic (use) value of the expensive water, but also provide much-needed recharge to groundwater in these hard rock regions. The study also showed that nearly 16% of the command area would be susceptible to water logging, which, according to the study, could be averted by growing paddy in the Kharif and winter seasons. Nevertheless, it is important to ascertain the storage potential of hard rock aquifers to absorb the seepage from the canal during Kharif seasons, as natural recharge from rainfall also takes place during the season.

7.12 Conclusion

The assessment of groundwater over-exploitation in erstwhile Andhra Pradesh based on simplistic considerations of annual recharge from various sources and abstraction provides highly misleading outcomes. The problems of groundwater depletion in the state are far more serious than what the official estimates show them to be, as evident from the growing incidence of well failures and the decline in the average area irrigated by a well. Indiscriminate drilling of wells, with their aggregate capacity far exceeding the utilizable resources in the shallow aquifers, is causing well interference and the fast drying of shallow wells. The investments, to the tune of 1648 crore, had been sunk by poor farmers in the state in drilling new wells without adding an extra hectare of well-irrigated area in just 3–4 years from 2000 to 2001. It also affected tank irrigation adversely. On the other hand, groundwater irrigation is unlikely to expand in the groundwater abundant regions, as the land in these regions is fully under irrigated production.

Given the existing cropping pattern dominated by paddy, very little can be achieved through drip systems, which too would require huge amounts of public expenditure in the form of subsidies for the capital equipment. Local water harvesting and groundwater recharge interventions can achieve too little given the hydrological regime, geological setting and soil conditions of the basins viz., the Krishna and Pennar, which are experiencing acute water scarcity and groundwater depletion. There is no evidence of community regulation and management of groundwater in a meaningful way.

Pro rata pricing of electricity in the farm sector appears to be the only major option the state is left with to tackle the problem of inefficient energy and water use in the short and medium terms. In the long term, establishing tradable property rights in groundwater based on physical sustainability and equity considerations might be a viable option if one considers the opportunity cost of not having them. A detailed discussion on establishing a functional water rights system is provided in Chapter 8. Such institutional interventions would also put a break on wasteful expenditure on competitive well drilling. Only that can protect the investments of farmers, which run into several billions of rupees. This is an arduous exercise given the scientific and institutional challenges of correctly assessing groundwater resources and instituting and enforcing water rights of individual users. But establishment and enforcement of tradable rights in groundwater in volumetric terms would ensure equity in access to groundwater and its efficient use, therefore achieving groundwater demand management (Hérivaux et al., 2020; Kumar, 2007; Rosegrant & Binswanger, 1994).

Inter-basin water transfers from water-rich basins (like the Godavari basin) to water-scarce basins also hold a big promise (Sharma et al., 2008). The state of Telangana has already embarked on a mammoth multi-purpose project that involves lifting around 215 TMC of water from the Godavari River, most of which is planned to be used for irrigation (details are in Chapter 10). But, the impacts of such interventions will be multiplied if we take the following measures. First: irrigation water is released for growing Kharif paddy in order to ensure sufficient return flows to the groundwater as recharge. Second: part of the excess flows diverted is used for reducing the environmental scarcity of the water-stressed river, apart from using it for recreational purposes. Third: caution is exercised in ensuring the conjunctive management of groundwater and surface water.

References

Aarnoudse, E., Qu, W., Bluemling, B., & Herzfeld, T. (2017). Groundwater quota versus tiered groundwater pricing: Two cases of groundwater management in north-west China. *International Journal of Water Resources Development, 33*(6), 917–934.

Ahmad, M. U. D., Masih, I., & Turral, H. (2004). Diagnostic analysis of spatial and temporal variations in crop water productivity: A field scale analysis of the rice-wheat cropping system of Punjab. *Journal of Applied Irrigation Science, 39*(1), 43–63.

Allen, R. G., Willardson, L. S., & Frederiksen, H. (1997). Water use definitions and their use for assessing the impacts of water conservation. In J. M. deJager, L. P. Vermes, & R. Rageb (Eds.),

Proceedings ICID Workshop on Sustainable Irrigation in Areas of Water Scarcity and Drought, Oxford, England, September 11–12 (pp. 72–82).

Amarasinghe, U., Anand, B. K., Samad, M., & Narayanamoorthy, A. (2007, August 30). *Irrigation in Andhra Pradesh: Trends and turning points* [Paper presentation]. In *The Strategic Analyses of India's River Linking Project: A Case Study of the Polavaram-Vijayawada link, C Fred Bentley Conference Center, Building # 212, ICRISAT Campus*, Patancheru, Hyderabad.

Athawale, R. N., (2003). Water harvesting for sustainable water supply in India, Centre for Environment Education, Ahmedabad, India.

Bassi, N., & Kumar, M. D. (2010, October). *NREGA and rural water management in India: Improving the welfare effects* [Occasional Paper # 3]. Institute for Resource Analysis and Policy.

Biggs, T. W., Gaur, A., Scott, C. A., Thenkabail, P., Rao, P. G., Gumma, M. K., Acharya, S., Turral, H., (2007). Closing of the Krishna Basin: Irrigation, streamflow depletion and macroscale hydrology, research report 111, International Water Management Institute, Colombo, Sri Lanka.

Central Ground Water Board (CGWB). (2019). *National Compilation on Dynamic Ground Water Resources of India 2017*. Faridabad: CGWB, Ministry of Water Resources, River Development and Ganga Rejuvenation. Government of India.

Central Water Commission. (2017, October). *Reassessment of water availability in India using space inputs, basin planning and management organization*. Central Water Commission.

Chand, R. (2018, November 15). Innovative policy interventions for transformation of farm sector. In *Presidential Address at the 26th Annual Conference of Agricultural Economics Research Association*, India.

Custodio, E. (2000). *The complex concept of over-exploited aquifer* (Secunda ed.). Uso Intensivo de Las Agua Subterráneas.

Dhawan, B. D. (2000). Drip irrigation: Evaluating returns. *Economic and Political Weekly, 35*(42), 3775–3780. http://eprints.icrisat.ac.in/13160/.

Dickenson, J. M., & Bachman, S. B. (1994). The optimization of spreading groundwater operations. In A. Ivan Johnson, R. David, & G. Pyne (Eds.), *Proceedings of Second International Conference on Artificial Recharge of Groundwater* (pp. 630–639). ASCE Publications.

Directorate of Economics and Statistics. (2013). *Season and Crop report 2012-13*. Directorate of Economics & Statistics Government of Andhra Pradesh. http://ecostat.telangana.gov. in/PDF/PUBLICATIONS/Season_crop_2012-13.pdf.

Frederick, K. D. (1992). *Balancing water demand with supplies: The role of management in a world of increasing scarcity* [Technical Paper 189]. World Bank.

Government of India. (2002, May). *Annual report (2001–02) of the working of state electricity boards and electricity departments 2001-2002*. Planning Commission-Power & Energy Division.

Government of India (GoI). (2005). Dynamic Ground Water Resources of India, Central Ground Water Board, Ministry of Water Resources, Government of India, August.

Hérivaux, C., Rinaudo, J.-D., & Montginoul, M. (2020). Exploring the potential of groundwater markets in agriculture: Results of a participatory evaluation in five French case studies. *Water Economics and Policy, 6*(1), 195009. https://doi.org/10.1142/S2382624×19500097.

Howe, C. W., Schurmeier, D. R., & Shaw, W. D. (1986). Innovative approaches to water allocation: The potential for water markets. *Water Resources Research, 22*(4), 439–445. https://doi.org/10.1029/wr022i004p00439.

Jadeja, Y., Maheshwari, B., Packham, R., Jadeja, Y., Maheshwari, B., Packham, R., Hakimuddin, Purohit, R., Thaker, B., Goradiya, V., Oza, S., Dave, S., Soni, P., Dashora, Y., Dashora, R., Shah, T., Gorsiya, J., Katara, P., Ward, J., Kookana, R., Dillon, P., Prathapar, S., & Varua, M. (2015, November 3–5). *Participatory groundwater management at village level in India: Empowering communities with science for effective decision making* [Paper presentation]. In *Australian Groundwater Conference 2015*, Canberra, Australia.

Jain, A. K. (2008, June 24). Challenges facing groundwater sector in Andhra Pradesh. In *Presentation by the Special Secretary to Government of AP.*

Kulkarni, H., Shah, M., & Vijay Shankar, P. (2015). Shaping the contours of groundwater governance in India. *Journal of Hydrology: Regional Studies, 4,* 172–192. https://doi.org/10.1016/j.ejrh.2014.11.004.

Kulkarni, H., & Shankar, P. S. V. (2009). Groundwater: Towards an aquifer management framework. *Economic and Political Weekly, 44*(6), 13–17.

Kumar, M. D. (2005). Impact of electricity prices and volumetric water allocation on energy and groundwater demand management. *Energy Policy, 33*(1), 39–51. https://doi.org/10.1016/s0301-4215(03)00196-4.

Kumar, M. D. (2007). *Groundwater management in India: Physical, institutional and policy alternatives.* Sage Publications.

Kumar, M. D. (2009). *Water management in India: What works, what doesn't.* Gyan Books.

Kumar, M. D., & Amarasinghe, U. (Eds.). (2009). *Water productivity improvements in Indian Agriculture: Potentials, constraints and prospects, strategic analysis of National River Linking Project of India Series 4.* International Water Management Institute.

Kumar, M. D., & Bassi, N. (2011). Maximizing the Social and Economic Returns from Sardar Sarovar Project: Thinking beyond the Convention. In R. Parthasarathy, & R. Dholakia (Eds.), *Sardar Sarovar Project on River Narmada: Impacts So Far and Ways Forward* (pp. 747–776). New Delhi: Academic Publishers.

Kumar, M. D., Scott, C. A., & Singh, O. (2011). Inducing the shift from flat-rate or free agricultural power to metered supply: Implications for groundwater depletion and power sector viability in India. *Journal of Hydrology, 409*(1–2), 382–394.

Kumar, M. D., Scott, C. A., & Singh, O. (2013). Can India raise agricultural productivity while reducing groundwater and energy use? *International Journal of Water Resources Development, 29*(4), 557–573.

Kumar, M. D., Narayanamoorthy, A., & Sivamohan, M. V. K. (2012). Key issues in Indian irrigation. In M. D. Kumar, M. V. K. Sivamohan, & N. Bassi (Eds.), *Water management, food security and sustainable agriculture in developing economies.* Routledge/Earthscan.

Kumar, M. D., & Singh, O. P. (2001). Market instruments for demand management in the face of scarcity and overuse of water in Gujarat, Western India. *Water Policy, 3*(5), 387–403.

Kumar, M. D., & Singh, O. P. (2008). How serious are groundwater over-exploitation problems in India? Fresh investigations into an old issue. In *Proceedings of the 7th Annual Partners' Meet of IWMI-Tata Water Policy Research Program "Managing Water in the Face of Growing Scarcity, Inequity and Declining Returns: Exploring Fresh Approaches,"* ICRISAT Campus, Patancheru, Hyderabad, India, April 2–4, 2008.

Kumar, M. D., Singh, O. P., Samad, M., Turral, H., & Purohit, C. (2008a). Water saving and yield enhancing micro irrigation technologies in India: When do they become best bet technologies? *Proceedings of the 7th Annual Partners' Meet of IWMI-Tata Water Policy Research Program "Managing Water in the Face of Growing Scarcity, Inequity and Declining Returns: Exploring Fresh Approaches,"* International Water Management Institute, South Asia Regional Office, ICRISAT Campus, Hyderabad, April 2-4, 2008.

Kumar, M. D., Sivamohan, M. V. K., & Narayanamoorthy, A. (2008b, October 24). *Irrigation water management for food security in India: The forgotten realities* [Paper presentation]. International Seminar on Food Crisis and Environmental Degradation-Lessons for India, Greenpeace.

Kumar, M. D., Singh, O. P., & Singh, K. (2001). *Groundwater depletion and its socioeconomic and ecological consequences in Sabarmati River Basin* [Monograph # 2]. India Natural Resources Economics and Management Foundation.

Kumar, M. D., Trivedi, K., & Singh, O. P. (2009). Analyzing the impact of quality and reliability of irrigation water on crop water productivity using an irrigation quality index. In M. D. Kumar, & U. Amarasinghe (Eds.), *Water productivity improvements in Indian Agriculture: Potentials, constraints and prospects, strategic analysis of National River Linking Project of India Series 4* (pp. 55–72). International Water Management Institute.

Kumar, M. D., & van Dam, J. C. (2009). Improving water productivity in agriculture in India: Beyond "more crop per drop". In M. D. Kumar, & U. Amarasinghe (Eds.), *Water productivity improvements in Indian Agriculture: Potentials, constraints and prospects, strategic analysis of National River Linking Project of India Series 4* (pp. 99–120). International Water Management Institute.

Marechal, J. C., Galeazzi, L., Dewandel, B., & Ahmed, S. (2003). Importance of irrigation return flow on the groundwater budget of a rural basin in India. In E. Servat, W. Najem, C. Leduc, & A. Shakeel (Eds.), *Hydrology of the Mediterranean* and semi-arid regions*, Proceedings of an International Conference held at Montpellier, France, April 2003* (pp. 62–67). IAHS Press.

Ministry of Agriculture and Farmer Welfare. (2018). *Horticulture statistics at a glance 2018*. Department of Horticulture, Ministry of Agriculture and Farmer Welfare, Government of India.

Monari, L. (2002, April). *Power subsidies* [Note Number 244]. Public Policy for the Private Sector, World Bank Group Private Sector and Infrastructure Network.

Mukherjee, M. (2021, October 19). Nationalisation of water is not an option for India: Mihir Shah [Interview]. *The Wire*. https://thewire.in/environment/interview-national-water-policy-management-mihir-shah.

Muralidharan, D., & Athawale, R. N. (1998). *Artificial recharge in India, base paper prepared for Rajiv Gandhi National Drinking Water Mission*. Ministry of Rural Areas and Development, National Geophysical Research Institute.

Murthy, K. V. S. R., & Raju, M. R. (2009). Analysis of electrical energy consumption of agriculture sector in Indian context. *ARPN Journal of Engineering and Applied Sciences, 4*(2), 6–9.

Narayanamoorthy, A. (2009). Drip and Sprinkler Irrigation in India: Benefits, Potential and Future Directions. In R. M. Saleth (Ed.), *Promoting Irrigation Demand Managment in India: Potentials, Problems, and Prospects* (pp. 253–266). Colombo, Sri Lanka: International Water Management Institute.

Narayanamoorthy, A. (2014). Groundwater depletion and water extraction cost: Some evidence from South India. *International Journal of Water Resources Development, 31*(4), 604–617. https://doi.org/10.1080/07900627.2014.935302.

Pfeiffer, L., & Lin, C. C. (2014). The effects of energy prices on agricultural groundwater extraction from the high plains aquifer. *American Journal of Agricultural Economics, 96*(5), 1349–1362. https://doi.org/10.1093/ajae/aau020.

Rama Mohan, R. V., & Sreekumar, N. (2009). *Evolving an integrated approach for improving efficiency of ground water pumping for agriculture using electricity: A few pointers from the field*. Centre for World Solidarity.

Reddy, M. D. (2007). *Water management: Andhra Pradesh*. Water Technology Centre, ANGRAU.

Reddy, M. D., & Kumari, V. R. (2007, March 5–9). *Water management in Andhra Pradesh, India*. Indo US Workshop on Innovative E-Technologies for Distance Education and Extension/Outreach for Efficient Water Management, 2007, ICRISAT, Patancheru, Hyderabad.

Reddy, V. R. (2005). Costs of resource depletion externalities: A study of groundwater overexploitation in Andhra Pradesh, India. *Environment and Development Economics, 10*(4), 533–556. https://doi.org/10.1017/s1355770X05002329.

Rosegrant, M. W., & Binswanger, H. P. (1994). Markets in tradable water rights: Potential for efficiency gains in developing country water resource allocation. *World Development, 22*(11), 1613–1625. https://doi.org/10.1016/0305-750x(94)00075-1.

Saleth, R. M. (1997). Power tariff policy for groundwater regulation: Efficiency, equity and sustainability. *Artha Vijnana, 39*(3), 312–322.

Sarkar, A. (2011). Socio-economic implications of depleting groundwater resource in Punjab: A comparative analysis of different irrigation systems. *Economic and Political Weekly, 46*(7), 59–66.

Shah, T., Gulati, A., Padhiary, H., Sreedhar, G., & Jain, R. C. (2009). Secret of Gujarat's agrarian miracle after 2000. *Economic and Political Weekly, 44*(52), 45–55.

Shah, M., Vijayshankar, P. S., & Harris, F. (2021). Water and Agricultural Transformation: A symbiotic relationship I. *Economic and Political Weekly, 56*(29), 17 July.

Shah, M., Banerji, D., Vijayshankar, P. S., & Ambasta, P. (1998). *India's drylands: Tribal societies and development through environmental regeneration.* Oxford University Press.

Shah, T. (2008). Groundwater management and ownership: Rejoinder. *Economic and Political Weekly, 43*(17), 116.

Shah, T., Scott, C. A., Kishore, A., & Sharma, A. (2004). *Energy irrigation nexus in South Asia: Improving groundwater conservation and power sector viability* [Research report # 70]. IWMI.

Shah, T., & Verma, S. (2008). Co-management of electricity and groundwater: An assessment of Gujarat's Jyotigram scheme. *Economic and Political Weekly, 43*(17), 59–66.

Sharma, B. R., Rao, K. V. G. K., & Massuel, S. (2008). Groundwater externalities of surface irrigation transfers under National River Linking Project: Polavaram Vijayawada link. In U. Amarasinghe, & B. R. Sharma (Eds.), *Strategic Analyses of the National River Linking Project (NRLP) of India, Series 2: Proceedings of the Workshop on Analyses of Hydrological, Social and Ecological Issues of the NRLP, New Delhi, India, 9-10 October 2007* (pp. 271–288). International Water Management Institute (IWMI).

Singh, R. (2005). *Water productivity analysis from field to regional scale: Integration of crop and soil modelling, remote sensing and geographical information* [Doctoral dissertation]. Wageningen University.

Watt, J. (2008). *The effect of irrigation on surface-ground water interactions: Quantifying time dependent spatial dynamics in irrigation systems* [Doctoral dissertation]. School of Environmental Sciences, Faculty of Sciences, Charles Sturt University.

World Food Programme/M S Swaminathan Research Foundation (WFP/MSSRF). (2001). *Report of the State of Food Security in Rural India*, Chennai, India.

Zekri, S. (2008). Using economic incentives and regulations to reduce seawater intrusion in the Batinah coastal area of Oman. *Agricultural Water Management, 95*(3), 243–252. https://doi.org/10.1016/j.agwat.2007.10.006.

Chapter 8

Institutions and policies for sustainable groundwater management in alluvial areas

8.1 Introduction

The Indo-Gangetic Plain (IGP) is a 700,000 km^2 fertile plain encompassing northern regions of the Indian subcontinent, including most of northern and eastern India, around half of Pakistan, a major part of Bangladesh, and the southern plains of Nepal. The IGP, also known as the "Great Plains," is a large floodplain of the Indus, Ganga, and Brahmaputra River systems. They run parallel to the Himalayan mountains, from Jammu and Kashmir and Khyber Pakhtunkhwa in the west to Assam in the east, and drain most of northern and eastern India. The IGP has the largest and probably the richest alluvial aquifer system in the world, covering the northern and eastern parts of India, large parts of Pakistan, and large parts of Bangladesh. It is historically well-known for settled agriculture and has experienced a virtual explosion in well irrigation during the past five decades (Pandey, 2014; Pant, 2004, 2005, for India; Qureshi, 2020, for Pakistan; and Qureshi et al., 2014, for Bangladesh). There has been a manifold increase in the number of groundwater abstraction structures in the form of wells and pump sets in this heavily populated and vast region. As per the data available from the 5th minor irrigation census, the number of wells (including dug wells, shallow tube wells, and deep tube wells) in the region (from the western Indian plains and the eastern Indian plains) stood at 9,023,402. (Source: based on data presented in GoI, 2017). The alluvial plains of the two basins, viz., the Indus and the Ganges, which fall within the Indian part of the sub-continent, however, display distinct characteristics in terms of the hydrological regime, the agro climate and the socio-economic characteristics, which largely determine the access to and use of groundwater (Kumar et al., 2012; Taneja et al., 2014). The Indus Basin part of the IGP has low rainfall and high aridity. But the region is economically prosperous; cropping and irrigation intensities are very high; and the agricultural productivity is one of the highest in the world, with very high yields of cereals such as wheat and paddy. Due to the high intensity of cropping, low rainfall, and high aridity, the groundwater use intensity is very high (Pandey, 2014), with

Groundwater Economics and Policy in South Asia. DOI: https://doi.org/10.1016/B978-0-443-14011-2.00016-4
177

problems of over-exploitation and groundwater quality deterioration. The major issues here are falling water levels, the increasing cost of construction of wells, and the constantly rising cost of pumping water from higher depths.

Contrasting conditions exist on the eastern part of the Gangetic plains. Though groundwater resources are abundant, the aggregate demand for water for irrigation is low, owing to high to very high rainfall, a sub-humid to humid climate, and abundant surface water resources available in wetlands (ponds and lakes). Yet, only a small percentage of the farm households enjoy direct access to groundwater through wells, owing to poor access to electricity and the rising cost of diesel, lack of access to finance for well construction, the poor economic conditions of the marginal holders who dominate the agricultural landscape and the high degree of land fragmentation. Millions of farmers depend on pump rental water markets to access water for irrigation, and this indirect access to groundwater is governed by the monopoly power of the well owners who offer irrigation services to the non–well-owning farmers (Pant, 2004, 2005).

As a clear indication of the intensification of groundwater use, the number of deep tube wells increased from 0.59 million to 1.04 million during a very short time span of seven years in the western part of the IGP, followed by a slight reduction in the number of open wells, while a similar trend was not seen in the eastern parts. In fact, there was no significant change in the number of deep tube wells in that region (source: based on GoI, 2017). It is therefore evident that while sustainability of resource use is an issue in the Indus Basin part of the IGP, inequity in access and financial scarcity of the resource are the key issues in the eastern part of the IGP. The governance and management challenges are different in the two situations and therefore are the institutional and policy measures required for improving the management of groundwater in the two distinct regions within the IGP. In the former case, the institutional and policy interventions should promote efficient and sustainable use, whereas in the latter case, they should be designed to promote distributional equity.

8.2   Groundwater development, use, and management in the Indo-Gangetic Plain: understanding the institutional landscape

8.2.1   Interventions by state agencies for developing well irrigation

In the 1980s, when access to drilling and pumping technologies was extremely limited to ordinary smallholder farmers in rural areas, the public tube well program was a major institutional intervention in the groundwater irrigation sector. At that time, public policies supported state-owned tube wells for irrigation in Uttar Pradesh, which received support from the intellectuals who believed that private tube owners could create short-run diseconomies on traditional, open-dug wells owned by small and marginal farmers (Pant, 2005). It has been argued that the performance of the state-owned tube wells had been poor, mainly attributed to poor electricity supply conditions in rural areas. In the 80s, there was overwhelming evidence of a correlation between caste and ownership of water abstraction devices, with land holding size. However, by the mid-2000s, while the average size of land holdings among the high-caste people was much higher than that of the low-caste people, the supremacy of this class in owning modern water abstraction devices started waning.

Pant (2004) reported on the basis of a survey carried out in 2002 and the early 80s that a higher proportion of farm households depended on public irrigation schemes such as canals and public tube wells as compared to the early 80s, while water markets had also become a dominant factor in irrigation amongst smallholders, with the percentage of farmers depending on both increasing over the two-decade period (Pant, 2004). Even though many of those who have their own private sources such as wells also depend on purchased water for meeting their irrigation requirements, this could be mainly due to the high degree of land fragmentation. Though researchers have downplayed the impact of public tube well irrigation programs on improving access equity in groundwater in the eastern Gangetic plains and eulogized private water markets (Mukherji, 2008; Pant, 2004, 2005; Shah, 2001), the analyses used to support such arguments are rather skewed and do not use any sound criteria for analyzing and comparing the equity impacts of water markets and public tube well programs. Parameters such as the monopoly price ratio and the cost that the water-buying farmers would have otherwise incurred to obtain the water through investment in their own wells were never part of the analytical framework.

8.2.2   Groundwater markets for irrigation for small and marginal farmers

Pump rental markets, also known as groundwater markets, exist in the IGP (Malik et al., 2008; Sarkar, 2011), but are more prevalent in the eastern Gangetic plains (Pant, 2004, 2005). The access to groundwater for irrigation and the degree of equity in access are heavily dependent on the way these markets function. A large proportion of the rural people in the eastern Gangetic plains, especially in eastern Uttar Pradesh, Bihar, and West Bengal, are very poor, with very small and fragmented holdings. As one can see from Table 8.1, in Bihar, the proportion of farmers who belong to the marginal category (less than one ha holding) is 91%, followed by 82.8% for West Bengal and 80.2% for Uttar Pradesh. It is only in Punjab where the marginal farmers make up a small percentage of the operational holders. The average holding of the marginal farmer is lowest in Bihar (0.25 ha), followed by Uttar Pradesh (0.38 ha).

The problem of small-sized holdings is compounded by land fragmentation. These factors reduce their ability to invest in individual wells and pump sets and purchase water from resource-rich, diesel, and electric well-owning farmers at prohibitive prices. Studies have shown that these markets are highly monopolistic in the sense that the price that well owners charge the buyers is far higher

TABLE 8.1 Proportion of marginal farmers in the operational holders in the five states of IGP.

Name of the state	Total no. of marginal farmers	Total no. of farmers	Percentage of marginal farmers	Average size of operational holdings of marginal farmers (ha)
Assam	1,868,000	2,742,000	68.00	0.42
Bihar	14,971,000	16,413,000	91.20	0.25
Punjab	154,000	1,093,000	14.10	0.60
Uttar Pradesh	19,100,000	23,822,000	80.17	0.38
West Bengal	5,998,000	7,243,000	82.81	0.49

From: The data presented in MOA and FW (2019).

than the actual cost incurred by them for abstracting water. Further, the prices are largely decided by the market conditions, and not the actual cost incurred by the well owners (Kumar et al., 2013). The electric well owners who incur a very low cost for abstracting water owing to heavy energy subsidies charge as much as what a diesel pump owner charges, and therefore the monopoly price of electric well owners was found to be very high (Kumar et al., 2013). Over a period of time, the monopoly prices charged by these well owners have however reduced, with a greater proportion of the farmers owing wells and pump sets, reducing the monopoly power of the well owners (Kishore, 2004). In the western IGP, especially in Punjab, there is no widespread incidence of water markets, though kinship partners owning wells jointly is reported by scholars.

8.2.3 Legal and regulatory instruments for checking over-development

There are some sporadic attempts by state governments to control and regulate groundwater use through top-down legislative measures, irrespective of whether such controls are required or whether they would be effective in checking over-draft of groundwater. In 1993, the government of West Bengal imposed restrictions on the drilling of low-duty tube wells in some districts through an executive order, though this step was not effective in controlling groundwater development, as evident from the Minor Irrigation Census of 2002, which showed a vast increase in the number of tube wells in the state (Kumar, 2018).

The government then introduced a check on the issuing of new electricity connections for farmers who want to install electrified pump sets for their wells. The state groundwater department also started issuing licenses for drilling agro wells, and the same was used as a mechanism for control in situations where such developments posed a threat to groundwater sustainability (Mukherji, 2006).

Some researchers criticized this as a "knee-jerk" reaction from officialdom to the impending threat of groundwater mining in the state and an avenue for rent-seeking, as according to them the official assessment of the resource does not show any sign of groundwater over-development. They also strongly argued for deregulating the process of drilling wells in the state (Mukherji et al., 2012). However, the point that is largely ignored is that the groundwater resource assessment methodology currently followed is not robust enough to capture all negative consequences of groundwater intensive use, such as its impact on wetlands, which is a very crucial factor for states like West Bengal. The government of Bihar had already framed a groundwater act for the state; however, it has not yet been enacted.

In Punjab, the government had asked farmers to delay the sowing time for Kharif paddy to reduce the dependence on groundwater for raising the paddy crop through an ordinance in 2008, which became law in 2009 and was called the "Punjab Preservation of Sub-Soil Water Act 2009." The Act prohibits the farmers from raising paddy nurseries before 10th May and transplanting paddy saplings before 10th June, and those who are found violating the Act would be fined Rs. 10,000 per ha and their crops would be destroyed in the field at their own expense (Singh, 2009).

The farmers in the region generally do early nursery and transplantation of monsoon paddy by early April and early May, respectively, which are the hottest months for north-western India for obtaining the best yield, and use water from their wells to raise the paddy nursery and also give 1–2 waterings to the transplanted paddy saplings. The penalty is used to discourage farmers from going for pre-monsoon sowing, which increases not only the crop water requirement but also the irrigation water requirement. Such measures can actually help reduce groundwater pumping. This is because, on the one hand, the crop water requirement for paddy would drop after the onset of monsoon due to a reduction in solar energy, temperature, and wind speed, and an increase in relative humidity, while on the other hand, the effective availability of rainwater for crop production would increase (Singh, 2009). Theoretically, this should work in the field if monitoring and enforcement are strict. Some reports and studies have suggested large-scale adoption of this practice by Punjab farmers and its positive impact on groundwater conservation (Singh, 2009; Tripathy et al., 2016), electricity consumption, and the environment, while reducing the revenue losses for the Punjab State Electricity Board (Singh, 2009). However, the effect of delayed transplanting on paddy yields needs to be studied.

8.2.4   The mechanism of minimum support price for cereals

It has long been argued that the minimum support price being offered to farmers in Punjab and Haryana for paddy and wheat had come at a huge cost to the society in the form of over-exploitation of groundwater resources in these states, as it is argued that paddy is not suitable for the agro-ecology of Punjab and the crop depletes large amounts of water. An immediate response has been to remove the minimum support price for wheat and paddy in the state, which encourages farmers to continue with the cropping system. However, the impact of paddy-wheat on the water ecology of Punjab is largely a perception that has originated from the fact that the irrigation water dosage to Kharif paddy in a semi-arid region like Punjab with sandy and sandy loam soils is very high. But what is equally important but often ignored is the fact that a large part of the water applied in paddy fields for inundation percolates down to reach the groundwater table. However, there has been no proper water balance study done in Punjab to substantiate this claim. Studies in the Indus basin in Punjab, which has more or less the same climate as that of the Indian part of the Indus basin, have shown that the evapotranspiration (ET) from the paddy-wheat cropping system is much less in comparison to the total amount of water applied in the form of irrigation and rainfall. Therefore, more than the water intensity of the crops, what matters is the high intensity of irrigated crop production in the state, with 95% of the land under cultivation and 90% of the same under irrigation.

8.3 Groundwater, energy, and food sector policies influencing groundwater development and use

8.3.1 Policies for providing power connections in the farm sector

All the Indian states had adopted very friendly policies towards the farm sector when it comes to issuing power connections. However, the outcomes of this policy have not been uniform across states. States such as Punjab and Haryana made enormous progress long ago in providing agricultural power connections and therefore water markets are not common. As noted by Malik et al. (2008), in Punjab, wells are often owned by several farmers, who allocate water among them on the basis of their ownership holding in the water abstraction structure. So are diesel engines, which are quite rare. This has been the result of the rapid pace of rural electrification that happened in these states to support tube well irrigation, which was the backbone of the Green Revolution in the early 70s. The strong demand from the rural sector, good state finances and good overall workings of the power sector in the states enabled this change. The relatively large operational holdings of farmers also generated demand for farm power connections in these agriculturally prosperous states.

Contrary to this, in states such as Bihar, Uttar Pradesh, and West Bengal, the pace of rural electrification has been very slow, owing to poor state finances, poor overall governance and management of the power sector, and the poor economic condition of the population in rural areas, which suppressed the demand for electricity. These states are also power-starved, with a major gap between demand and supply, widening over time.

In Uttar Pradesh and Bihar, owing to poor quality power supply, de-electrification has been happening in rural areas for some time, with some improvements in the recent past (Pant, 2004, 2005). The high cost of pumping water for irrigation using diesel-powered engines had adversely affected both well owners and water buyers. Based on past empirical studies, it could be safely argued that the delay in providing power connections to farmers would have had a negative impact on access equity in groundwater, as monopoly prices are found to be higher in situations of a few sellers against many potential buyers (Kumar et al., 2013).

8.3.2 Electricity supply policies for the farm sector

Most states in the IGP provide free power to the farmers for agricultural use, which is mainly used for pumping groundwater (Badiani et al., 2012; Gulati & Pahuja, 2015). In Punjab, the state power utility started supplying electricity to the farmers free of charge, abandoning connected load-based charging (Gupta, 2021). The only exception to this is West Bengal, which introduced metered pro rata tariffs for agricultural users in 2006. While Uttarakhand introduced the policy of metering agricultural power connections and later revoked it, West Bengal has been continuing with the policy of metering power supply to agriculture. Since the state subsidy to electricity consumed by the farm sector is very high and most of the revenue to the utilities comes from the commercial and industrial sectors in these states, the utilities have not been very keen to provide quality power supply to the sector and have also tried to restrict electricity use through rationing of power supply in order to reduce the subsidy burden. In none of the states, the power supply is round the clock, and in most cases, it is now available for a maximum of 6 hours.

Over the years, with the financial workings of the state power utilities deteriorating, the quality of power supply to agriculture has been on the decline in Uttar Pradesh (Pant, 2004, 2005) and Bihar (Kishore, 2004). Poor quality and reliability of power supply influence the quality of irrigation water supply from wells. This in turn also forced farmers to shift to diesel engines in shallow groundwater areas (Pant, 2004, 2005). This is evident from the latest statistics. According to the 5th Minor Irrigation Census report (2017), the total number of diesel engines (for wells and for surface lift) in the eastern Indian plains went up from 3.11 million during a time span of 7 years from 2006/07 to 2013/14, while there has been a significant reduction in the number of diesel engines in the country as a whole from 6.47 million to 5.14 million (GoI, 2017, p. 19). In the eastern Indian plains, the intensity of use of electric pumps is very low, especially in eastern Uttar Pradesh, Bihar, West Bengal, Odisha, and Jharkhand (Fig. 8.1). However, it is the relatively shallow water table in this region (CGWB, 2019) that allows farmers to shift to diesel engines that run on expensive diesel, an option that is not available in the deep water table areas of the western Indian plains and the peninsular region.

The poor reliability and quality of power supply force farmers to go for low-risk and low-value crops such as cereals (paddy and wheat), which are not

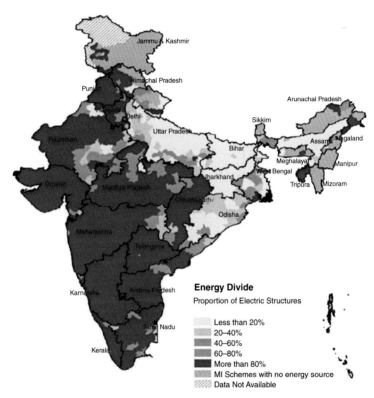

FIGURE 8.1 Intensity of electrification of minor irrigation (MI) schemes in India. (*From: GoI, 2017*).

very sensitive to the timing of irrigation. On the other hand, the zero marginal cost of using electricity for pumping water creates no incentive to use water and electricity efficiently. As a result, the returns are very low. The diesel well owners, who have greater control over irrigation water, on the other hand, are found to be allocating comparatively more water for growing vegetables and cash crops (Kumar et al., 2010).

8.3.3 Electricity pricing policies and subsidies

The argument by scholars from economics and public policy circles that free electricity for groundwater pumping, which results in almost zero marginal cost of pumping, has resulted in over-exploitation of groundwater in semi-arid and arid regions of India is at least a quarter century old (Badiani et al., 2012; Dubash, 2007; Saleth, 1997; Singh et al., 2022). While metering of electricity supplied to the farm sector and charging a pro-rata tariff was suggested as a measure to check this (Dubash, 2007; Kumar, 2005; Narayanamoorthy, 1997), the counter-argument was that the cost of metering electricity used for running pumps by well irrigators would be so high that if transferred to the users, it would force farmers to reduce groundwater use for irrigation, thereby reducing the welfare effects of groundwater irrigation, including negative impacts on access equity in groundwater use, as a result of the increasing price of water in the market (Shah, 1993; Shah et al., 2004). Another counterview is that the price at which the demand for electricity for pumping water responds to price changes would be so high that irrigation would become unviable for the farmers (Saleth, 1997).

The first argument does not consider the fact that there is scope for farmers to improve the efficiency of electricity and water use in agriculture so that with reduced energy use, the same amount of farm output could be generated with no welfare losses. As regards the second argument, with a positive marginal cost of using electricity and water and with high energy and water costs, farmers would take greater risk and shift to high-value crops that yield a higher income return per unit of water and electricity. Empirical studies done in alluvial north Gujarat (Kumar, 2005), the eastern Gangetic plains (south Bihar and eastern Uttar Pradesh) and alluvial north Gujarat (Kumar et al., 2011, 2013) validated this argument.

Nevertheless, the electricity utilities in many Indian states, including Uttar Pradesh, Bihar, and Punjab, follow the practice of providing free electricity to the farm sector for groundwater pumping or charging for power on the basis of the connected load. One major justification offered by the utilities for this has been that the transaction cost of metering electricity use in the farm sector is sometimes more than the revenue from the sale of electricity to that sector, in view of the fact that the wells are located in remote rural areas and electricity anyway will have to be supplied at heavily subsidized rates to the farmers given the poor affordability among the farming community. However, the transaction cost argument is just used as an excuse for rent-seeking by field-level officials of the state power utilities.

8.3.4 Provisions of subsidized diesel engines and fuel

Millions of farmers in the eastern Gangetic plains (in Bihar, Uttar Pradesh, West Bengal, and Assam) use diesel engines for pumping groundwater in lieu of the fact that obtaining power connections for running electric pumps is extremely difficult in the rural areas (GoI, 2017). The access to groundwater at low depths keeps the energy used for pumping a unit volume of irrigation water low (Kumar et al., 2010). These states were also offering capital subsidies for the purchase of diesel engines (Kumar, 2018). Over and above this, diesel prices were kept very low in India for many decades through subsidies. Hence, the costs incurred by diesel well owners for pumping groundwater are not high (Kumar et al., 2010). Kumar et al. (2010) showed that given the shallow water table conditions and the subsidized diesel, the farmers in the eastern Gangetic plains were not very cautious about diesel use, with the result that the energy use efficiency is quite low, though higher as compared to electric well commands.

The study also showed that there is enormous scope for increasing diesel prices in the farm sector, which can result in improved energy use efficiency through improved water use efficiency in irrigated crop production without adversely affecting farm incomes. The empirical analysis that supported this inference showed that the farmers who purchased water from diesel well owners and paid a very high price for water were found to have higher water use efficiency in their cropping system as compared to their diesel well-owning counterparts, who incurred a much lower cost for pumping water from their own wells. More importantly, in spite of the high cost of irrigation water, the water buyers were getting almost the same net income per unit area of land as compared to the diesel well owners, as they allocated more water and land for high-value crops (especially vegetables) that give a higher return per unit volume of water and unit area of land (Kumar et al., 2010).

8.3.5 Procurement policies

It has long been argued that the agricultural procurement policy adopted in the Green Revolution states such as Punjab and Haryana, which offers a minimum support price higher than the market price for the cereals, viz., paddy and wheat, combined with free electricity for agricultural groundwater pumping and the storage facilities offered by Food Corporation of India (FCI), keeps the rice-wheat system of farming popular among the farmers, depleting the groundwater resources in the region (Gulati et al., 2017). However, more than the good procurement price offered for paddy and wheat, what keeps the farmers hooked to these crops is the absence of a good market for crops such as maize and ground nut, which are grown in smaller areas in these states (Shergill, 2005). Nevertheless, the discussion on the appropriateness of agricultural procurement prices revolved around resource conservation. In the process, what is being ignored is the fact that the two states of Punjab and Haryana contributed significantly to the nation's food basket, with large stocks of wheat and rice from these states going to the buffer stock of grains for a long time. Any large-scale shift in the cropping pattern of the cereal to oil seeds and other cash crops would have posed a major threat to national food security. However, these constraints would no longer be existent. In recent times, states such as Madhya Pradesh, which have a large amount of cultivable land, have become a major supplier of wheat and paddy to the nation's buffer stock, with its irrigated area expanding remarkably during the past 10–12 years.

However, even if we leave aside the problem of food security, there are significant challenges to crop diversification in states like Punjab. This is because in highly productive farms like those in Punjab, crop diversification is not desirable as improving productivity warrants that farmers specialize in a few crops (Shergill, 2005). Further, economic incentives for shifting to less water-intensive crops are lacking in the absence of inefficient electricity and water pricing (Srivastava et al., 2015).

8.4   Resource sustainability: impact of institutional and policy interventions

8.4.1   Resource depletion in the Indo-Gangetic Plain

The alluvial IGP had a large groundwater stock (source: based on GoI, 1999; Hussain & Abbas, 2019; Lytton et al., 2021; Shamsudduha et al., 2011). This unique characteristic of the aquifer systems underlying this region poses a unique threat to aquifer mining. In the Indian part of the IGP, the free electricity for groundwater pumping available in the states and the good procurement prices for paddy and wheat encourage over-use of the resources for irrigating these water-intensive crops. It is possible to abstract water from the aquifers even when there is no recharge from rainfall in a particular year, unlike the hard rock areas where the aquifers have to receive replenishment from annual precipitation to yield water. In fact, the groundwater abstraction in the western part of the IGP goes up during drought years, with a sharp annual decline in water levels as farmers in the intensively cropped region go for tube well irrigation of transplanted paddy in the wake of delayed monsoon (Singh, 2009).

With potential evapotranspiration (PET) far exceeding annual rainfall, intensive cropping of paddy-wheat, which is heavily dependent on irrigation, causes negative groundwater balance, as a large share of the irrigation demand is met from wells. In recent decades, the groundwater table in the state has been declining at an alarming rate. The rate of fall in the water table was 18 cm during 1982–87, which increased to 42 cm during 1997–2002 (Hira et al., 2004) and further to 75 cm during 2002–06 (Singh, 2009). Annual groundwater extraction in Punjab is 31.16 billion m^3 as opposed to 21.44 billion m^3 renewable groundwater resources (CGWB, 2014). As pointed out by Kumar (2007), had it not been for the water release from canals, which provide additional recharge from irrigation return flows and canal seepage, the groundwater balance in alluvial Punjab would have been even more precarious.

In the eastern part of the IGP, rainfall is much higher than in the western parts, and reference evapotranspiration is much lower. Also, the effective renewable water resource (including soil profile) is far higher than the reference evapotranspiration in the eastern IGP, making it water-rich (Kumar et al., 2012). Permanent depletion of groundwater is very unlikely in this region through agricultural intensification, except in certain pockets of western Uttar Pradesh due to low rainfall and high aridity (Kumar, 2018).

8.4.2   Groundwater quality deterioration

A major threat facing the Indian part of the IGP is the contamination of groundwater resources. The most serious among them is arsenic contamination of groundwater in the Eastern IGP, with an increasing incidence of arsenic in the groundwater pumped from shallow aquifers, spatially (Shankar et al., 2014). This is reported to be caused by the oxidation of arsenic-containing minerals such as arsenopyrite or pyrite in the formations bearing groundwater in the deep strata, due to abstraction of water and consequent geochemical processes (of reducing redox conditions) resulting from the arsenic-containing ores, which lie in anaerobic conditions, coming in contact with air. The presence of arsenic in groundwater causes a major threat to human health in the form of arsenicosis due to excessive and long-term (such as 5–10 years) human intake of toxic inorganic arsenic from drinking water and food. These health problems include skin disorders, skin cancers, internal cancers (bladder, kidney, and lung), diseases of the blood vessels of the legs and feet, possibly diabetes, increased blood pressure, and reproductive disorders. If groundwater containing arsenic is used for irrigating crops, residues of arsenic can enter the crops, posing a threat to human health.

Arsenic contamination of groundwater poses a serious human health risk to people living in several districts of West Bengal and a few districts of Uttar Pradesh, Bihar, and Assam, falling in the alluvial IGP, given the fact that groundwater is the major source of drinking water and irrigation[1]. Studies also report that arsenic also enters the food chain through rice, which is cultivated on a large scale in the eastern Gangetic plains using groundwater from deep tube wells contaminated with arsenic, through the irrigation water–soil–crop–human, and irrigation water–soil–crop–animal–human pathways (Ahmed, 2009).

Another major groundwater contamination problem is increasing salinity levels in the groundwater. With lowering water levels due to groundwater overdraft, water from deep aquifers is being abstracted for irrigation and other uses. As the groundwater in the deeper strata of alluvial areas contains several minerals, the total dissolved solids (TDS) levels in groundwater have increased over the years, rendering it unsuitable for drinking purposes. There are large pockets in Punjab and Haryana where the salt concentration in groundwater (TDS) is above permissible levels for use in irrigation. High levels of nitrate in groundwater are reported from both the western IGP and eastern IGP (GoI, 2010). While in the western IGP, intensive irrigation and excessive use of nitrogenous fertilizers for paddy cultivation cause nitrate contamination of groundwater, in the eastern IGP, high rainfall, humidity, and shallow water table conditions create a favorable environment for nitrate leaching from cultivated soils and animal waste.

8.4.3 Drying up of wetlands in the eastern Gangetic Plains

In the past, many argued for exploiting the abundant groundwater in West Bengal and Bihar to bring about agrarian change and poverty reduction (Mukherji, 2003,

---

1. As of 2008, West Bengal, Jharkhand, Bihar, and Uttar Pradesh in the flood plain of the Ganga River; Assam and Manipur in the flood plain of the Brahmaputra and Imphal rivers; and Rajnandgaon village in Chhattisgarh state had been exposed to drinking arsenic-contaminated hand tube-wells' water. A total area of 88,688 km^2 and approximately 50 million people have been estimated to be vulnerable to groundwater arsenic contamination.

2005, 2008; Mukherji et al., 2012; Shah, 2001), and that limiting exploitation of groundwater has a huge opportunity cost (Mukherji et al., 2012). However, the climatic conditions (with respect to rainfall and PET), the amount of land available for crop intensification and flooding problems induce major constraints on increasing the use of groundwater for irrigated agriculture (Kumar et al., 2012). Interestingly though it may appear, in a recent empirical work involving field surveys, Mukherji et al. (2020) showed that in West Bengal, in spite of the state having favorable policies for promoting groundwater development (in the form of new electricity connections for tube wells) and with 200,000 new electric tube wells installed after the introduction of the policy, the cropping intensity had not increased. One of the factors identified by the researchers as the reason for this phenomenon was that there is not much scope for increasing cropping intensity in the state in lieu of the lack of arable land that can be brought under cultivation (Mukherji et al., 2020).

Yet, future exploitation of groundwater in West Bengal should be done with utmost care. The reason is that the region has vast wetland resources that provide water for domestic and livestock use and irrigation and support inland fisheries (Mukherjee and Kumar, 2011; Varghese et al., 2008). A study carried out by Varghese et al. (2008) using geospatial data in the EGP covering five districts of Uttar Pradesh showed a reduction in wetland area from a lowest of 45% to a highest of 90% over a period of 32 years from 1972 to 2004. A small increase in groundwater pumping can play havoc with the region's water ecology, as thousands of wetlands receive their inflows from shallow aquifers. Such changes are possible if the farmers introduce perennial tree crops (like mango, guava, sapota, and litchi) and annual crops such as banana and papaya, increasing the evapotranspiration per unit of land. Lowering of the water table of shallow aquifers during winter-summer seasons, when agricultural water demand actually picks up, can result in the temporary drying up of the shallow wetlands. This will have a huge impact on very poor families. The reason is that the poor farmers depend on these water bodies not only for domestic water supplies and irrigation but also for fish, which is a major source of protein. Clearly, the negative externalities of such approaches were ignored (Mukherjee and Kumar, 2011).

Arguments about recharging the aquifers during monsoon using schemes such as the National Rural Employment Guarantee Scheme (IWMI, 2012, p. 2) do not hold much water, as the aquifers are fully replenished and do not have much storage space during the monsoon months. In a region that is endowed with a large number of small water bodies (wetlands) that function as natural recharge systems, investing resources for exploiting groundwater and then recharging does not make hydrological or economic sense. Instead, if these wetlands are well managed, the poor farmers can tap water from these small water bodies for irrigation at a much lower cost. The obsession with the idea of achieving 100% utilization of natural recharge can only be attributed to a poor understanding of how groundwater systems interact with wetlands.

As is now understood (also discussed in detail in Chapter 3), "groundwater over-development" is a complex concept and concerns various "undesirable consequences" that are physical, social, economic, ecological, environmental, and ethical in nature (Custodio, 2000). Therefore, an assessment of groundwater over-development involves hydrological, hydro-dynamic, economic, social, and ethical considerations (Kumar and Singh, 2008; Kumar et al., 2012).

8.5 Institutional measures for sustainable groundwater management

8.5.1 Refining resource assessment methodologies

Developing a nuanced understanding of the resource dynamic is critical to evolving strategies for managing groundwater in the IGP. In the case of the western IGP, understanding the various factors influencing the groundwater balance of the region is important. Particularly, knowing the extent to which imported surface water used for canal irrigation augments groundwater recharge in the alluvial plains of Punjab and Haryana, and the extent to which resource depletion occurs through paddy cultivation in the form of evapotranspiration, is crucial.

However, the current groundwater assessment methodology does not consider these complex hydrological processes. Therefore, the discussion on arresting groundwater depletion in this part of the IGP has largely focused on reducing water withdrawal for paddy production and artificial recharge of groundwater. There is little recognition of the fact that a significant proportion of the water applied in paddy is available for recharge through irrigation return flows (Tripathy et al., 2016). A proper quantification of the various components of the water balance of the region would help evolve sound strategies for groundwater management. To improve groundwater balance in areas with negative groundwater balance, the surface water contribution for meeting the crop ET requirements needs to be enhanced. A poor understanding of the region's hydrology led policy makers and practitioners to push for water harvesting and groundwater recharge schemes to arrest groundwater depletion in the region.

In the case of the eastern IGP, understanding groundwater–wetland interactions is important to ascertain to what extent groundwater use could be intensified in the region. This is urgent, as the proposal for intensive groundwater pumping to reduce floods in the eastern part of the Gangetic basin has gained momentum in the recent past in the wake of this knowledge gap (see, for instance, Amarasinghe et al., 2016). Blanket use of the concept of intensive groundwater pumping during summer to reduce monsoon flooding in the Ganges, anywhere in the Ganga basin, can cause havoc in areas of the basin that are not flood-prone. Refining the current resource evaluation methodologies to suit the unique hydrological conditions of an area is going to be an institutional challenge.

8.5.2 Subsidy removal and correct pricing of electricity

Foster et al. (2017) presented the results of a laboratory experiment conducted in the city of León, Guanajuato, México, to study the groundwater extraction decisions of stakeholders under alternative subsidy structures. They analyzed the performance of two traditional policy interventions, i.e., the elimination and reduction of subsidy—and then a novel policy: decoupling the subsidy from the electricity rate by replacing it with a lump sum transfer. Their results suggested that the rate of water extraction and the level of water in the aquifer may be improved significantly by altering the subsidy structure. An important finding for policy makers is that the decoupling leads to outcomes similar to those of eliminating the subsidy, but with fewer political economy conflicts.

In the Indian context, two research studies carried out, one in north Gujarat (Kumar, 2005), and the other covering north Gujarat, eastern Uttar Pradesh, and south Bihar (Kumar et al., 2011, 2013) had shown that the energy prices at which the farmers started responding to tariff changes in terms of reducing the demand for these inputs would be socio-economically viable.

Kumar et al. (2011) showed that farmers who were confronted with a marginal cost of using electricity and water, which included farmers who paid for electricity on the basis of consumption (in north Gujarat), buyers of water from electric well owners who paid for irrigation services on an hourly basis (in eastern Uttar Pradesh and south Bihar), diesel well owners who incurred positive marginal cost for every hour of pumping, and water purchasers from diesel well owners who paid for irrigation services on an hourly basis (in eastern Uttar Pradesh and south Bihar) secured not only higher water productivity in economic terms (Rs/m^3 of water), but also higher net income per unit area of land (Rs/ha) than farmers who paid for electricity on the basis of flat rate or got free electricity supply in these regions (columns 4 and 6, respectively, of Table 8.2). The rate of groundwater pumping per unit of irrigated land (hour/ha) was low for the farmers who were confronted with a positive marginal cost of using electricity/water (column 7 of Table 8.2). The extent of the reduction in irrigation water demand that is possible without causing any reduction in net income is roughly 20%.

This is because they are able to improve the efficiency of the use of irrigation water and other farm inputs and also modify their farming systems when confronted with a positive marginal cost of electricity and a higher unit tariff, with the result that they get not only higher water productivity but also higher net income per unit area of land. Therefore, one should arrive at the ideal unit price for electricity for different regions that would be most efficient in terms of demand reduction, affordable for farmers and also improving the viability of the power sector (Kumar et al., 2011, 2013). The studies also showed that the shift from free power or flat rate (based on connected load) to pro rata pricing would not have any effect on the price at which water is traded in the market (Kumar et al., 2013).

TABLE 8.2 Impact of electricity prices and volumetric water charges on water productivity and net income from farming and groundwater use per unit area.

Region	Type of well command	Type of farmer	Water productivity in economic terms (Rs/m³)	Total farm-level income (Rs) with crops and dairying	Farm-level income per unit land (Rs/ha)	Average irrigation pumping per ha (hours)
Eastern Uttar Pradesh	Electric well	Well owner	10.98	131,740	24,880	175.00
		Water buyer	11.18	60,803	27,570	184.00
	Diesel well	Well owner	8.67	82,194	14,528	222.00
		Water buyer	12.89	68,584	18,075	148.00
North Gujarat	Electric well	Flat rate pricing	6.30	768,287	57,531	444.00
		Pro rata pricing	7.90	669,250	56,882	304.00
South Bihar	Electric well	Well owner	9.28	130,770	210,345	330.00
		Water buyer	10.13	76,024	190,031	250.00
	Diesel well	Well owner	11.97	150,064	191,387	231.00
		Water buyer	12.43	84,043	197,895	198.00

From: Kumar et al. (2011, 2013).

As regards the impact on the aggregate groundwater draft, results of the modeling study by Badiani et al. (2012) indicated that electricity subsidies increased groundwater draft. They suggest that a 10% reduction in the average electricity subsidy, which amounts to roughly a 50% increase in the price of electricity, would lead to a 6.6% reduction in groundwater draft. The estimated elasticity was −0.13.

That said, the argument of increasing inequity in access to groundwater due to a rise in the market price of water as a likely impact of pro rata pricing (Mukherji, 2008; Mukherji et al., 2012) missed the point that under flat rate pricing of electricity, large well owners, whose implicit cost of irrigation is very low and return from crop production is high, may enjoy monopoly power and decide the price at which water should be sold in the market. Conversely, the smallholders owning wells, whose implicit cost of irrigating their own farms is high, are left without much choice but to look for buyers to whom they could sell water to earn extra income, at a price decided by the market, as they will have limited bargaining power. Such prices obviously offer high-profit margins for the large farmers, but not for the small farmers (Kumar, 2018).

If electricity is charged on a pro-rata basis, both small and large farmers from the same locality will incur the same unit cost of pumping water, and the large farmers will not enjoy a comparative advantage over the small farmers. In view of the fact that there are no fixed costs of keeping pump sets, the small landholders are not under pressure to sell water at a very low price to stay in the market. Hence, both large and small farmers would have equal incentives to invest in wells and sell water and have equal opportunities to make profits in the shallow groundwater areas. This will lower the price at which water would be sold in the market (Kumar, 2018).

But given the sharp variations in water table conditions across India (CGWB, 2017), electricity tariff fixation has to be done carefully in different regions. The pro rata tariff should be an inverse function of the depth to groundwater table to make sure that the cost of electricity to abstract a unit volume of groundwater remains more or less the same across geo-hydrological environments (Kumar, 2018).

However, the price for electricity would also have to be a function of what farmers would be willing to pay. The WTP would be a function of the economic surplus generated from the use of water for irrigation or the value of water use in irrigation (Dinar et al., 1997; Young, 1996). This is a function of the cropping system that is feasible under the given climate, the wherewithal available to the farmers to adopt the technologies needed to raise the crops, and the market demand for the crop. These factors need to be considered while fixing the price of electricity.

The question is what should the actual unit charge for electricity be to affect a reduction in demand for groundwater without causing adverse effects on the economic prospects of farming? The experience in West Bengal and Gujarat shows that within the same locality, a uniform tariff (unit charges) for electricity can be followed, irrespective of the total electricity consumption by individual farmers. In the case of north Gujarat, where the water table is very deep (often 400–500 feet below the ground), the charge levied by the electricity utility on farmers is Rs. 0.70/kWh, whereas in West Bengal, where the groundwater table is very shallow, it is Rs. 4.10/kWh. In both states, this mode of charging electricity has been in use for many years.

8.5.3 Reducing the transaction costs of metering

Some Indian researchers who do not like the idea of metering electricity supply in the farm sector have been continuously arguing for the past three decades that charging for farm power on the basis of its actual consumption is next to impossible in the Indian context, and political parties have lost power because of such measures initiated by their governments. Another convenient argument put forth by them against metering is the heavy transaction cost of metering. The state electricity boards and policymakers recognize the importance of metering electricity in the farm sector. But they were also struggling with the idea of carrying out metering in a way that made it foolproof as well as cost-effective (Kumar, 2018). Today, technologies exist not only for metering but also for controlling energy consumption by farmers. The pre-paid electronic meters, which are operated through scratch cards and can work on satellite and internet technology, are ideal for remote areas to monitor energy use and control groundwater use online from a centralized station. Over the past 15 years, there has been a remarkable improvement in the quality of services provided by internet and mobile (satellite) phone services, especially in rural areas (Kumar et al., 2011).

Pre-paid meters prevent electricity pilferage through the manipulation of pump capacity. They can be operated through tokens, scratch cards, magnetic cards, or recharged digitally through the internet and SMS, and can be used by an electricity company to restrict the use of electricity. The company can decide on the "energy quota" for each farmer on the basis of the reported connected load and total hours of power supply per unit of irrigated land. Farmers can pay and obtain an activation code through mobile SMS (Zekri, 2008). Alternatively, automatic metering infrastructure (AMI), which are used to meter electricity and, in many cases, water supply, can be used (Aarnoudse et al., 2016). All these will reduce costs and theft (Zekri, 2008). Prepaid meters are now used for agro wells in some countries, including China (Aarnoudse et al., 2016) and Bangladesh (Schmidt-Rosen, 2016).

8.5.4 Improving the direct groundwater access of the poor small and marginal farmers in EGP

As we have seen from the earlier discussions, pro rata pricing would benefit the poor non–well-owning farmers by reducing the monopoly power of rich well owners. Though the cost of pumping water would go up under the metered tariff system, the price at which water is traded in the market would not increase as the monopoly power of the well owner, which determines the market conditions, does not change. That said, with the subsidized power connections, the profit margins of the old diesel well owners would go up substantially. This would be a windfall gain for the well owners.

The state governments in the eastern IGP states, viz., Uttar Pradesh, Bihar, West Bengal, and Assam, must formulate schemes that offer micro diesel engines to small and marginal farmers at subsidized rates. It would not only be economically viable but also financially feasible for resource-poor farmers, with very small holdings. The financial burden on the state would also be much lower as compared to the results from the new policy. But proper targeting of the subsidy would be essential to prevent pilferage and misappropriation by the rural elite. Simultaneously, managing the hundreds of thousands of small and large wetlands in those states through judicious management of groundwater would be crucial in improving the rural livelihoods and nutritional security of its people. Periodic monitoring of wetlands and groundwater in the region is crucial for the management of the two interconnected resources.

8.5.5 Evolving a functional water rights system and its enforcement in the "over-exploited" areas

Researchers have long been arguing for the use of tradable water rights to ensure the sustainability of water use (Lohmar et al., 2007; Rosegrant and Binswanger, 1994; Saleth, 1996). Tradable property rights for groundwater exist in many countries, including the United States, Australia, Mexico, and Chile. A variety of projects that examine the ways to allocate water rights and promote water conservation are being tried in China, though there are institutional barriers to the adoption of these models (Lohmar et al., 2007). Analyses show that tradable water rights can serve as a viable instrument to improve allocative efficiency in water use (Haisman, 2005; NWC, 2010; Rosegrant & Sinclair, 1994; Thobani, 1997). There are legal, institutional, and technological challenges in establishing and enforcing property rights for groundwater, given the invisible, dynamic nature of the resource, the pattern of its use, and the legal status vis-à-vis its ownership. Newly introduced water rights, if designed to address concerns about the current inequity in access to the resource, can take away some of the existing rights. With many millions of farmers in remote rural areas accessing groundwater, monitoring volumetric resource use by individual farmers required for tracking water allocations, however, would require new institutions with the associated transaction costs (Kumar, 2018).

The process of establishing water rights and restricting water use by farmers is feasible in areas where farmers use electricity supplied by the state electricity utility because energy consumption can be used as a proxy for groundwater abstraction. Restricting farmers' energy use for pumping groundwater is analogous to rationing groundwater withdrawal for irrigation volumetrically (Aarnoudse et al., 2016; Kumar, 2018; Zekri, 2008). This can be done through pre-paid electricity meters (Kumar et al., 2011). As studies have shown, when water allocation is volumetrically rationed, farmers would allocate the water to economically more efficient crops (Kumar, 2005). Hence, restricting energy use will have a positive impact on groundwater use efficiency by all categories of farmers. More importantly, it will help achieve sustainable groundwater use if the energy quota is fixed by taking into account the sustainable yield of the aquifer. But, in such cases, it is important that the consumers are informed about their energy quota, and the approximate number of hours for which they could pump water from their wells using this quota, well in advance. Such information would help them choose the crops. The energy quota will have to be decided on the basis of the geo-hydrological environment prevailing in the area and the optimum irrigation requirements (Kumar, 2018).

As regards the cost of introducing such a system, the modern electricity metering technology with farm-level metering combined with the establishment of ICT would cost around US $250 per well (Gulati & Pahuja, 2015). The cost of a prepaid meter (including the cost of installation) works out to be only US $90 approximately (Singh, 2020).

Here again, the energy quota will have to be decided on the basis of the yield characteristics of the aquifers and water table conditions prevailing in the area and the optimum irrigation requirements. However, there are governance challenges in implementing this idea. First, the utility has to frame rules/norms regarding the allocation of energy quota amongst the farmers, which will have to be based on groundwater rights/entitlements and the energy required to abstract a unit volume of groundwater. The decision on these entitlements should use sound criteria based on principles of equity and resource sustainability. The process can be politically sensitive because the decisions would ideally result in the allocation of water rights/entitlements to many farmers who currently do not enjoy direct access to groundwater, and limits to the use of groundwater by many large and medium farmers.

Surveys done in Haryana and Punjab suggest that farmers would be willing to pay a higher tariff for electricity, provided that a good quality and reliable power supply is guaranteed (World Bank, 2001). As mentioned previously, farmers bear significant economic costs of power rationing and poor quality of electricity supply due to voltage fluctuations, low voltage, frequent interruptions, and phase imbalances, which increase their actual cost of irrigation by 25–30% (Gulati & Pahuja, 2015). If good quality power is supplied, they will spend less for the repair of pump sets. Further, as studies by the World Bank in erstwhile Andhra Pradesh and Punjab suggest, under the flat rate system of pricing electricity and fee power supply, a major share of the electricity subsidy to the farm sector went to large and medium farmers (Vashishtha, 2006a, b), who are very few in number.

Vashishtha (2006a) showed that in AP, large and very large farmers together received 73% of the subsidy benefits in 2003–04, whereas small and marginal farmers received only 5.1%. The corresponding figures for Punjab were 73.8% and 6.3%, respectively (Vashishtha, 2006b). While a large majority of the farming community does not gain from such policies, the politicians are largely ignorant of this. Because of this reason, the many millions of small and marginal farmers will be happy to shift to a pro-rata tariff.

Understanding the nuances of this would go a long way in convincing the political leaders about the need to get away with such perverse policies, which neither benefit the large section of the rural masses nor help improve the water and energy economies of the states concerned. Hence the political economy problems of raising the electricity tariff can be overcome (Kumar et al., 2013). Moreover, with metering and a mechanism for energy rationing in place, no restrictions on the duration of farm power supply would be required, and electricity can be made available to the farm round the clock, thereby substantially increasing the quality of the power supply.

8.5.6 Framing appropriate rules for groundwater abstraction in the EGP for managing groundwater quality

As per the latest estimates of the Central Ground Water Board, no district or block located in the eastern Gangetic plains experiences groundwater over-development (CGWB, 2019). However, the criterion currently used for assessing the status of groundwater development in India is only hydrological in nature, considering the annual recharge against abstraction (Kumar & Singh, 2008; also see Chapter 3 of this book). Such simplistic criteria used in the method for assessing the stage of groundwater development will have disastrous consequences for fragile regions such as the eastern IGP. Therefore, the methodology for assessing the status of groundwater development needs to be refined by integrating environmental considerations in the decision-making, with hydrological ones. The criteria should consider the changes in contribution of groundwater outflows to wetlands at different levels of groundwater stress and the negative impacts of groundwater over-exploitation on water quality, such as the dissolution of arsenic in groundwater. Along with the newly defined criteria and methodology, rules will have to be framed for deciding when to classify the groundwater resources in an area as "over-exploited" or safe for future exploitation.

Studies indicate that the vulnerability of the aquifers in the IGP to nitrate pollution of groundwater due to intensive rice farming with the use of chemical fertilizers is high (Sihi et al., 2020), given the shallow water table, permeable soils, heavy rainfall, and flat topography (Kumar et al., 2010). There are no regulatory instruments or incentive mechanisms in place to bring about behavioral change among farmers to prevent the widespread use of nitrogenous fertilizers in irrigated paddy fields in intensively cropped and irrigated areas of Punjab and Haryana, which cause groundwater pollution. Rules will have to be framed to affect changes in agricultural practices that can lead to reduced use of nitrogenous fertilizers, which are built on either economic incentives for reducing fertilizer use or disincentives for excessive use.

8.6 Conclusion

The western part of the IGP and the eastern part of the IGP in India characterize two distinctly different situations with respect to the groundwater environment and the governance and management challenges (Taneja et al., 2014). Public policies in water, agriculture, food security, and energy have influenced the way groundwater resources are developed, used and managed in the eastern and western parts of the IGP. In the western part of the IGP, rapid rural electrification, free or subsidized power to the farm sector, and attractive procurement prices for major cereals had led to intensive use of groundwater, with low water and energy use efficiency and unsustainable resource use emerging as major challenges. Lack of well-defined property rights in groundwater is another factor driving over-exploitation of the resource and inefficient resource use.

In the eastern IGP, groundwater is relatively abundant, and demand for the resource for irrigation is comparatively much lower. But inequity in access to groundwater is a major challenge in the region, with a high monopoly price for the water traded in the market for irrigation. At the same time, resource use efficiency is higher among the millions of diesel well owners and buyers of water from diesel and electric wells. As the analysis presented in the chapter showed, free electric power connections or flat rate power are very unlikely to result in improved access equity in groundwater or intensification of well irrigation, as the benefits of energy subsidy would not be passed on to the water buyers.

To improve the sustainability of groundwater use in the western IGP, two policy measures are needed. They are pro-rata pricing of electricity and a functional water rights system. Pre-paid energy meters or automatic metering infrastructure can be used to introduce a pro rata pricing system and monitor the energy use of individual farmers by the electricity utility. This can be combined with energy rationing. A modern electricity metering technology with farm-level metering combined with the establishment of ICT would cost around US $250 per well (Gulati & Pahuja, 2015). The cost of a prepaid meter (including the cost of installation) works out to be only US $90 approximately (Singh, 2020). The electricity prices that would affect the efficient use of water but also be affordable to the farmers need to be worked out for different water table conditions. Energy rationing can also become a proxy for a functional water rights system, with the energy quota of individual farmers decided on the basis of their volumetric water rights or entitlements (Nabavi, 2018; Zekri, 2008).

On the knowledge front, it is important to evolve methodologies for assessing the stage of groundwater development that integrate a variety of environmental considerations, such as the protection of wetlands and preservation of ground-water quality, particularly in the context of the eastern IGP, where the interactions between shallow aquifers and wetlands are quite significant. Also, rules need to be framed on the use of nitrogenous fertilizers in intensively cropped regions like Punjab, supported by the design of incentives for reducing the use of fertilizers and disincentives for their excessive use.

Some of the measures for improving the sustainability of groundwater use in the western IGP (such as metered tariffs and energy rationing) apply to large parts of the Indus basin area in Pakistan, especially the areas falling in Punjab province that are experiencing groundwater over-draft. However, a large proportion of the well-owning farmers there (about 84%) use diesel engines (Qureshi, 2020). The farmers using diesel engines for pumping groundwater are incurring a cost three times higher than their counterparts using electric motors. The energy and water use efficiencies of diesel pump owners are expected to be quite high as they are confronted with high marginal costs of using energy and water. For instance, the cost of pumping groundwater using shallow tube wells run by diesel was reported to be 2.20 US cents/m^3 of water, whereas it was only 0.70 US cents/m^3 of water for electric tube wells (Qureshi et al., 2003). Ideally, most of the shallow tube wells are diesel engine-operated and all the deep tube wells are driven by submersible pumps that run on electricity. The focus of energy metering and rationing should therefore be on the deep tube wells as they pump large volumes of water, and the option of shifting to diesel engines for running their wells is extremely limited for them. However, this does not mean that the shallow tube wells that run on electricity do not need to be covered. They should also be brought under the purview of any such scheme. But their effect on groundwater abstraction will be less.

In Bangladesh, the areas facing groundwater depletion will have to reduce the consumptive use of water pumped from the aquifers for crop production if over-draft is to be checked. As groundwater depletion is already forcing farmers to shift to surface water bodies as sources of irrigation in the wake of increasing costs of pumping (Qureshi et al., 2014), the wetlands have to be protected if agricultural growth is to be sustained. For this too, net groundwater withdrawal from the shallow aquifers will have to be regulated, as shallow groundwater pumping can adversely impact the hydrological integrity of the wetlands. Therefore, a mere reduction in pumping will not help, as a reduced dosage of irrigation water (to paddy) will only lead to a reduction in return flows to the shallow aquifer. What is important is a sustained reduction in the consumptive use of water for irrigated crops through a reduction in evapotranspiration per unit cropped area.

References

Aarnoudse, E., Qu, W., Bluemling, B., & Herzfeld, T. (2016). Groundwater quota versus tiered groundwater pricing: Two cases of groundwater management in north-west China. *International Journal of Water Resources Development, 33*(6), 917–934.

Ahmed, B. (2009). *Arsenic in food chain through irrigation water-soil-crop pathway: Risk assessment for sustainable agriculture of Bangladesh (Ex0523)* [Master's thesis]. Dept. of Urban and Rural Development, Swedish University of Agricultural Sciences.

Amarasinghe, U., Muthuwatta, L., Surinaidu, L., Anand, S., & Jain, S. K. (2016). Reviving the Ganges water machine: Potential. *Hydrology and Earth Systems Sciences, 20*, 1085–1101.

Badiani, R., Jessoe, K., & Plant, S. (2012). Development and the environment: The implications of agricultural electricity subsidies in India. *Journal of Environment and Development, 21*(2), 244–262.

Central Ground Water Board (CGWB). (2014). *Ground water year book 2013–14*. Central Ground Water Board, Ministry of Water Resources, Government of India.

Central Ground Water Board (CGWB). (2017). *Dynamic ground water resources of India (as on 31st March 2013)*. Central Ground Water Board, Ministry of Water Resources, River Development & Ganga Rejuvenation, Government of India.

Central Ground Water Board (CGWB). (2019). *National compilation on dynamic ground water resources of India 2017*. CGWB, Ministry of Water Resources, River Development and Ganga Rejuvenation. Government of India.

Custodio, E. (2000). *The Complex Concept of Over-exploited Aquifer, Secunda Edicion*. Madrid: Uso Intensivo de Las Agua Subterráneas.

Dinar, A., Rosegrant, M. W., & Meinzen-Dick, R. S. (1997). Water Allocation Mechanisms: Principles and Examples. Available at SSRN: https://ssrn.com/abstract=615000.

Dubash, N. K. (2007). The electricity-groundwater conundrum: Case for a political solution to a political problem. *Economic and Political Weekly, 42*(52), 45–55.

Foster, E. T., Rapoport, A., & Dinar, A. (2017). Groundwater and electricity consumption under alternative subsidies: Evidence from laboratory experiments. *Journal of Behavioral and Experimental Economics, 68*, 41–52. https://doi.org/10.1016/j.socec.2017.03.003.

Government of India. (1999). *Integrated water resource development a plan for action*. National Commission for Integrated Water Resource Management, Ministry of Water Resources, Government of India, New Delhi.

Government of India. (2010). Groundwater quality in the shallow aquifers of India, Central Ground Water Board, Ministry of Water Resources, Government of India, Faridabad.

Government of India (GoI). (2017). *5th census of minor irrigation schemes report*. Minor Irrigation (Statistics Wing), Ministry of Water Resources, River Development and Ganga Rejuvenation, Government of India.

Gulati, M., & Pahuja, S. (2015). Direct delivery of power subsidy to manage energy–ground water–agriculture nexus. *Aquatic Procedia, 5*, 22–30. https://doi.org/10.1016/j.aqpro.2015.10.005.

Gulati, A., Roy, R., & Hussain, S. (2017). *Getting Punjab agriculture back on high growth path: Sources, drivers and policy lessons*. Indian Council for Research on International Economic Relation.

Gupta, D. (2021). *Free power, irrigation and groundwater depletion: Impact of the farm electricity policy of Punjab, India* [Working paper no. 316]. Centre for Development Economics, Delhi School of Economics.

Haisman, B. (2005). Impacts of Water rights reform in Australia. In B. Bruns, C. Ringler, & R. S. Meinzen-Dick (Eds.), *Water rights reform: lessons for institutional design* (pp. 113–152). Washington D. C.: International Food Policy Research Institute.

Hira, G. S., Jalota, S. K., & Arora, V. K. (2004). *Efficient management of water resources for sustainable cropping in Punjab*. Department of Soils, Punjab Agricultural University.

Hussain, A., & Abbas, H. (2019, September 27). To save Pakistan, look under its rivers. *The Third Pole*. https://www.thethirdpole.net/en/climate/pakistans-riverine-aquifers-may-save-its-future/.

International Water Management Institute (IWMI). (2012, April). *Agricultural water management learning and discussion brief*. AGWAT Solutions, Improved Livelihood for Small Holder Farmers.

Kishore, A. (2004). Understanding agrarian impasse in Bihar. *Economic and Political Weekly,* *39*(31), 3484–3491. https://scholar.harvard.edu/avinashkishore/publications/understanding-agrarian-impasse-bihar.

Kumar, M. D. (2005). Impact of electricity prices and volumetric water allocation on energy and groundwater demand management: Analysis from Western India. *Energy Policy, 33*(1), 39–51.

Kumar, M. D. (2007). *Groundwater management in India: Physical, institutional and policy alternatives.* Sage Publications.

Kumar, M. D., & Singh, O. P. (2008). How serious are groundwater over-exploitation problems in India? A fresh investigation into an old issue. In M. D. Kumar (Ed.), Managing water in the face of growing scarcity, inequity and declining returns: Exploring fresh approaches. *7th Annual Partners' meet of IWMI-Tata Water Policy Research Program, ICRISAT, Patancheru, AP, 2–4 April 2008.*

Kumar, A., Pal, R. N., Sharma, H. C., & Singh, R. (2010). Assessment of non-point source nitrate pollution of groundwater in Gangetic alluvial plain using Bayesian model. *Indian Journal of Soil Conservation, 38*(2), 75–79. http://www.indianjournals.com/ijor.aspx?target=ijor:ijsc&volume=38&issue=2&article=003.

Kumar, M. D., Singh, O. P., & Sivamohan, M. (2010). Have diesel price hikes actually led to farmer distress in India? *Water International, 35*(3), 270–284.

Kumar, M. D., Scott, C. A., & Singh, O. (2011). Inducing the shift from flat-rate or free agricultural power to metered supply: Implications for groundwater depletion and power sector viability in India. *Journal of Hydrology, 409*(1–2), 382–394.

Kumar, M. D., Sivamohan, M. V. K., & Narayanamoorthy, A. (2012). The food security challenge of the food-land-water nexus in India. *Food Security, 4*(4), 539–556.

Kumar, M. D., Scott, C. A., & Singh, O. (2013). Can India raise agricultural productivity while reducing groundwater and energy use? *International Journal of Water Resources Development, 29*(4), 557–573.

Kumar, M. D. (2018). Institutions and policies governing groundwater development, use and management in the Indo-Gangetic Plains of India. In K. G. Vilholth, E. L. Gunn, K. Conti, A. Garrido, & J. van der Gun (Eds.), *Advances in groundwater governance.* CRS Press/Balkema.

Lohmar, B., Huang, Q., Lei, B., & Gao, Z. (2007). Water pricing policies and recent reforms in China: The conflict between conservation and other policy goals. In F. Molle, & J. Berkoff (Eds.), *Irrigation water pricing: Theory and practice* (pp. 277–294). CABI Publishing.

Lytton, L., Ali, A., Garthwaite, B., Punthakey, J. F., & Saeed, B. (2021). *Groundwater in Pakistan's Indus Basin: Present and Future Prospects.* World Bank, Washington, DC. https://openknowledge.worldbank.org/handle/10986/35065.

Malik, A. K., Junaid, M., Tiwari, R., & Kumar, M. D. (2008). Towards evolving groundwater rights: The case of shared well irrigation in Punjab. In M. D. Kumar (Ed.), Managing water in the face of growing scarcity, inequity and declining returns: Exploring fresh approaches. *Proceedings of the 7th Annual Partners Meet, IWMI TATA Water Policy Research Program, Vol. 1. Hyderabad, India* (pp. 439–451). International Water Management Institute (IWMI), South Asia Sub Regional Office.

Ministry of Agriculture and Farmer Welfare (MOA and FW). (2019). *Agriculture census 2015–16 (phase-I): All India report on number and area of operational holdings.* Agriculture Census Division, Dept. of Agriculture, Cooperation & Farmer Welfare, Ministry of Agriculture and Farmer Welfare, Govt. of India.

Mukherjee, S., & Kumar, M. D. (2011). Economic valuation of a multiple use wetland water system: A case study from India. *Water Policy, 14*(1), 80–98.

Mukherji, A. (2003). Groundwater development and agrarian change in Eastern India. IWMI-Tata Comment # 9), based on Vishwa Ballabh, Kameshwar Chaudhary, Sushil Pandey and Sudhakar Mishra, IWMI-Tata Water Policy Research Program.

Mukherji, A., & Shah, T. (2005). Socio-ecology of groundwater irrigation in south Asia: An overview of issues and evidence. In A. Sahuquillo, J. Capilla, L. Martinez-Cortina, & X. Sánchez-Vila (Eds.), *Groundwater intensive use: IAH selected papers on hydrogeology 7* (pp. 67–92). Taylor & Francis.

Mukherji, A. (2006). Political ecology of groundwater: the contrasting case of water-abundant West Bengal and water-scarce Gujarat, India. *Hydrogeology Journal, 14*(3), 392–406. https://doi.org/10.1007/s10040-005-0007-y.

Mukherji, A. (2008, January 29-31). The paradox of groundwater scarcity amidst plenty and its implications for food security and poverty alleviation in West Bengal, India: What can be done to ameliorate the crisis? [Paper presentation]. In *9th Annual Global Development Network Conference*, Brisbane, Australia.

Mukherji, A., Shah, T., & Banerjee, P. (2012). Kick-starting a second Green Revolution in Bengal. *Economic and Political Weekly, 47*(18), 27–30.

Mukherji, A., Buisson, M.-C., Mitra, A., Banerjee, P. S., & Chowdhury, S. D. (2020, May 31). *Does increased access to groundwater irrigation through electricity reforms affect groundwater and agricultural outcomes? Evidence from West Bengal*. Final report submitted to Australian Center for International Agricultural Research, International Water Management Institute.

Nabavi, E. (2018). Failed policies, falling aquifers: unpacking groundwater overabstraction in Iran. *Water Alternatives, 11*(3), 699.

Narayanamoorthy, A. (1997). Impact of electricity tariff policies on the use of electricity and groundwater: arguments and facts. *Artha Vijnana, 39*(3), 323–340.

National Water Commission (NWC). (2010). *The impacts of water trading in the southern Murray–Darling Basin: An economic, social and environmental assessment.* NWC.

Pandey, R. (2014, August). *Groundwater irrigation in Punjab: Some issues and way forward* [Working paper no. 2014-140]. National Institute of Public Finance and Policy. https://www.nipfp.org.in/media/medialibrary/2014/09/WP_2014_140.pdf.

Pant, N. (2004). Trends in groundwater irrigation in eastern and western UP. *Economic and Political Weekly, 39*(31), 3463–3468.

Pant, N. (2005). Control of and access to groundwater in UP. *Economic and Political Weekly, 40*(26), 2672–2680. http://www.jstor.org/stable/4416815.

Qureshi, A. S., Akhtar, M. J., & Sarwar, A. (2003). Effect of electricity pricing policies on groundwater management in Pakistan. *Pakistan Journal of Water Resources, 7*(2), 1–9.

Qureshi, A. S., Ahmed, Z., & Krupnik, T. J. (2014). *Groundwater management in Bangladesh: An analysis of problems and opportunities* [Research report no. 2]. Cereal Systems Initiative for South Asia Mechanization and Irrigation (CSISA-MI) Project. CIMMYT.

Qureshi, A. S. (2020). Groundwater Governance in Pakistan: From Colossal Development to Neglected Management. *Water, 12*(11), 3017. https://doi.org/10.3390/w12113017.

Rosegrant, M. W., & Binswanger, H. P. (1994). Markets in tradable water rights: Potential for efficiency gains in developing country water resource allocation. *World Development, 22*(11), 1613–1625. https://doi.org/10.1016/0305-750x(94)00075-1.

Rosegrant, M. W., & Synclair, R. G. (1994). Reforming Water Allocation Policy through Markets in Tradable Water Rights: Experience from Chile, Mexico and California [Paper presentation]. In *DSE/IFPRI/ISISI workshop on Agricultural Sustainability, Growth and Poverty alleviation in East and South East Asia*. https://doi.org/10.22004/ag.econ.42822.

Saleth, R. M. (1996). *Water institutions in India: Economics, law and policy*. Commonwealth Publishers.

Saleth, R. M. (1997). Power tariff policy for groundwater regulation: Efficiency, equity and sustainability. *Artha Vijnana, 39*(3), 312–322.

Sarkar, A. (2011). Socio-economic implications of depleting groundwater resource in Punjab: A comparative analysis of different irrigation systems. *Economic and Political Weekly, 46*(7), 59–66.

Schmidt-Rosen, M. (2016, August). *Prepayment metering in Bangladesh how to improve electricity delivery and eliminate theft* [Case study]. Global Delivery Initiative, German Development Cooperation.

Shah, T. (1993). *Groundwater Markets and Irrigation Development: Political Economy and Practical Policy*. Oxford University Press.

Shah, T. (2001). *Wells and welfare in Ganga Basin: Public policy and private initiative in eastern Uttar Pradesh, India* [Research Report 54]. International Water Management Institute.

Shah, T., Scott, C. A., Kishore, A., & Sharma, A. (2004). *Energy Irrigation Nexus in South Asia: Improving Groundwater Conservation and Power Sector Viability, Research Report # 70*. IWMI, Colombo, Sri Lanka.

Shamsudduha, M., Taylor, R. G., Ahmed, K. M., & Zahid, A. (2011). The impact of intensive groundwater abstraction on recharge to a shallow regional aquifer system: evidence from Bangladesh. *Hydrogeology Journal, 19*(4), 901–916. https://doi.org/10.1007/s10040-011-0723-4.

Shankar, S., Shanker, U., & Shikha (2014). Arsenic Contamination of Groundwater: A Review of Sources, Prevalence, Health Risks, and Strategies for Mitigation. *The World Scientific Journal, 2014*, 304524. http://dx.doi.org/10.1155/2014/304524.

Shergill, H. S. (2005). Wheat and paddy cultivation and the question of optimal cropping pattern for Punjab. *Journal of Punjab Studies, 12*(2), 239–250.

Sihi, D., Dari, B., Yan, Z., Sharma, D. K., Pathak, H., Sharma, O. P., & Nain, L. (2020). Assessment of water quality in Indo-Gangetic Plain of South-Eastern Asia under organic vs. conventional rice farming. *Water, 12*, 960.

Singh, K. (2009). Act to save groundwater in Punjab: Its impact on water table, electricity subsidy and environment. *Agricultural Economics Research Review, 22*, 365–386.

Singh, R. K. (2020). *Rs 1.5 lakh crore: The bill India is set to pay for the coming power gamechanger*. The Economic Times. https://economictimes.indiatimes.com/industry/energy/power/rs-1-5-lakh-crore-the-bill-india-is-set-to-pay-for-the-coming-power-gamechanger/re_show/74329268.cms.

Singh, O., Kasana, A., & Bhardwaj, P. (2022). Understanding energy and groundwater irrigation nexus for sustainability over a highly irrigated ecosystem of north western India. *Applied Water Science, 12*(3). https://doi.org/10.1007/s13201-021-01543-w.

Srivastava, S. K., Chand, R., Raju, S. S., Jain, R., Kingsly, I., Sachdeva, J., Singh, J., & Kaur, A. P. (2015). Unsustainable groundwater use in Punjab agriculture: Insights from cost of cultivation survey. *Indian Journal of Agricultural Economics, 70*(3), 365–378.

Taneja, G., Pal, B. D., Joshi, P. K., Aggarwal, P. K., Tyagi, N. K. (2014). *Farmers' preference for climate-smart agriculture: An assessment in the Indo Gangetic Plain* [IFPRI discussion paper 01337]. International Food Policy Research Institute. https://www.ifpri.org/publication/farmers%E2%80%99-preferences-climate-smart-agriculture-assessment-indo-gangetic-plain.

Thobani, M. (1997). Formal Water Markets: Why, When, and How to Introduce Tradable Water Rights. *The World Bank Research Observer, 12*(2), 161–179. https://doi.org/10.1093/wbro/12.2.161.

Tripathy, A., Mishra, A. K., & Verma, G. (2016). Impact of preservation of sub-soil water on groundwater depletion: The case of Punjab, India. *Environmental Management, 58*(1), 48–59.

Varghese, S., Narayanan, S. P., Raj, V. M. P., Prasad, V. H., & Prasad, S. N. (2008). Analyses of wetland habitat changes and it's impacts on avifauna in select districts of the Indo-Gangetic Plains of Uttar Pradesh, India, Between 1972 and 2004. In M. Sengupta, & R. Dalwani (Eds.), *Proceedings of Taal 2007: The 12ᵗʰ World Lake Conference* (pp. 2045–2055).

Vashishtha, P. S. (2006a). *Input subsidy in Andhra Pradesh agriculture* [Unpublished manuscript]. International Food Policy Research Institute.

Vashishtha, P. S. (2006b). *Input subsidy in Punjab agriculture* [Unpublished manuscript]. International Food Policy Research Institute.

World Bank. (2001). *India power supply to agriculture, volume 1: Summary report (Report no. 22171-IN)*. New Delhi, India: Energy Sector Unit, South Asia Regional Office.

Young, R. A. (1996). *Measuring Economic Benefits for Water Investments and Policies* (pp. 1–119). Washington DC: World Bank Technical Paper 338, World Bank.

Zekri, S. (2008). Using economic incentives and regulations to reduce seawater intrusion in the Batinah coastal area of Oman. *Agricultural Water Management, 95*(3), 243–252. https://doi.org/10.1016/j.agwat.2007.10.006.

Chapter 9

Regulating groundwater use through economic incentives

9.1 Introduction

India's groundwater challenges are not limited to merely arresting over-exploitation of the resource. Over the years, the issue has become very complex, with the political economy taking dominance over hard science (Kumar, 2018a). Groundwater use in many "problem areas," which have deep water table conditions, is also linked to the power supply in agriculture. Agricultural electricity subsidies are found to increase the farmers' incentive to increase groundwater abstraction for expanding cropped areas and go for water-intensive crops. While they help increase the value of agricultural outputs, they induce huge negative externalities on the environment, and in the long run, agricultural production itself might get affected adversely (Badiani et al., 2012). In India, the states are free to decide the price for the electricity supplied to various consumers through the utilities owned by them. Power supply is heavily subsidized in most Indian states, with annual subsidies running into several hundreds of billions of rupees (Chand, 2018). Scholars had begun to explore how the mode of pricing supplied to agriculture could be changed to manipulate groundwater abstraction to achieve efficiency, equity, and sustainability goals in the early 90s. The most frequently suggested instrument for controlling groundwater draft was metering and pro rata pricing of electricity (Bassi, 2014; Kumar, 2005; Kumar et al., 2011; Narayanamoorthy, 1997; Pfeiffer & Lin, 2014). Realizing the limited role of energy pricing as a tool only to achieve higher efficiency of use of groundwater and not to reduce abstraction (as farmers tend to use the saved groundwater for the increasing area under irrigation if extra land is available), the scholars had also explored the possibility of using energy rationing (fixing energy quota) in agriculture (Kumar, 2018b; Kumar et al., 2011). These ideas also emerged from China (Aarnoudse et al., 2017) and some of the countries in the Middle East such as Oman (Zekri, 2008) and Iran (Nabavi, 2018).

Punjab, a state that has been infamous for groundwater depletion for several decades and also for free electricity for agricultural pumping in the recent past, had experimented with several methods to check over-abstraction. The latest is the direct delivery of power subsidies to agriculture introduced by

Groundwater Economics and Policy in South Asia. DOI: https://doi.org/10.1016/B978-0-443-14011-2.00010-3
205

the state electricity department. The approach works by incentivizing farmers who stick to using just the pre-determined quota of electricity fixed by the utility for each season through payment of cash. In this chapter, we critique this approach for its effectiveness in conserving electricity and groundwater and also explore alternative ways of improving the sustainability of groundwater use for agriculture in the state. For this, we first examine whether the criteria used for fixing the electricity quota for individual farmers is robust enough to encourage farmers to economize groundwater use for irrigation, and then examine whether such use of irrigation water for the cropping system there would result in actual water-saving.

9.2   Managing groundwater: a plethora of flawed ideas

While in Western countries and in China, governments are exploring various economic instruments for controlling groundwater use, such as pricing of electricity supplied, rationing of energy supplied to farms, and fixing groundwater quota (Aarnoudse et al., 2017; Lohmar et al., 2007), in India, some scholars are highly pessimistic about the idea of metering agro wells and charging of electricity on the basis of consumption (Mukherji et al., 2009; Shah et al., 2004; Shah & Verma, 2008); others claim that it is political *hara-kiri* for any government to even think about installing meters in farmers' fields and that whenever it was attempted, it led to a fall of the government (citing the experience of a particular state). For instance, Shah et al. (2004) argued that metering of agricultural power connections is not politically feasible in agriculturally prosperous states due to fierce opposition from the farm lobby and the lack of willingness of the state governments to take the risk, and whenever attempts were made to raise the electricity tariff in Gujarat, a state where farmers enjoyed heavily subsidized power for several decades, it led to the "fall" of the government. Beginning in 2004, a series of articles were published in popular journals and magazines singing the false narrative that farmers and politicians alike hate metering of agricultural power connections and thus energy pricing is not part of the solution, without an iota of evidence to substantiate this claim (and continued until around 2015) (Gulati & Pahuja, 2015; Mukherji et al., 2009, 2012; Shah et al., 2004; Shah & Verma, 2008).

For instance, a 2015 study by the International Institute for Sustainable Development, which explored various options for electricity subsidy reforms in Haryana, says: "Irrigation and electricity subsidies are a politically sensitive subject in India with far-reaching implications. Therefore, any efforts made by the utilities for metering agricultural connections, tariff hikes, etc. face stiff resistance not only from the agrarian community but the political diaspora as well" (Sharma et al., 2015, p. 38). The emptiness of this statement can be discerned from the fact that it was made in the context of introducing energy-efficient pump sets in the farmers' fields and not energy meters. The aim of such a tirade by some researchers against the idea of metering agro wells was to create a narrative by which they sought wider support for the "half-baked" alternatives that they wanted to propagate.

Even a study by the World Bank in 2001 (conducted in Andhra Pradesh and Haryana) showed that farmers are willing to pay for electricity if good quality power supply is assured (World Bank, 2001). The study in both locations showed farmers incurring substantial costs due to limited hours of power supply, unreliable supplies, poor quality of power (voltage fluctuations), and transformer burnouts. The study further found that in Haryana, farmers' willingness to pay for improved availability was quite high, especially among small (landholding 1 hectare or more but less than 2 hectares) and marginal (land holding less than 1 hectare) farmers. The econometric analysis of data from the recall survey showed that marginal and small farmers are willing to pay between Rs. 9400 and Rs. 9700 (US $125 to US $130) annually for an additional hour/day of increase in the availability of power in the short run (referred as the time period over which irrigation technology does not change). However, medium (landholding 2 hectares or more but less than 10 hectares) and large (landholding 10 hectares or more) farmers seem to have a zero valuation for power availability at the margin, indicating that, given their technology choices, the available power supply does not currently constrain them and greater availability is not likely to have any short-run effects on their net farm incomes (World Bank, 2001). This result implies that important resources like water and power have zero marginal valuation in the short run for around 60% of the electric pump-owning population in Haryana. To improve the conservation of these scarce resources, it is imperative to shift to energy metering and per-unit tariffs (World Bank, 2001).

As regards the reliability of the power supply, the willingness to pay for its improvements was quite high in the short run for medium and large farmers, small and marginal farmers attached no significance to improving reliability in the short run. In the medium run (the time period over which farmers are able to adjust their irrigation technologies), however, the effect was quite large. Since the small and marginal farmers had overinvested in electric pumps as a way to cope with the unreliable supply, when reliability improves, these farmers could shift to lower capacity pump sets and thus lower their costs. The willingness to pay for a reduction in days lost due to transformer burn-outs was also quite high for medium and large farmers, suggesting that, in general, farmers value improvements in reliability and quality much more than increases in availability (World Bank, 2001).

While the tirade against "electricity metering" continued by a group of researchers, the government in West Bengal introduced pro-rata pricing of electricity in the agriculture sector in 2006, with a charge almost close to the cost of production of electricity (Mukherji et al., 2009). Mukherji et al. (2009) found that, with a charge of Rs. 3.75/kWh paid by the well owners, energy metering promoted reduced use of irrigation water for paddy among the electric well owners as compared to the situation when electricity charges were levied on the basis of the connected load of the pump-set, though the study was critical of the impact of such a pricing policy on water buyers, arguing that pro rata pricing of electricity increased the price of water in the market. Gujarat introduced metering of agricultural power in 2012, and around 75% of the wells in that state are currently metered, with farmers paying a price of 0.70/kWh of electricity consumed (Kumar & Perry, 2019). Here again, the West Bengal government's move to charge the farmers for the electricity supplied was criticized as "anti-poor," arguing that a switch from connected load-based pricing of electricity to pro rata pricing actually led to an increase in the price of irrigation water in the market, adversely affecting the non-well owning farmers (Mukherji et al., 2009), though the analysis based on which such an argument was made was found completely wrong (Kumar, 2018b). Clearly, the motivated and relentless attack on metering and pro-rata pricing was to "create" a market for "innovative solutions" for controlling groundwater abstraction in India that they wanted to push.

Of the many such solutions dished out, the "big ticket" one was the "separation of feeder line for agriculture from that of domestic supply" (Gupta, 2012; Shah et al., 2004; Shah & Verma, 2008). A pilot project started by govt. of Gujarat to provide 24×7 three-phase power to the domestic sector (which was later on implemented in the entire state as *Jyotigram Yojna*) was misinterpreted and touted by some of these scholars as an innovative scheme to ration power supply to agriculture, to conserve groundwater and to reduce the subsidy burden of the state power utility with no social and political risk! In fact, even before the feeder line separation was taken up, the power supply to agriculture was rationed (supplied for only 8 hours a day). The reason why the Gujarat state government decided to separate the feeder line for agriculture from that of the domestic one was that the Utility feared that with a 24-hour three-phase power supply going through a common feeder, farmers would easily pilfer power and use it for running wells. However, contrary to its intended objective, this intervention was paraded for nearly a decade since 2004 as a successful intervention for groundwater management. Later on, this program was implemented in a few other Indian states (Kumar & Perry, 2019).

Nevertheless, later on, studies showed that this intervention had no positive effect on controlling groundwater abstraction and electricity use (Kumar & Perry, 2019). For instance, the World Bank conducted a study of the feeder segregation program implemented in Haryana during 2005–10. It found a significant variance between utility records and actual connections (World Bank, 2013, p. 56). Specifically, the World Bank study found "a relatively large proportion of consumers without working meters; significant variation in actual connected load vis-à-vis utility records; higher hours of three-phase supply made available ranging between 8 and 14 hours; and finally, peak load was higher than connected load, indicating the presence of unauthorized load" (World Bank, 2013, p. 57). The anomalies between the reported and actual number of electricity connections, the reported load and actual connected load, the large number of malfunctioning meters, and higher electricity supply hours indicated high electricity pilferage, meaning feeder line separation is unlikely to have any desirable impact on farmers' pumping behavior.

In the case of Gujarat, the analysis showed that electricity consumption in the agriculture sector increased after the implementation of the scheme (Kumar & Perry, 2019). As noted by Kumar and Perry (2019), the increase in power consumption with a proportionate increase in state subsidy and the increasing raids conducted in the rural areas by the power utilities after the implementation of *Jyotigram Yojna* suggest that the scheme had huge social and political repercussions. It was also clear that what is required to regulate groundwater draft is energy rationing and not rationing of power supply (please see the discussion in Chapter 6).

9.3   Depletion and current status of groundwater in Punjab

The depletion problem in Punjab is unique in the sense that the decline in the water table is consistent and has been witnessed for a long time period, i.e., from 2000 onwards (source: based on CGWB, 2011). This is despite the large amount of surface water resources that the region receives for irrigation. One major factor driving over-exploitation is the presence of deep alluvial strata in the region with a large groundwater stock (over and above the dynamic groundwater resources from annual recharge from rainfall, irrigation return flows, and canal seepage), which allows farmers to tap water from a higher depth as water table drops (Hira & Khera, 2000). As discussed in Chapter 3, the aquifers in Punjab get mined even when the recharge is very poor due to monsoon failure, with the result that the water level fluctuation during the monsoon in bad rainfall years is negative.

The average depth to water table was 7.32 m in 1998 and 12.79 m in 2012, thus indicating an annual fall of 41.6 cm/year. This corresponds to a negative water balance of around 50 mm per year if we assume the average specific yield of the alluvial aquifers of the region to be 0.13. Though groundwater levels have declined across the state (Gulati et al., 2017), the rate of decline has not been uniform, both spatially and temporally (Baweja et al., 2017; Srivastava et al., 2015).

As per the analysis by Baweja et al. (2017), about 2.04 million hectares (m.ha) of area had experienced a fall between 0.0 and 3 m, about 1.24 m.ha of area experienced a fall between 3 m and 10 m, and about 1.10 m.ha of area experienced a fall greater than 10 m in various parts of the state. A rise in the water table was experienced at around 0.64 m.ha, mainly in the southwest region besides marginal areas of the northeast region of the state. The rise in water levels in these areas was attributed to the continuous seepage of water from the network of unlined canals and distributaries and due to the negligible draft from groundwater in the area. The areas where the water level has not lowered much happened to be in the vicinity of rivers or in those districts where the canal irrigation network exists for agriculture purposes. The rate of decline was observed to be the maximum in the central zone with 50 cm/year, 36 cm/year in the southwest zone, and 33 cm/year in the northeast zone. A major portion of the central zone registered a fall of more than 10 m over a span of 14 years from 1998–2012 (Baweja et al., 2017).

Similar trends are available from other studies as well. As per the estimates provided by Srivastava et al. (2015) based on data from CGWB (2014), around 18% of the monitored wells in the state had groundwater levels within 5 m below ground; another 18% of the monitored wells had groundwater levels in the range of 5–10 m; 35% had groundwater levels in the range of 10–15 m; and 28% had it in the range of 20–40 m.

As regards the present situation, as per CGWB (2019a), the depth to water level during May 2019 lies between 0.65 m below ground level (bgl) at Kotla in Rupnagar district and 50.70 m bgl at Pedol pz in SAS Nagar district. Very shallow water levels of 0–2 m cover nearly 1% area of the geographical area in isolated patches in Muktsar, Fazilka, Faridkot, Gurdas, and Pathankot districts. Shallow water levels of 2–5 m cover about 10% of the total area in the south Western parts of Muktsar, Fazilka, Faridkot, and in the northern parts of Gurdaspur and Pathankot districts, and a few isolated patches in the northeastern parts. These are mainly canal command areas. Depth to water levels in the range of 5–10 m is observed in the northern parts, south and southwestern parts, and eastern parts of the Ropar district, covering about 17% of the state's geographical area. Moderately deep water levels (10–20 m) are observed in 32% area of the state in central and western parts. Deep water levels (20–40 m) are also observed in about 39% area of the state in the central and southeastern parts. Very deep water levels (>40 m) are observed as patches in Patiala, SAS Nagar, and SBS Nagar districts, covering 2% area of the state (CGWB, 2019a). From this description, it is clear that nearly 60% of the state's geographical area has a water table within 20 m below ground level. The pre-monsoon depth to water levels in Punjab is shown in Fig. 9.1.

9.4 Punjab's direct power subsidy model

As the idea of feeder line separation (in the lines of the Jyotigram model) was put to dust, "direct delivery of power subsidy to farmers" was an idea that came from a prominent research group recently to manage groundwater and improve the viability of the power utility (Gulati & Pahuja, 2015; Prasad, 2018). This model conforms to the earlier narrative that any intervention that makes farmers pay for the electricity they consume will not be acceptable to both the political class and the farming community (Mukherji et al., 2012; Shah et al., 2004; Shah & Verma, 2008). It is based on the premise that farmers get free power, but the government/power utility reduces its subsidy burden gradually by incentivizing the farmers to use less electricity, thereby saving both groundwater and electricity. As a result, it involves the metering of agricultural power connections but no metered tariff. This proposal considered the segregation of electricity feeders as a prerequisite for implementing the direct subsidy model.

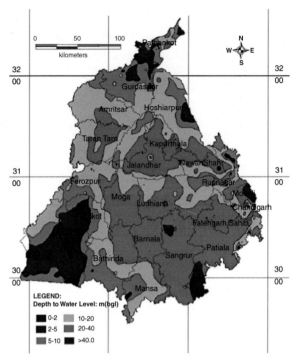

FIGURE 9.1 Depth to water levels in Punjab and union territory of Chandigarh (May 2018). (*From: CGWB, 2019a*).

This model, adopted by the power utility of Punjab, involves offering cash incentives to well irrigator farmers in Punjab who use less than a designated quota of electricity each season. The quota for each farmer is decided on the basis of the connected load the farmer had and the season. For one HP of connected load, a farmer is entitled to 200 units of power per month during the Kharif season and 50 units during the winter season (Prasad, 2018). The assumption here is that paddy is four times more water-guzzling than wheat in Punjab. This assumption appears to be wrong when we compare the actual water consumption by the two crops. While winter wheat consumes a little less than 400 mm of water in Punjab, paddy consumes only 480–500 mm of water as the ET in the region. That said, if a farmer manages to limit the energy consumption to a level that is 1000 units less than the quota, the farmer would be entitled to get a cash incentive of Rs. 4000 (Rs. 4 per unit of electricity saved). Consumption above this quota would not invite any penalty though (Chaba, 2019). The basic argument is that the farmer had saved 4000 units of electricity and the corresponding groundwater equivalent. But how far is this claim real? To know this, we should first know the rationale behind fixing the quota based on the connected load.

9.5   Ad-hoc basis for fixing electricity quota

Rationing of energy supply has been discussed by researchers worldwide as a tool for controlling groundwater withdrawal for agriculture (Kumar, 2005; Nabavi, 2018; Zekri, 2008). In a given year, in a given season and in a given locality, the power demand will be a function of the cropped area and cropping pattern, which would determine the aggregate irrigation water demand of the farmer. However, the electricity quota is not fixed on the basis of these considerations but instead on the basis of the connected load of the pumps. The rationale of fixing the electricity quota on the basis of connected load can be said to be "sound" only if the farmers have correctly chosen the pump capacity, taking into consideration the actual quantum of energy required for irrigating their farm and the number of hours for which power supply is available. But this is not the case in reality. Many resource-rich farmers have chosen oversized pumps and own more wells. Both the number of operational tube wells and average depth is much higher for tube wells of large farmers, especially in villages facing depletion problems. The limited funds prevent small farmers from overexploiting groundwater resources. In a village experiencing groundwater depletion, the average number of tube wells owned by large farmers was 1.8, compared with 0.41 for marginal farmers, 1.0 for small farmers, and 0.91 for medium farmers (Sarkar, 2011). Further, the average depth of a tube well was 109 m for large farmers and 58 m for small farmers (Sarkar, 2011), a clear indication of the greater wherewithal that large farmers have to exploit groundwater. By virtue of having more tube wells and higher capacity wells (which clearly suggests the use of higher capacity pump sets), the energy quota available for large farmers will be much higher than what is required to irrigate the plot even at the current excessive levels of dosage.

Even with the cash incentive, these large farmers might maintain a high level of irrigation dosage. One reason could be that it can be rewarding from an economic point of view as it might result in yield improvement. Ideally, the farmer will continue irrigating the plot till the gross marginal return from a one-kilowatt hour of electricity use becomes Rs. 4.0.

If we consider an average water table depth of 12 m in Punjab and a pump efficiency of 50%, a unit (kilowatt hour) of electricity can pump out 15 m^3 of water. Hence, every unit of water saved will attract a cash incentive of Rs. 0.27 (Rs. 4.0/15). This means the gross marginal return from crop production has to reduce to become Rs. 0.27 for every extra cubic meter of water applied for the farmer to stop irrigation. Such low marginal returns will obviously correspond to very high levels of dosage of irrigation. Studies using data for 2008–09 showed that an additional one cubic meter of irrigation water in Punjab (without considering the effect of fertilizers, labor, and pesticides) generated a marginal return of Rs. 1.2 in the case of wheat and Rs. 0.4 in the case of paddy (Srivastava et al., 2015). Adjusting it for inflation at the current prices, these marginal returns will be roughly equal to Rs. 2.4 for wheat and Rs. 0.80 for paddy for every additional cubic meter of water applied. These estimates show that the cash incentive offered by the government will not encourage the efficient use of groundwater by the farmers under the current cropping system.

Hence, these resource-rich farmers will be able to keep their power consumption much below the allocation or "quota," while continuing with inefficient irrigation, and yet be able to claim the cash incentive from the utility. Whereas a resource-poor farmer, who has a low-capacity pump that is just sufficient to irrigate the land with the restricted power supply, might end up using the full quota of energy or even more and therefore will not be entitled to any cash subsidy. Therefore, in addition to cornering a large share of the electricity subsidy benefits that they appropriate under the flat rate system of electricity pricing and free power (Vashishtha, 2006a, b), the resource-rich farmers would eventually also corner the cash incentives provided by the power utility for energy conservation.

Now the pilot project, which was done on 135 farmers across Punjab, has shown reduced electricity consumption by around 60% of the farmers, while nearly 33% of the farmers had increased electricity consumption even after accepting the scheme. It is claimed that in the Kharif season of 2018, a total of 1.75 lac units of electricity were saved by the participating farmers (Chaba, 2019). The analysis presented above shows the agency's claim about energy conservation is invalid. Ideally, the "energy quota" for deciding on the incentive should have been fixed on the basis of the actual land holding cultivated by the farmer during a particular season, which determines the water and energy requirements for irrigation. The foregoing analysis basically leads us to the point that the current subsidy structure under direct delivery of power subsidy to farmers may not create any special incentive to save either electricity or groundwater in Punjab.

9.6 False claims of groundwater conservation

The importance of understanding the difference between notional water saving (or "dry water saving") and real water saving (or "wet water saving") in water resources management has long been talked about. In regions where irrigation produces significant return flows, the application of such concepts is extremely crucial for deciding on strategies for water resource conservation (Perry, 2007; Seckler, 1996). Now, for the time being, let us assume that the farmers are really saving energy by reducing groundwater pumping through efficient irrigation methods (such as micro-irrigation systems). Will this help conserve groundwater in Punjab, as claimed by the proponents of this model? The answer to this question lies in a nuanced understanding of Punjab's groundwater balance.

Punjab has a flat alluvial plain (which holds most of the groundwater) with a thin belt of mountains along the north eastern border and stable dunes in its south western parts (Baweja et al., 2017). As per the March 2017 estimates (CGWB, 2019b), the annual replenishable groundwater resources in the state are 23.98 billion cubic meters (BCM). With a natural discharge of 2.35 BCM during non-monsoon months, the net annual groundwater availability was 21.60 BCM. It is quite well known that the irrigated paddy fields, as well as rainfall, contribute to the recharge of shallow groundwater during the monsoon season in alluvial Punjab (CGWB, 2014; Hira & Khera, 2000; Srivastava et al., 2015). Further, the recharge from canals and irrigation return flows contributes 71% (17.08 BCM of the total annual replenishment of 23.98 BCM) in the state (CGWB, 2019b). Nevertheless, the annual groundwater draft according to CGWB is 35.78 BCM (about 98% of it is for irrigation), which makes groundwater "overexploited" in the state as per their methodology (stage of groundwater development is 171%). Out of the 138 assessment units, 80% (109) are categorized as "over-exploited" and only 15% (22 nos.) are "safe." Safe units are those where groundwater development is less than 70% (CGWB, 2019a).

Table 9.1 provides the names of the districts and their status with regard to groundwater development, i.e., whether "over-exploited," "critical," "semi-critical," or "safe." It also shows the number of over-developed assessment units in each district. Over-exploited districts are those where the net annual groundwater abstraction exceeds the total annual recharge.

The groundwater storage change is the net effect of gross abstraction/draft and the "total recharge," the latter being the sum of rainfall infiltration and irrigation return flows (including that from pumped well water). The gross abstraction equals the water consumed by crop + soil moisture depletion after crop harvest + soil moisture storage + the deep percolation or irrigation return flows (Kumar, 2007). Therefore, reducing gross draft might only result in the reduction in irrigation return flows (which are a major contributor to groundwater recharge in the alluvial plains of Punjab), so long as the farmers do not change the crop to reduce evapotranspiration (ET). In the case of paddy and wheat, reduction in ET will be practically impossible unless shorter-duration varieties are chosen. In the present situation, without any research on plant genetics to come up with new varieties producing higher transpiration efficiency (kg/T), such shifts can come at a cost in terms of reduced yield and income. Hence, they would not prefer that option.

But the people who designed the project seem to be completely unaware of this. Only a fraction of the total water applied (around 1200 mm) to the field for Kharif paddy actually gets consumed by the crop. The ET for paddy in Punjab is around 450–480 mm. The rest of the water is available as return flows to the aquifer underlying the irrigated land. For instance, analysis of soil water balance in two rice-wheat fields in the Sirsa district of Haryana (which had a similar climate) using the soil, water, atmosphere & plant (SWAP) model showed that the total water applied for the paddy-wheat system of cropping (1239 mm and 1427 mm, respectively) was in excess of the estimated ET (949 mm and 858 m) in the order of 290 mm and 561 mm. This excess water was available as recharge to groundwater and soil moisture storage change. Interestingly, the actual ET value was higher for the field that had a lower dosage of irrigation (Singh, 2005).

TABLE 9.1 Block-wise groundwater development in Indian Punjab.

Name of the district	Stage of groundwater development (%)	Categorization	No. of over-exploited blocks in the district
Fazilka	99.21	Critical	1
Hoshiarpur	107.20	Over-exploited	4
Amritsar	147.61	Do	8
Barnala	210.80	Do	3
Bathinda	97.50	Critical	3
Faridkot	167.55	Over-exploited	2
Fatehgarh Saheb	207.55	Do	5
Jalandhar	239.16	Do	10
Kapurthala	223.75	Do	5
Ludhiana	182.94	Do	12
Ferozepur	163.98	Do	6
Gurdaspur	134.24	Do	8
Mohali	119.50	Do	2
Moga	229.47	Do	5
Mansa	141.44	Do	4
Patiala	216.82	Do	8
Nawan Shahr	116.90	Do	3
Pathankot	76.03	Safe	0
Ropar	211.70	Over-exploited	3
Sangrur	260.20	Do	9
Tarn Taran	152.90	Do	8
Total			109

From: CGWB (2019b).

However, this hard science is least understood by the people who champion this model, including the mass media. Anecdotes such as the following statement from an adopter farmer are used to promote it. "The techniques taught by World Bank teams have massively reduced water consumption. We keep the fields moist and reduce water used for irrigation during rains. Hence, water input has been reduced from 70 to 80 days to 15 to 20 days now." In return, the government has disbursed subsidies amounting to Rs. 42,757,23 to Doaba farmers under the scheme, which was launched in June 2018. In all, 175 farmers in Doaba are enrolled under the scheme (Banerjee, 2020).

One might argue that while the reduction in soil moisture depletion is one avenue for saving groundwater, this fraction of water use is unlikely to be significant (as farmers in Punjab go for wheat cultivation soon after harvesting paddy, some of the residual moisture might get beneficially used). Hence, to sum up, even if the farmers reduce the amount of water applied to irrigated paddy fields to reduce electricity consumption, there won't be any groundwater saving as it would only result in reduced return flows.

Even though subsidies are available for the purchase of efficient irrigation technologies (drips and sprinklers) under the scheme, the paddy-wheat cropping system is not amenable to these technologies. More importantly, even if farmers save irrigation water through the use of such technologies (assuming a different cropping system), that saving will not result in any gain in net income for them due to the zero-marginal cost of water and electricity. It is also quite obvious that the small cash incentive will not offset the high capital cost of the micro irrigation system.

Lastly, the heavy cash incentive of Rs. 4 offered to farmers for saving a unit of electricity is also not justified, as the actual cost of supplying it (Rs. 5.16/unit) is only a little more. Does this incentive reflect the actual value of the water saved? The foregoing analysis shows that the electricity and water "saving," as claimed by the proponents of the scheme, are unreal.

The analysis also exposes the fallacy behind the unrelenting criticism of the paddy-wheat cropping system in Punjab. The groundwater depletion in Punjab is due to intensive crop cultivation with 95% of the land area under irrigation and a cropping intensity of 195%, which increases the total water depletion (outflow) per unit area of land against a limited inflow in the form of annual precipitation and imported surface water, a point also made by Baweja et al. (2017). Therefore, reducing the total area under irrigated cropping would itself help reduce groundwater depletion to a great extent, though this option may not be socioeconomically viable. In the next section, we will explore various viable options for arresting groundwater depletion in the Indian Punjab.

9.7 The way to arrest groundwater depletion in Punjab

9.7.1 Better understanding of the groundwater balance

Solving the groundwater over-draft problem first requires understanding the groundwater balance, i.e., inflows, outflows, and the change in storage. Given the flat topography (in most parts) and the homogenous alluvial aquifers that make the groundwater discharge and lateral flows least significant, constructing the groundwater balance of Punjab is relatively simple as compared to many other regions (especially regions with hard rock formations and undulating topography) (source: based on CGWB, 2019b). Analysis of the agricultural water use in the state is also amenable to simplistic formulations if we understand the water use hydrology under the paddy-wheat system. Fortunately, the paddy-wheat system of regions with similar agro-ecological and geohydrological conditions has been well studied by Ahmad et al. (2004) and Singh (2005), who estimated different components of the field water balance for paddy-wheat irrigation, such as total water applied, evapotranspiration (ET), soil moisture storage change and recharge to groundwater. The analysis presented in Chapter 4 of this book also shows that return flows from irrigation are a major component of the groundwater balance in Punjab, amounting to 2570 MCM per annum in a wet year and 1760 MCM in one district (Ludhiana) alone. The root cause of groundwater over-exploitation in Punjab is the very high evapotranspiration per unit of land (contributed by the paddy-wheat cropping system, which is intensive) against the total amount of water available annually to meet this demand (Baweja et al., 2017).

The average evapotranspiration for the paddy-wheat system of cropping is around 950–1000 mm in the region (source: based on Ahmad et al., 2004; Singh, 2005). Let us consider this as the average water outflow from the land area in Punjab, as 95% of the land is under crop cultivation. The spatial average of the rainfall in the state is around 500 mm. On average, the northeast, central, and southwest zones received 710 mm, 544 mm, and 295 mm of annual rainfall from 1998 to 2012 (Baweja et al., 2017). If we assume that all the precipitation falling on the cropland is available for crop production (through recharge to groundwater and soil moisture with no surface runoff), there is a deficit of around 450–500 mm. The imported canal water (at the headworks) is 17,950 MCM per annum (Jain & Kumar, 2007). This is equivalent to spreading water at a depth of 0.44 m (440 mm) over the entire irrigated crop land of 4.072 m.ha in Punjab. This amount is just sufficient to meet the deficit. Hence, in the most ideal condition, there does not seem to be any shortfall.

The problem, however, occurs when the monsoon rains depart from the normal values significantly downwards, and this is quite common in Punjab. For instance, the average annual rainfall of the state was 418 mm in 2006, 438 mm in 2007, 529 mm in 2008, and 384 mm in 2009 (a drought year). As evident from these data, in drought years, the deficit can be very large. Since the cropping pattern by and large remains the same, the high irrigation water deficit is met by mining groundwater. Further, the rainfall is not uniform across Punjab. South and southwestern Punjab receives around 350 mm of mean annual rainfall, and northeastern Punjab receives around 900 mm. The high-rainfall regions are unlikely to experience depletion, whereas the low-rainfall regions will experience depletion even during normal rainfall years if they follow the paddy-wheat cropping system.

9.7.2 Getting additional surface water for irrigation through imports

The foregoing analysis suggests that one way to arrest groundwater depletion in Punjab is to get additional exogenous water for irrigation. This can be roughly 4080 MCM of water annually if we assume that the extent of depletion is around 100 mm. The amount required will change according to rainfall conditions.

The greater the rainfall deficit of the region, the higher will be the additional allocation of surface water required. Currently, the contribution of canal water to Punjab's irrigated area is only 28% in terms of area irrigated (Jain & Kumar, 2007). To implement the idea, the canal network will have to be extended to new, unserved areas, especially those areas that are experiencing groundwater overdraft problems. Simultaneously, water allocation to areas currently served by canals but experiencing groundwater over-draft (or areas where the sum of surface water allocation and renewable groundwater resources is less than the overall water demand for irrigation) needs to be increased.

9.7.3   Encouraging crop shift through fixing of energy quota

The second option to arrest depletion is to go for cropping systems, which would consume less amount of water (Srivastava et al., 2015) than rice-wheat. Attempts were made to persuade farmers in Punjab to go for a shift in cropping patterns to low–water-consuming maize and high-value fruit and vegetable crops with micro-irrigation. For instance, Gulati et al. (2017) recommend encouraging crop diversification away from common rice in the Kharif season by moving on to maize and promoting fruits and vegetables to at least 10 percent of the GCA, including their protected cultivation through micro irrigation, etc. However, this has not been met with much success, as the farmers are not convinced about the assumed price for maize. Shah et al. (2021) recently advocated a major shift in cereal crops from paddy and wheat to sorghum and pearl millet as a way to transform agriculture and water sectors in regions like Punjab that burn a lot of water for crop production. But, the economic incentive for crop shift is absent, with no marginal cost of using water or electricity, which essentially means that water and electricity saving will not result in input cost saving for the farmers (Srivastava et al., 2015).

More importantly, recent analysis shows that crops such as sorghum and pearl millet that are less water-intensive (in terms of ET requirements) have a much greater water footprint than wheat and paddy, because of their disproportionately lower yields (in the range of 1–1.2 ton/ha against 4–5 ton/ha for paddy and wheat). This essentially means that shifting to such crops would have serious implications for aggregate cereal production in the country (Perry & Kumar, 2022), with much less cereal output from the available cropped area in Punjab. For the farmers, shifting to these crops also means compromising on the income from a unit area of cultivated land.

Since the state power utility does not have domain knowledge of crop sciences and groundwater hydrology, it should take the help of the department of agriculture and the department of water resources in Punjab to start working on establishing (volumetric) water use rights amongst the groundwater users, fix an equivalent energy quota, and then start monitoring groundwater and electricity use (both abstraction and consumptive use). As empirical studies have shown, under pro rata pricing of electricity, they will prefer crops that are less water-consuming as compared to rice + wheat or varieties of the same crops, or use water more efficiently for some of the existing ones, to manage the input costs so that the net income does not reduce (Kumar, 2005; Kumar & van Dam, 2013). As there is no scope for expanding the area under irrigation almost anywhere in the state, the pricing of water/energy will lead to some reduction in water use at the aggregate level. Further, under rationed energy/water allocation combined with unit pricing, farmers will be concerned with maximizing the net return per unit of water rather than land as water becomes a limiting factor for irrigated crop production, and hence will choose crops that are water-efficient in economic terms (Aarnoudse et al., 2017; Kumar, 2000, 2005).

There are around 9.96 lac tube wells in Punjab that are run by electricity (source: Statistical Abstract, Government of Punjab, 2009), and the use of this instrument will influence water use by these farmers.

9.7.4 The criteria for fixing energy quota

Since the sustainability of groundwater use is a major concern for Punjab, at the macro level, energy allocation decisions should be based on the sustainable level of abstraction from the aquifer, which is largely decided by the average renewable recharge. The quota of groundwater for individual farmers shall be decided on the basis of the sustainable level of abstraction from the aquifer and the total amount of land that is currently irrigated (source: based on Aarnoudse et al., 2017; Hérivaux et al., 2019; Zekri, 2008). The net annual groundwater replenishment in Punjab is around 21,600 MCM (CGWB, 2019b). We can consider 80% of this (i.e., 17,280 MCM per annum) as a sustainable yield.

Though the annual renewable groundwater would keep fluctuating in Punjab due to rainfall variation and changes in canal water allocation, as the region has a large groundwater stock (over and above the replenishable resources available from rainfall and irrigation return flow), the approach of using an average figure will be feasible. If we consider the land holding under irrigation (4.07 m.ha) as the only criterion for water allocation, that gives each farmer around 4245 m^3 of groundwater per ha of (net) irrigated land (i.e., 17280/4.07 = 4245). However, the actual gross water allocation per ha for the farmer will be far greater (8657 m^3/ha), which would include the surface water allocation through canals. A more complex consideration for water allocation, wherein we consider the family size of the agriculturist along with the land holding as one of the criteria, can also be thought about.

The energy equivalent of this quota of water can be worked out (on the basis of the electricity required to pump out a unit volume of groundwater), and the energy quota for each farmer shall be fixed on the basis of this formula (Kumar, 2018b; Zekri, 2008). If a farmer's water entitlement per annum is 20,000 m^3 of water, and if it requires one unit of electricity to pump out 15 m^3 of water from a locality, then the electricity quota of the farmer will be 1350 units per annum (source: author's own estimates). This model of allocating energy quotas to individual farmers can be implemented using pre-paid energy meters that work on scratch cards or coupons (Aarnoudse et al., 2017; Zekri, 2008). China had already piloted the use of smart card machines to regulate and monitor groundwater use in north China plains through different instruments such as groundwater quota and tiered groundwater pricing, and studies have shown the first one to be more effective in regulating groundwater draft by farmers (Aarnoudse et al., 2017).

However, there is a considerable difference between water applied and consumptive water use in the case of paddy (Ahmad et al., 2004; Perry, 2007; Seckler, 1996; Singh, 2005). Only a fraction of the water applied to paddy in this alluvial area actually gets consumed by the crop, and a major portion of the rest is available as return flows to groundwater (Ahmad et al., 2004; Kim et al., 2009; Singh, 2005). Hence, there are significant challenges involved in water allocation for this crop, which is grown extensively during Kharif in Punjab. The actual allocation of water and therefore energy has to be much higher than that, which corresponds to the amount of water consumed by the crop and shall include the allowance for return flows to the aquifer. For instance, Srivastava et al. (2015) found that the total irrigation water applied for paddy in Punjab was 12,150 m^3 per ha. But going by the estimates provided by Singh. (2005), the ET demand of paddy that is met from irrigation could be in the range of 7000 to 7800 m^3. Nevertheless, the actual water allocation (that can be used for crop production) shall only consider the consumptive use part. In that case, providing an energy quota to the farmers growing paddy would require proper monitoring to arrive at their actual water consumption.

9.7.5 Macro-level implications of energy quota fixing

The macro-level implications of the planned restrictions on energy supplies, which directly impact groundwater access, are expected to be large. The state accounts for a major proportion of the wheat and paddy procurement in the country (Gulati et al., 2017; Singh & Bhangoo, 2013). Hence, if the farmers of the state replace paddy and wheat with an alternative cropping system that can help them earn a higher income, it will have some significant implications for India's food security, at least in the near future (Shergill, 2005). Shergill (2005) argued that in highly productive farms like those in Punjab, crop diversification is not an option as improving productivity warrants that farmers specialize in a few crops. Further, he points out that crop diversification reduces the chances of creating a market surplus. Dairy farming will also get impacted by crop diversification, as residues from paddy and wheat are extensively used for milk production in the state (Kumar et al., 2008; Kumar & van Dam, 2013).

The fixing of the energy quota will also have an impact on the functioning of groundwater markets in Punjab. Though groundwater markets are not very extensive in the state, as reported by Malik et al. (2008), there are many private tube wells in the state under shared ownership whose water is shared by kinship partners. With energy/water rationing, water markets are likely to become more widespread, with well owners selling water to other well owners in addition to non–well-owning farmers, depending on the relative scarcity of land and water. With restrictions on the volume of groundwater that can be pumped by a well, the price at which water is traded in the market will reflect the opportunity cost of pumping groundwater under the new regime. Hence, the well owning farmers would use their quota for irrigating their own crops only if the net returns per unit of water were higher than the price at which water is traded in the market. If they are not able to grow high-value crops that yield high returns per unit volume of water ($/m^3), they might sell the water to their neighbors, who are willing to pay a high price for the water.

9.8 Conclusion and the way forward

Data available from extensive well monitoring by the CGWB and the scientific studies carried out by research institutions are unequivocal on the point that groundwater resources have been depleting for a long time in Punjab. In nearly 40% of the geographical area of the state, the depth to water table is above 20 meters, as per the data available from recent monitoring by the CGWB. While well irrigation is the major factor contributing to the over-draft of groundwater, studies clearly indicate that a significant proportion of the water pumped from underground and the water delivered from canals to the cropland returns to the aquifers. There were various attempts by state governments (such as in Gujarat, Karnataka, and Rajasthan) in the recent past to intervene in the groundwater sector with popular schemes, with the presumption that they would solve the multiple problems of poor power sector viability, unsustainable groundwater use and inefficient energy use. A careful analysis will show that most of these schemes were designed without a proper understanding of the system's behavior and were on a weak conceptual footing. Instead of solving any problem facing the sector, these schemes will only create more problems, as the rich farmers would be addicted to pervasive subsidies (Bassi, 2018; Kumar & Perry, 2019; Sahasranaman et al., 2018).

Our analysis shows that the model of "direct delivery of power subsidy to farmers" being implemented by the power utility in Punjab is one such scheme. This intervention is on a weak conceptual footing and is unlikely to result in any real benefits to society in terms of efficient use of electricity supplied to agriculture and groundwater conservation.

There are two reasons for this. First, the limits kept by the power utility for power consumption by individual farmers are ad hoc and do not consider the water requirements of the crops cultivated and the electricity required for pumping that water. Since the state power utility does not have the expertise to estimate crop water requirements and assess groundwater availability, the situation demanded the agency's technical collaboration with the state agriculture department (to get information on crop water requirements), and the state water resources department (to get information on groundwater in different regions of Punjab). However, such efforts were not made.

Second, the heavy return flows from irrigated fields, mainly during the paddy season, which contributes to the annual groundwater replenishment, are the key component of Punjab's groundwater balance. As per CGWB estimates, the total recharge from sources other than rainfall in the region is 17.08 BCM, most of which is during the paddy-growing season (CGWB, 2019b). Some reduction in water application for paddy is unlikely to result in reduced consumptive use of water for that crop. The model will instead benefit large farmers, having deep wells installed with disproportionately higher capacity pumps, who would be able to corner most of the cash incentives along with the power subsidy benefits without any substantial reduction in groundwater abstraction and use.

Offering a cash incentive for the fictitious power saving and notional water saving will only lead to the draining of the state exchequer with no welfare gains. Instead of doing this, the state power utility, in collaboration with the agriculture department and water resources department of Punjab, should start working on establishing (volumetric) water use rights amongst the farmers who run their tube wells with electricity, fix an equivalent energy quota, and then start monitoring groundwater and electricity use (both abstraction and consumptive use). Smart meters can be introduced for this under a prepaid system and energy quota can be decided on the basis of socio-economic parameters such as land holding to be irrigated, the water quota per unit area of land, and the energy required to lift a unit volume of water. The establishment of a water/energy quota will promote the allocation of water to economically efficient crops and extensive water trading. No doubt, these measures are extremely difficult and would require a great degree of coordination among various departments—water resources, electricity, agriculture—apart from sound technical knowledge about agricultural water use and the electricity-groundwater nexus. But once they are initiated, there would not be any need to promote efficient irrigation technologies and water-efficient crops. Farmers will take up these measures on their own, without any subsidy support.

References

Aarnoudse, E., Qu, W., Bluemling, B., & Herzfeld, T. (2017). Groundwater quota versus tiered groundwater pricing: Two cases of groundwater management in north-west China. *International Journal of Water Resources Development, 33*(6), 917–934.

Ahmad, M. U. D., Masih, I., & Turral, H. (2004). Diagnostic analysis of spatial and temporal variations in crop water productivity: A field scale analysis of the rice-wheat cropping system of Punjab. *Journal of Applied Irrigation Science, 39*(1), 43–63.

Badiani, R., Jessoe, K., & Plant, S. (2012). Development and the environment: The implications of agricultural electricity subsidies in India. *Journal of Environment and Development, 21*(2), 244–262.

Banerjee, A. (2020, February 14). Water consumption cut to half, 175 Doaba farmers reap benefit. *The Tribune News Service*. https://www.tribuneindia.com/news/punjab/water-consumption-cut-to-half-175-doaba-farmers-reap-benefit-41769.

Bassi, N. (2014). Assessing potential of water rights and energy pricing in making groundwater use for irrigation sustainable in India. *Water Policy, 16*(3), 442–453.

Bassi, N. (2018). Solarizing groundwater irrigation in India: A growing debate. *International Journal of Water Resources Development, 34*(1), 132–145.

Baweja, S., Aggarwal, R., Brar, M., & Lal, R. (2017). Groundwater depletion in Punjab, India. In *Encyclopedia of soil science, Third edition: Three volume set*, p. 5. Taylor & Francis. doi:10.1081/E-ESS3-120052901.

Central Ground Water Board (CGWB). (2011). *Dynamic groundwater resources of India (as on 31st March 2009)*. CGWB, Ministry of Water Resources, Government of India.

Central Ground Water Board (CGWB). (2014). Dynamic groundwater resources of India (As on 31st March, 2011). Ministry of Water Resources, River Development and Ganga Rejuvenation. Government of India.

Central Ground Water Board (CGWB). (2019a). *Ground water year book Punjab and Chandigarh (UT) 2018–19* September. Central Ground Water Board North Western Region, Ministry of Water Resources, River Development and Ganga Rejuvenation.

Central Ground Water Board (CGWB). (2019b). *National compilation on dynamic ground water resources of India 2017*. CGWB, Ministry of Water Resources, River Development and Ganga Rejuvenation. Government of India.

Chaba, A. A. (2019, January 4). Reforming agriculture: When power is worth saving. *The Indian Express*. Retrieved July 30, 2019, from https://indianexpress.com/article/india/reforming-agriculture-when-power-is-worth-saving-5520945/.

Chand, R. (2018). *Innovative policy interventions for transformation of farm sector* [Presidential Address] 26th *Annual Conference of Agricultural Economics Research Association (India)*. November 15.

Government of Punjab. (2009). *Statistical Abstract of Punjab*. Economic and Statistical Organisation, Government of Punjab.

Gulati, A., Roy, R., & Hussain, S. (2017). *Getting Punjab agriculture back on high growth path: Sources, drivers and policy lessons*. Indian Council for Research on International Economic Relation.

Gulati, M., & Pahuja, S. (2015). Direct delivery of power subsidy to manage energy–ground water–agriculture nexus. *Aquatic Procedia, 5*, 22–30.

Gupta, R. K. (2012). The Role of Water Technology in Development: A Case Study in Gujarat, India. In R. Ardakanian, & D. Jaeger (Eds.), *Water and the green economy: Capacity development aspects*. UNW-DPA.

Hérivaux, C., Rinaudo, J. D., & Montginoul, M. (2019). Exploring the potential of groundwater markets in agriculture: results of a participatory evaluation in five French case studies. *Water Economics and Policy, 6*(1), 1950009. https://doi.org/10.1142/s2382624×19500097.

Hira, G. S., & Khera, K. L. (2000). *Water resource management in Punjab under rice-wheat production system* [Research Bulletin 1]. Department of Soil and Water Conservation, Punjab Agricultural University.

Jain, A. K., & Kumar, R. (2007). Water management issues–Punjab, North-West India. In *Proceedings Indo-US Workshop on Innovative E-technologies for Distance Education and Extension/Outreach for Efficient Water Management, March, 5–9, Patancheru/Hyderabad, India, ICRISAT*.

Kim, H., Jang, T., Im, S., & Park, S. (2009). Estimation of irrigation return flow from paddy fields considering the soil moisture. *Agricultural Water Management, 96*(5), 875–882. https://doi. org/10.1016/j.agwat.2008.11.009.

Kumar, M. D. (2000). Institutional framework for managing groundwater: A case study of community organisations in Gujarat, India. *Water Policy, 2*(6), 423–432.

Kumar, M. D. (2005). Impact of electricity prices and volumetric water allocation on energy and groundwater demand management: Analysis from Western India. *Energy Policy, 33*(1), 39–51.

Kumar, M. D. (2007). *Groundwater management in India: Physical, institutional and policy alternatives.* Sage Publications.

Kumar, M. D. (2018a). *Water policy science and politics: An Indian perspective.* Elsevier.

Kumar, M. D. (2018b). Institutions and policies governing groundwater development, use and management in the Indo-Gangetic Plains of India. In K. G. Villholth, E. Lopez-Gunn, K. Conti, A. Garrido, & J. van Der Gun (Eds.), *Advances in groundwater governance.* CRC Press.

Kumar, M. D., Malla, A. K., & Tripathy, S. K. (2008). Economic value of water in agriculture: Comparative analysis of a water-scarce and a water-rich region in India. *Water International, 33*(2), 214–230. https://doi.org/10.1080/02508060802108750.

Kumar, M. D., & Perry, C. J. (2019). What can explain groundwater rejuvenation in Gujarat in recent years? *International Journal of Water Resources Development, 35*(5), 891–906.

Kumar, M. D., Scott, C. A., & Singh, O. P. (2011). Inducing the shift from flat-rate or free agricultural power to metered supply: Implications for groundwater depletion and power sector viability in India. *Journal of Hydrology, 409*(1-2), 382–394.

Kumar, M. D., & van Dam, J. C. (2013). Drivers of change in agricultural water productivity and its improvement at basin scale in developing economies. *Water International, 38*(3), 312–325.

Lohmar, B., Huang, Q., Lei, B., & Gao, Z. (2007). Water pricing policies and recent reforms in China: The conflict between conservation and other policy goals. In F. Molle, & J. Berkoff (Eds.), *Irrigation water pricing: Theory and practice.* CABI Publishing.

Malik, A. K., Junaid, M., Tiwari, R., & Kumar, M. D. (2008). Towards evolving groundwater rights: The case of shared well irrigation in Punjab. In M. D Kumar (Ed.), *Managing water in the face of growing scarcity, inequity and declining returns: Exploring fresh approaches. Proceedings of the 7th Annual Partners Meet, IWMI TATA Water Policy Research Program, Vol. 1* (pp. 439–451). International Water Management Institute (IWMI), South Asia Sub Regional Office.

Mukherji, A., Das, B., Majumdar, N., Nayak, N. C., Sethi, R. R., & Sharma, B. R. (2009). Metering of agricultural power supply in West Bengal, India: Who gains and who loses? *Energy Policy, 37*(12), 5530–5539.

Mukherji, A., Shah, T., & Banerjee, P. S. (2012). Kick-starting a second green revolution in Bengal. *Economic and Political Weekly, 47*(18), 27–30.

Nabavi, E. (2018). Failed policies, falling aquifers: unpacking groundwater overabstraction in Iran. *Water Alternatives, 11*(3), 699.

Narayanamoorthy, A. (1997). Impact of electricity tariff policies on the use of electricity and groundwater: Arguments and facts. *Artha Vijnana, 39*(3), 323–340.

Perry, C. J. (2007). Efficient irrigation; inefficient communication; flawed recommendations. *Irrigation and Drainage, 56*(4), 367–378. https://doi.org/10.1002/ird.323.

Perry, C. J., & Kumar, M. D. (2022). Agricultural transformation or compromising food security. *Economic and Political Weekly, 57*(2), 58–60.

Pfeiffer, L., & Lin, C.-Y. C. (2014). The effects of energy prices on agricultural groundwater extraction from the high plains aquifer. *American Journal of Agricultural Economics, 96*(5), 1349–1362. https://doi.org/10.1093/ajae/aau020.

Prasad, A. V. (2018). India: Direct delivery of electricity subsidy for agriculture, an innovative pilot in Punjab. In *Presented at Geneva, October 30, 2018.* http://esmap.org/sites/default/files/ KEF%20Geneva%202018/Energy%20Subsidy%20Reform%20Forum%20%202018_Punjab_ Final%20-%2025.10.2018.pdf.

Sahasranaman, M., Kumar, M. D., Bassi, N., Singh, M., & Ganguly, A. (2018). Solar irrigation cooperatives: creating the Frankenstein's monster for India's groundwater. *Economic & Political Weekly, 53*(21), 65–68.

Seckler, D. (1996). *The new era of water resources management: From "dry" to "wet" water savings* [Research Report 1]. International Irrigation Management Institute (IIMI).

Sarkar, A. (2011). Socio-economic implications of depleting groundwater resource in Punjab: A comparative analysis of different irrigation systems. *Economic and Political Weekly, 46*(7), 59– 66.

Shah, M., Vijayshankar, P. S., & Harris, F. (2021, July 2021). Water and agricultural transformation in India: A symbiotic relationship—I. *Economic and Political Weekly, 56*(29), 43–55.

Shah, T., Scott, C., Kishore, A., & Sharma, A. (2004). *Energy-irrigation nexus in South Asia: Improving groundwater conservation and power sector viability: Vol. 70.* International Water Management Institute.

Shah, T., & Verma, S. (2008). Co-management of electricity and groundwater: An assessment of Gujarat's Jyotirgram scheme. *Economic and Political Weekly, 43*(7), 59–66.

Sharma, S., Tripathi, S., & Moerenhout, T. (2015). *Rationalizing energy subsidies in agriculture: A scoping study of agricultural subsidies in Haryana, India.* International Institute for Sustainable Development.

Shergill, H. S. (2005). Wheat and paddy cultivation and the question of optimal cropping pattern for Punjab. *Journal of Punjab Studies, 12*(2), 239–250.

Singh, I., & Bhangoo, K. S. (2013). *Irrigation system in Indian Punjab.* Patiala, India: Centre for Advanced Studies, Economics Department, Punjabi University.

Singh, R. (2005). Water productivity analysis from field to regional scale: integration of crop and soil modelling, remote sensing and geographical information *[Doctoral thesis].* Wageningen University, The Netherlands. https://edepot.wur.nl/121640.

Srivastava, S. K., Chand, R., Raju, S. S., Jain, R., Kingsly, I., Sachdeva, J., Singh, J., & Kaur, A. P. (2015). Unsustainable groundwater use in Punjab agriculture: Insights from cost of cultivation survey. *Indian Journal of Agricultural Economics, 70*(3), 365–378.

Vashishtha, P. S. (2006a). *Input subsidy in Andhra Pradesh agriculture* [Unpublished manuscript]. International Food Policy Research Institute.

Vashishtha, P. S. (2006b). *Input subsidy in Punjab agriculture* [Un-published manuscript]. International Food Policy Research Institute.

World Bank. (2001). *India power supply to agriculture, volume 1: Summary report* [Report no. 22171-IN]. Energy Sector Unit, South Asia Regional Office.

World Bank. (2013). *Lighting rural India: Experience of rural load segregation schemes in states.* Open Knowledge Repository Handle/10986/16690. World Bank.

Zekri, S. (2008). Using economic incentives and regulations to reduce seawater intrusion in the Batinah coastal area of Oman. *Agricultural Water Management, 95*(3), 243–252.

Chapter 10

Market instruments and institutions for managing groundwater

10.1 Introduction

Groundwater use in agriculture produces significant economic returns in South Asia. For instance, Kumar (2007) estimated that the surplus value product generated from the use of groundwater in India's irrigated agriculture is enormous (for 15 major states), accounting for nearly 5% of its gross domestic product. In Pakistan, around 14.3 m.ha of cropped areas are irrigated by wells, with around 60 BCM of groundwater pumped annually (Qureshi, 2020). The contribution of these irrigated crops to the country's agricultural GDP ranges from 20% to 24%, with an estimated value of these crop outputs in the range of 15–18.0 billion US dollars[1] against the total agricultural GDP of around US $75 billion. The negative externalities induced by groundwater over-exploitation by individual users on the society at large on the social, economic, and environmental fronts are so great that institutional interventions are required to manage the resource (Burke & Moench, 2000). The focus of this chapter is on the institutional alternatives for groundwater management. The institutional alternatives that have been debated elsewhere in the world so far to arrest groundwater over-draft problems like the one being faced in India are centralized legal and regulatory measures, the establishment of property rights in groundwater with traceability, well registration and well permits, groundwater taxes, and electricity rationing (Kemper, 2007; Kumar, 2007; Schiffler, 1998).

While several researchers in the past three decades have mooted the idea of local, community-based management of groundwater and participatory aquifer management, the discussion has not advanced in the absence of any clarity on the management structures and institutional arrangements needed. One of the inherent weaknesses of such proposals is the overemphasis on one role of local institutions such as the Gram Panchayats and village community organizations

---

1. Based on an average economic water productivity of 0.25–0.30 US $/m^3 for cereal, cotton, and sugarcane dominated cropping patterns there.

Groundwater Economics and Policy in South Asia. DOI: https://doi.org/10.1016/B978-0-443-14011-2.00014-0
227

in groundwater management and the tendency to over-estimate the capacity of these institutions to manage groundwater locally without having a mechanism that creates a credible incentive for the same. While aquifer-level management of groundwater is advocated by some scholars, a lack of information (about aquifer conditions, flows, etc.) is considered the only barrier to self regulation. The need for support from higher-order institutions and enabling legal regimes for these local institutions to play their role effectively is never discussed.

We first argue that top-down regulatory approaches to manage groundwater are less likely to be effective with numerous tiny users in rural areas and very high heterogeneity in the resource conditions. At the same time, local management of groundwater by user groups is subject to several externalities on the demand side as well as the supply side. The chapter discusses the institutional structure for local groundwater management, consisting of village institutions, watershed committees, and aquifer management committees, and identifies and determines the externalities. It recognizes the role of markets in encouraging economically efficient use of water, to be promoted through the establishment of tradable property rights in water, which would also integrate concerns of equity and sustainability. The chapter discusses the role of these institutions in the overall institutional arrangements for groundwater management at the aquifer level to internalize these externalities, including the enforcement of tradable property rights in groundwater.

10.2 Legal and regulatory approaches to manage groundwater

De jure rights to groundwater in South Asian countries such as India, Pakistan, and Bangladesh (which were all parts of British India before 1947) are not clearly defined. But *de facto*, groundwater belongs to all those who have land overlaying it. These three countries follow the English common law with regard to rights over groundwater. Anyone who has a piece of land has access to the resource underlying. A landowner can legally abstract any amount of water unless the geo-hydrology limits. The lack of well-defined property rights, invisibility, indivisibility, and the complex flow characteristics of groundwater makes it difficult to monitor its use (Cullet, 2014). On the other hand, the landless do not enjoy ownership rights in groundwater. In a socialistic society, such a legal framework is inappropriate as it does not suit the interests of all sections of society (Sharma, 1995).

The most dominant institutional approach to check the over-development of groundwater and mitigate the environmental consequences has been regulatory in nature. The oldest in India is the Gujarat groundwater legislation of 1976, which was an amendment to the Bombay Irrigation Act. The legislation could never be enacted due to a lack of popular support. Nevertheless, the legislation passed by the Government of Maharashtra for the protection of public drinking water sources is found to be effective, the reason being that the number of drinking water sources is very small.

Regulatory approaches also include control of institutional financing for well development by the National Bank for Agriculture and Rural Development in "dark" and "grey" talukas (blocks). However, this was not very effective due to a large amount of private financing in the sector, as clearly shown by the 4th and 5th Minor Irrigation Census reports (see GoI, 2014, 2017). The state electricity departments use power connections as leverage to control groundwater extraction in over-exploited and critically developed areas. However, this has not been effective as farmers use connections of the wells no longer in use for newly drilled wells (Kumar, 2007). Also, there are a large number of illegal connections, and electricity theft is predominant in rural areas.

Shah et al. (2004a) argue that any direct regulations to manage groundwater would be meaningless in the South Asian context, and they attribute that to the sheer number of agricultural pumps and their distribution in the vast countryside, and according to them, the pessimism is justifiable given the past unsuccessful experiences of the governments with direct regulations (Burke & Moench, 2000). Qureshi et al. (2010) and Qureshi (2020) also express similar concerns about the effectiveness of state regulations in the context of Pakistan, citing the same reasons as Shah et al. (2004a). However, with a lot of advancements in information technology and widespread use of mobile phones in rural areas, which enable the agencies to track each and every water user who is connected to an electricity grid (Zekri, 2008), the argument downplaying the importance of such regulations merely by citing the well numbers is perverted and does not carry much substance.

Top-down legal and regulatory approaches to check over-development are most likely to be less effective in the Indian context due to many reasons. *First*: legislation is generally applied to large regions within which significant variations in the characteristics of groundwater exist. Even within north Gujarat, both alluvial and hard rock (basalt and granite) conditions exist. The quality of groundwater also varies from fresh to brackish. Both recharge/outcrop areas and discharge areas are found in the region (Kumar, 2007). While recharge interventions and aquifer protection from contamination should receive attention in outcrop areas, preventing a further rise in water levels should be the management concern in discharge areas due to salinity (Central Ground Water Board, 2007).

It is very unlikely that legislation will reflect local specific problems and the interests of the communities at large, and hence could face strong opposition during enforcement. *Second*: there are millions of wells located in rural areas, which suffer from poor road networks and are inaccessible. It will be extremely difficult to get a monitoring mechanism established to ensure that a particular regulation is enforced. *Third*: the farming lobby, which is strong enough to get political patronage and can mobilize rural masses against the legislation. The farmers' movement against the proposed hike in electricity tariffs in agriculture in Gujarat is an example. Such moves are sufficient to cause political instability. Governments require great political will to strictly enforce legislation.

There are international experiences in legal and regulatory approaches. Jordan adopted a Water Policy Framework, which established general principles for future water resources management. The framework commits to public ownership of water, formulates water allocation policies, and sets planning and investment priorities. As per the allocation policy, domestic consumption gets the highest priority, followed by industry, and agriculture, the largest user of water in the country, gets the lowest priority (Schiffler, 1998). Hence, the reallocation of water from the existing uses was to happen voluntarily. To arrest or slow down depletion, sustainable management plans were to be worked out for each groundwater basin, including the installation of volumetric meters in each well. But reallocation of water from agriculture to municipal and industrial uses still does not take place. The rural masses, accounting for 22% of the country's population, receive political patronage and economic privilege from the King. Schiffler (1998) argues that reallocation would happen if some of the benefits of water transfer from agriculture to urban areas are transferred to the farmers.

Intelligent energy rationing, advocated by some (see, for instance, Bhanja et al., 2017; Gupta, 2012; Shah & Verma, 2008; Shah et al., 2004a), though an indirect approach to managing groundwater, is also regulatory in nature. This approach has a few basic premises. (1) Farmers in areas facing groundwater depletion are heavily dependent on electricity supply. (2) The power supply farmers get during most of the year is in excess of the actual demand. (3) Good quality power for a longer duration, which is required during the critical stages of crop growth, is needed for a few days in a year; but is currently unavailable. These conditions, according to them, lead to inefficient use of electricity and precious groundwater, given the fact that there is no marginal cost of using electricity under the flat rate system, and there are very few days in a crop year when farmers need the power supply for long hours (Kumar, 2007).

There is a major problem with this proposition on the theoretical front. In most Indian situations, power demand during a particular crop season is almost evenly distributed due to the following reasons: (1) major time lags in the time of sowing between farmers; (2) adoption of different types of crops (having different seasonality and duration), and also crop varieties having different maturity periods; (3) variation in the energy required to pump a unit volume of groundwater with changing geo-hydrological environments; and (4) the pump capacity per unit of irrigable land and land holding size are not uniform across farmers creating differential power supply requirements. The other problem is on the practical front. Even if rationing of power supply is done in accordance with actual power demand, farmers can still manipulate their energy/water use by choosing over-sized pump sets, which is done to make management of their farms easy (Kumar, 2007; Kumar & Perry, 2019).

10.3 Institutional reforms for groundwater management

10.3.1 Establishing tradable water use rights

The absence of well-defined property rights regimes is a major source of un-certainty about the negative environmental impacts of resource use, leading to inefficient and sustainable use (Kay et al., 1997; Pearce & Warford, 1993). This is applicable to groundwater resources as well. Equity in access to the resource and the physical sustainability of the resource base are important considerations for establishing institutional regimes for sustainable water use like property rights (Kumar, 2000a; Kumar & Singh, 2001; Saleth, 1996). These property rights should be legally enforceable and tradable. But a realistic assessment of total water resource availability, renewable freshwater, and economically accessible water resource in different regions/basins will be the cornerstone of establishing tradable property rights in water (Hérivaux et al., 2020). The concept of tradable property rights in groundwater resources has been used in several western states of the United States and Australia (Brozovic & Young, 2014; Lawson et al., 2010; Skurray et al., 2012; Thompson et al., 2009; Wheeler et al., 2016).

In deep alluvial areas, the total groundwater storage is many times greater than the groundwater that is accessible economically (CGWB, 2019). On the other hand, economically accessible groundwater is much greater than the replenishable groundwater. Since sustainability considerations over-ride economics, only the renewable portion of water should be brought under the property rights regimes. This criterion will then be applicable to north Gujarat. But, there are certain practical considerations in evolving the criteria, which will differ for north Gujarat and south and central Gujarat parts of the basin, as per the prevailing conditions of the resources.

In the alluvial areas of north Gujarat and central Punjab, in view of the problem of overdraft, the total volumetric use rights to be allocated every year should be much less than the renewable groundwater. This will help reverse the process of mining over a period of time. In hard rock areas of north Gujarat, Saurashtra, Karnataka (Kolar district), Tamil Nadu, Maharashtra, and Madhya Pradesh (Dewas, for instance), the economically accessible groundwater will be less than the renewable groundwater resources as these regions have very little static groundwater. Another factor that needs to be considered while establishing private property rights systems is the effect of the phenomenon called "well interference" on the individual's ability to use his/her rights. Here, the actual "abstraction quota" that can be permitted to any farmer and that does not deprive the neighbors of their rights will be much less (for details, see Kumar, 2002). In order to internalize this phenomenon of well interference and make the rights "non-correlative" or mutually exclusive, the communities can devise mechanisms.

In the absence of proper legal and institutional regimes under which water can be allocated among the competing uses, rights will be politically contested. As urban areas in India are economically and politically more powerful than rural areas, it is very likely that they manage the additional supplies required from the rural areas, depriving the socially and economically weaker groups of their share of water. Already, there are conflicts in the water-scarce regions of Gujarat and Rajasthan over the diversion of water from irrigation reservoirs to urban areas for drinking during scarcity (Kumar, 2003). Unregulated transfer of water from rural areas to urban areas through the mining of aquifers for cash gains is a widespread phenomenon in many areas, including the water-scarce areas surrounding Chennai city in Tamil Nadu (Palanisami, 1994), as cited in Rosegrant & Ringler, 2000). Such water transfers will have both short-term and long-term impacts on agriculture and rural livelihoods. The high prices that water can fetch would encourage a few rich and enterprising farmers to deplete the aquifers at the cost of the poor. The negative equity and sustainability effects of such transfers can be mitigated if water allocation is done under properly instituted water rights, which take into account aquifer yield constraints.

Different viewpoints exist on the characteristics of tradable property rights in groundwater. For instance, Easter et al. (1999) argue that given the South Asian situation of having a very large number of small land holdings, community ownership of groundwater would only be feasible encompassing one or two villages whereas Saleth (1996) argues that such ownership regimes could raise issues about accountability. It would also enhance the capacity of individual users to take action to protect his/her water rights in the case of free riders. In our view, property rights in groundwater have to be not only tradable but also privately owned. Many of the hydraulic assets owned by the communities in villages are never managed due to a lack of accountability.

Legally well-established water rights systems exist in some parts of developed countries for groundwater. Examples are real property rights in groundwater in California and Kansas in the western United States and private tradable property rights existing for groundwater in Mexico (source: Peck, 2003; Shah et al., 2004b). In recognition of groundwater mining problems, the Kansas legislature passed an enabling act for the creation of groundwater management districts (GMDs) for the purpose of conservation of groundwater resources.

The Kansas state applies the prior appropriation doctrine—"first in time is the first in right" to groundwater. The states in the western United States recognize rights to use water as property rights to be treated much like land. Under the Kansas Water Appropriation Act (1945), from the effective date of the law, any person seeking the right to use water had to apply for a permit. The "junior water rights" were to be given away to "senior water rights" in times of conflict. However, these water appropriation rights were not expressly stated as property rights but were recognized as usufructuary rights. The Act is administered by the chief engineer of the Division of Water Resources in the State Board of Agriculture. But in 1957, the Kansas legislature amended the Act to redefine water rights as "real property rights." However, these water rights differed from land ownership rights in the sense that a water right is a usufruct, i.e., a right to use water and not absolute ownership. Though several thousands of permits were granted under the Act, it led to the creation of several groundwater mining areas, which subsequently led the chief engineer to close many parts of Kansas to new appropriation permits.

The GMD Act led to the creation of groundwater management districts (which constituted 1/4th of Kansas State). Part of the GMD Act gave the chief engineer the power, after notice and hearing, to establish "intensive groundwater use control areas" (IGUCAs). Inside the IGUCAs, the chief engineers would have extraordinary power to reduce the permissible withdrawal of groundwater by any one or more appropriators there. However, such intensive groundwater use control areas still account for a small share of the Ogallala aquifer in the western United States that experiences serious mining problems (source: Peck, 2003).

Though the western United States examples with its establishment of well-defined water rights do not provide successful cases in checking or controlling groundwater mining problems[2], one should keep in mind that those property rights were based on water right laws framed in the 40s and 50s. They have not evolved in response to the problem of groundwater mining *per se*. They have evolved in response to a series of court cases to either protect the rights of prior appropriators or secure new rights, and at a time mining was not a social concern. In contrast to the western United States, India does not have water rights defined, but almost every land-owning farmer enjoys *de facto* rights to access groundwater underlying the farmer's piece of land, and over-draft and mining are problems that concern the society at large. It will be easier to define water rights as "property rights" and grant/allocate rights in such a manner that the total allocation does not go beyond the renewable recharge.

One should be cautious in drawing on the experiences of the western United States for making prescriptions for water rights reforms in India simply due to the reason that the socio-economic and legal environments in India are not comparable with that of western United States. First of all, the socioeconomic profile of farmers is very different. Whereas an average farmer in the western United States owns nearly 1000 ha of land, in India it is less than a hectare. The groundwater management districts in the western United States have to deal with a few hundred farmers. Whereas in India, there could be 100,000 farmers tapping water from the same aquifer. Large farmers, who run farming-like industries, also mean that the stakes on the water are high; so is the political clout and power to litigate to protect his/her rights. The legal system and the judiciary are also far more effective in the United States. These factors also point to the inference that the institutional framework tried in the United States ("groundwater management districts") may not be as much relevant for India, as the technological solutions, such as determining withdrawal limits for aquifers and monitoring draft. We would deal with the institutional framework separately in Section 10.5 of this chapter.

---

2. A few exceptions are the Walnut Creek IGUCA and Rattlesnake Creek Management program. There was also a recent proposal to extend the life of the Ogallala aquifer. The three main recommendations of the proposals were to delineate the aquifer into subunits to allow management decisions in areas of similar aquifer characteristics; for the GMDs and DWR to identify each aquifer subunit in decline or suspected decline and establish water use goals to extend and conserve the life of the Ogallala aquifer; and to identify priority subunits for action.

Leaving aside the West, there are excellent examples of water rights systems in many developing countries. Chile and Mexico were two countries that had introduced tradable property rights in both surface water and groundwater at the national level, while many of the western states of the United States and some states in Australia had them. In both Chile and Mexico, water rights are separate from land ownership rights (Thobani, 1997, p. 166). As regards the effect of introducing tradable water rights, Thobani writes: "Anecdotal evidences and studies showing the gains from trading water suggest, however, that water rights have facilitated economic growth. Agricultural GDP in Chile grew at a rate of 6% annually in the decade following the passage of the Water Law" (Thobani, 1997, p. 170).

Also, there are illuminating examples from small island countries. For instance, in Mauritius, a water law, which recognizes water rights in surface and groundwater, is part of the institutional setup for water resources management. A system of permits for drilling wells and withdrawing groundwater exists in this country. Licenses for agricultural withdrawal are renewed once every three years, and those for industrial uses are renewed every year. Water withdrawal by permit holders is closely monitored, license holders are regularly advised not to exceed their authorized abstraction rates, and punitive rates are applied for withdrawals in excess of the approved volumes. In sensitive zones, groundwater abstraction is closely monitored to prevent seawater intrusion. Twenty bore holes spread over the island are fitted with data loggers, which record water conductivity along with water levels. Against an average annual utilizable groundwater potential of 160 MCM, the current utilization in the country is 148 MCM (Sharma, 2005, pp. 290–292).

Defining property rights alone will not ensure equitable access to groundwater for all from a practical standpoint. The landed farmers who do not have the resources to invest in wells will either have to enter into agreements with the neighboring well owners for providing pumping services or promote their own organizations for water extraction (Kumar & Singh, 2001). This is actually the case in many parts of Gujarat, Rajasthan, Tamil Nadu, Karnataka, and Bihar, where many farmers currently have no direct access to groundwater. Several institutional models for groundwater development that operate on the principles of equity have been in existence in Gujarat for many decades (Kumar, 1996, 2000a; Shah, 1993; Shah & Bhattacharya, 1993).

That said, in developing countries like India, carrying out institutional reforms for water management, including well-established water rights, is going to be an arduous exercise, and the transaction cost is going to be enormous (Garduño, 2005; Kumar, 2000a; Kumar, 2003; Kumar, 2007) Referring to such transaction costs, Garduño (2005), while reviewing water rights administration in several countries from Latin America, Africa, and South Asia, cautions that when designing a water rights administration system and setting up a time frame for implementation, it is wise to be aware of the tools that may be required to implement it; otherwise, it might result in an unrealistic system.

In the context of groundwater, the transaction cost essentially comes from the technical and institutional processes involved in registering wells, monitoring water withdrawals, groundwater resource evaluation, and evolving norms for the allocation of volumetric water rights. It is important to moot the views of others from the field on this pertinent issue (Kumar, 2007). At least a few researchers have argued that such transaction costs would be too high in the Indian situation even for registering wells and users, given the many millions of dispersed and tiny users who are located in rural areas that are often inaccessible and therefore totally impractical (Moench, 1995; Shah et al., 2004b). It is to be kept in mind that at least some of the regions with intensive groundwater use have very few wells. For instance, the entire north Gujarat has only 17,000 plus tube wells. The regions that have millions of operational wells, such as eastern Uttar Pradesh and Bihar, anyway do not require such management interventions as those required in over-draft areas. The water used by these wells is tiny. Secondly, in the problem areas, the groundwater economy is becoming more and more formalized and it is technically easy for the government to make any interventions. The reason is that wells are electrified, and electricity departments possess details of the power users in the farm sector (Kumar, 2007).

Recent Indian experiences of efforts at introducing power tariff reforms in Punjab, Odisha, and West Bengal showed that such transaction costs are related to the political patronage and power enjoyed by the farm lobby, not the number of users to be dealt with. In Punjab, though the farmers are progressive, the government has failed to introduce a hike in electricity tariff in the farm sector due to the high political patronage enjoyed by the farm lobby. On the contrary, in Odisha, which is characterized by high illiteracy and rural backwardness, the state government had successfully introduced power tariff reforms, including the setting up of local institutions for the recovery of electricity bills from agricultural pumpers (Kumar, 2007). In West Bengal, the agro wells are metered and farmers are paying for electricity on the basis of consumption, and the unit charges are considerably high (Kumar, 2018).

But, as Saleth (1996) notes, there are major complexities involved in assessing resource availability and sustainable levels of abstraction. Over-estimation of the sustainable levels of abstraction of the resource and thereby over allocation of water rights might lead to communities not being able to use their entitlements. Further, as Kumar et al. (2006) established in the case of the Narmada River basin, groundwater outflows into surface streams were quite significant, and as a result, the utilizable groundwater resources could be much less than the recharge. However, the Central Ground Water Board continues to consider 85–90% of the annual recharge as utilizable groundwater. But, as Kumar et al. (2006) note, an increased draft could affect downstream surface water users in those basins where the groundwater system interacts with the river system. Also, as Saleth (1996) notes, water rights may not be mutually exclusive and could be correlative, especially in areas where interference with wells is very common. An individuals exercising their own rights could deprive their neighbors of their rights. Such phenomena can lead to future litigation. The only way to overcome such planning difficulties is to make the tenure of water rights short and keep modifying the volumetric rights that can be allocated based on actual observations of groundwater levels. Nevertheless, such issues would be non-existent in the case of deep alluvial areas of north Gujarat, which have significant groundwater stock.

While agencies such as the Central Ground Water Board and State Ground Water Departments have sufficient scientific and technical expertise for resource monitoring and evaluation, they lack adequate financial and human resources for maintaining a closer network of observation wells with monitoring instruments, which can result in proper data acquisition and processing. With the hydrology project implemented in 13 Indian states, the quality of data acquisition and processing had improved in many states, such as Gujarat, Karnataka, Maharashtra, and Kerala. Also, the methodologies for resource evaluation need to be modified to incorporate social concerns. One recent development is to capture the short-term and long-term trends in water levels in assessing the status of development of groundwater apart from recharge and abstraction.

Further, the opportunity costs of not investing in institutional reforms in the water sector are going to be enormous due to the growing water scarcity and can exceed the transaction cost (Saleth & Dinar, 1999). Therefore, in severely water-stressed regions like north Gujarat, Kolar in Karnataka, Coimbatore in Tamil Nadu, and Dewas in MP, investment in creating institutions could be justified to a great extent. Further, it is a phenomenon around the world that when water becomes scarce, communities are increasingly found to be evolving and enforcing social institutions.

10.3.2 Market instruments for groundwater demand management

The key to managing water in the face of growing scarcity and conflicts is demand management. It has two distinct components: the transfer of water among alternative use sectors and encouraging the conservation of water within a particular sector through efficiency improvements (Frederick, 1993). While fiscal instruments can bring about positive changes in the efficiency of use in a particular sector, say irrigation, they will not be adequate to serve the demands of other sectors, such as domestic uses and industry. Institutional changes are necessary to affect water transfers across sectors. Markets can serve as institutional arrangements for the transfer of water to economically efficient uses for demand management in most situations (Frederick, 1993; Howe et al., 1986). Property rights, when clearly defined, would make markets function efficiently (Frederick, 1993; Saleth, 1996; Thobani, 1997). Tradability and exchangeability are crucial to capture and reflect the scarcity value of water through price signals (Hérivaux et al., 2020; Saleth, 1996, p. 246).

Several barriers might impede the development or operation of groundwater markets (Bauer, 2012; Easter & Huang, 2014; Skurray et al., 2012; Wheeler et al., 2016), and therefore, water markets do not always function as predicted as a groundwater management instrument. In many groundwater basins where water markets were introduced, transactions remain limited and far below expected levels. This may be due to an inadequate institutional framework, high transaction costs, over-allocation of water rights, and problems with compliance and enforcement (Brozovic & Young, 2014).

In the over-exploited region of north Gujarat in India, access to the resource is still unlimited for those who have resources to invest in tube wells. In spite of the limited-hours power supply (for 10–12 hours a day), the tube well owners in the area can pump out 40–100 m^3 of water per hour. However, farmers here are growing less water-intensive crops such as wheat, castor, cotton, and mustard. But the well owners are pumping much more water than is required to irrigate their own fields and selling the excess water to the neighboring farmers to recover electricity charges. Though water markets are widespread, they are not efficient enough to create opportunity costs for growing water-intensive crops. A similar situation prevails in the water-abundant central Gujarat (Kumar & Singh, 2001, pp. 400–401).

The introduction of a unit pricing system, though encouraging farmers to make efficiency improvements in their pumps or use water more efficiently, may not result in positive impacts on economic efficiency (Kumar & Singh, 2001, p. 401). Eventually, the establishment of tradable property rights in groundwater will put restrictions on the volume of water that the tube well owners can pump, if the rights system is to ensure minimum water for those farmers, including the land-less, who are presently deprived. As a result, the large farmers will have to resort to buying water if they want to bring the entire farm under irrigation, while the smaller holders and land-less may have a little extra water, which they can sell. The price of water will reflect the opportunity cost of using it. The markets and market-determined prices could work in two ways: (1) farmers can shift to alternative uses that provide higher economic returns than the price of water; or (2) they could continue the existing uses with more efficient practices or resort to selling (Frederick, 1993). Such transfers can promote access equity and efficiency in use (Kumar, 2003).

The analyses presented in Chapter 8 showed that volumetric rationing motivates shareholders of tube well companies to go for highly water-efficient crops (Kumar et al., 2011, 2013). Well owners will be motivated to adopt crops that are not only economically efficient but also low water consuming, or to sell the saved water at prices determined by the market to industries or urban water utilities. With a growing water deficit in Chennai, the farmers from the surrounding region sell water from their wells, earlier allocated to growing paddy, for urban domestic uses and get high returns (Palanisami, 1994, as cited in Rosegrant & Ringler, 2000). The opportunities for the transfer of water from agriculture to urban and industrial uses for demand management are great in central Gujarat (see, for instance, Kumar & Bassi, 2011; Kumar & Singh, 2001).

While competition for water use is growing between sectors with increasing water scarcity (Kumar, 2003), the economic pressure to trade water for non-agricultural uses would be high. Large farmers are likely to take advantage of this situation. But, water rights would put restrictions on the amount of water large farmers could tap from aquifers for transfers to urban areas or other sectors, unless they are willing to make substantial cuts in their own farm-level use. At the same time, the high prices for water in the market would give opportunities to the resource-poor and landless farmers and farm laborers to trade their rights to more progressive farmers or for higher-valued uses (in urban areas or for industries) and earn profits.

In the case of north Gujarat, farmers can grow low water-intensive and high-value crops such as pomegranate, lemon, and gooseberry with drip irrigation. They could also go for crops such as castor and cotton with drip irrigation, thereby increasing both land and water productivity. Adoption of fruit crops is already happening in Banaskantha district in a reasonably big way without the introduction of the financial instruments mentioned above, but by highlighting the strong economic benefits that could be derived from such crops and offering marketing facilities for these crops. The argument here is that market interventions such as volumetric rationing with the pricing of water would force farmers to take a risk, become more enterprising, and it would explore full social ingenuity.

10.4 Management of groundwater by local user groups

The socio-ecology of groundwater in India is not uniform and is instead characterized by complexity and heterogeneity due to a varying physical environment affecting availability across regions and localities, millions of tiny users, a spatially varying socioeconomic environment affecting the nature and degree of use, the differential agricultural and energy policies across states, and changing institutional settings (Kumar, 2007). The nature of groundwater problems also varies from region to region and location to location (CGWB, 2019), and as a result, management solutions for one locality cannot be adopted for another. Also, there is a range of physical, social, economic, and institutional factors that affect the effectiveness of any groundwater management solution for a locality. These factors also vary from locality to locality. As a result, groundwater management solutions are to be locale specific and should incorporate the range of physical, social, economic, and institutional characteristics of the locality (Burke & Moench, 2000; Kumar, 2000a; Kumar, 2007; Moench, 1995, p. 3).

As regards regulations, as earlier indicated, top-down, centralized approaches to regulate the use of groundwater would be practically impossible. It is important to note that the benefits to individuals from collective action by local users are also not distinct and direct in the case of groundwater, as aquifer conditions are dependent upon the decisions of many thousands of dispersed users (Burke & Moench, 2000, p. 3). Also, lack of participation in the collective action doesn't involve any "opportunity cost" to the individuals. Hence, there could be resistance from individual users to joining the group initiatives. Therefore, establishing tradable property rights in groundwater in volumetric terms and enforcing them would be crucial to the success of local management. Once the property rights are made tradable, the price of water will increase in the market. Again when physical transfer of water is made fully possible, the price at which water is "traded" would be the opportunity cost of using water. As a result, the users would be encouraged to transfer water to economically more efficient uses to maximize the returns from every unit volume of water or sell it at the market price.

Over the past three and a half decades, India has seen large-scale NGO initiatives in experimenting with and implementing water management solutions for many water-scarce regions and localities, particularly in Gujarat, Rajasthan, Madhya Pradesh, Maharashtra, Tamil Nadu, and Karnataka. These interventions are mostly based on strategies for augmenting surface or groundwater supplies using approaches such as runoff water harvesting, artificial recharge of groundwater and watershed treatments. But the management regimes for these local interventions do not extend beyond village administrative or watershed boundaries (Glendenning & Vervoort, 2011; Kumar et al., 2006, 2008; Talati & Kumar, 2019).

In many areas, these interventions have resulted in temporary improvements in local water situations in terms of greater availability of surface water in local storages, rises in water levels in wells, and greater productivity of the rain-fed crops (Glendenning & Vervoort, 2011; Kumar et al., 2006, 2008; Ray & Mahesh Bijarnia, 2006). The management institutions, created by the NGOs, operate at the village or micro watershed level, implementing projects, and/or performing maintenance on structures. On the one hand, such approaches led to increased resource use by the communities that actually get the water benefits, owing to a lack of institutional capacities for regulation. On the other hand, there have been no institutional innovations worked out for sustainable scaling up of the activities for the local management regime to cover mega watersheds, sub-basins, or aquifers.

Sustainable scaling up does not merely refer to multiplying physical interventions to cover all villages/watersheds in the problem area but also means optimum planning of interventions in all villages/watersheds while keeping in view overall resource availability (the "uncommitted flows") in the basin and existing level of utilization from all the watersheds falling in the basin/aquifer for optimal impact in all treated watersheds. "Scale effects" are very significant in water management in a river basin due to the hydrological interdependence. It refers to the difference between the impact of interventions in a watershed/village carried out in isolation in that watershed/village and the impact of the same interventions carried out in all watersheds in the basin. This can reduce the local "impact" of the interventions when one moves from one micro watershed to a basin (Glendenning & Vervoort, 2011; Kumar et al., 2006, 2008). The Watershed Organization Trust of Ahmednagar district in Maharashtra is an exception, where the Trust took up the task of scaling up the watershed interventions in the district. But the organization also does not carry out mega-watershed level or basin-level planning.

10.4.1 The negative externalities in local groundwater management

The local community organizations generally operate at the village level. Institutions for managing groundwater are also proposed and created at the village level (Jadeja et al., 2015). But most of these proposals do not contain any mechanisms that create a credible incentive for self-regulation of groundwater use, without which the local institutions are bound to fail (Moench et al., 2013), as pointed out by Lopez-Gunn and Martinez Cortina (2006) in the context of Spain's groundwater user associations and Wester et al. (2011) in the context of Mexico's COTAS (*Consejos Técnicos de Aguas*). In cases where many villages tap the same aquifer, the actions of a single village to protect and manage their groundwater resource may not find success unless the neighboring communities cooperate. Understanding the scale at which interventions should occur is an important aspect of institutional analysis (Kumar et al., 1999; Moench et al., 2013). In this section, an attempt is made to determine the externalities in local groundwater management.

The first and most important aspect of designing any resource management institution is the identification of the resource boundaries (Ostrom, 1992). Inconsistency between hydrological boundaries and the jurisdiction of the management institution may lead to obstacles in broad planning, data collection, and regulatory activities (Frederiksen, 1998). In the case of groundwater, the resource boundaries could be demarcated on the basis of the physical system affecting groundwater recharge and the social systems affecting its use. The physical systems affecting groundwater availability are integrated at the level of the watershed. The social systems affecting their use are integrated at the aquifer level. North Gujarat's alluvial aquifer, for instance, encompasses many districts, viz., Ahmedabad, Gandhinagar, Patan, and most parts of Mehsana and Banaskantha districts. There are hundreds of thousands of farmers accessing the groundwater from this aquifer.

For any local supply-side approaches to augment the available groundwater supplies to be effective, the approach has to be watershed-based. This is one of the factors that induces a negative externality for local groundwater management efforts. Negative externality here refers to factors that prevent us from maximizing the local impacts. For the demand-side approaches for groundwater management to be effective from physical sustainability considerations, the total pumping across the aquifer should balance with the average annual replenishment. Hence, this is one of the externalities of local groundwater management initiatives. Again, in order to ensure sustainable use of the available groundwater, it is essential to maintain the quality. There are two sources of pollution in aquifers: (1) internal sources, and (2) external sources. The external sources of pollution can be further classified into point pollution and non-point pollution. The potential sources of external pollution are fertilizers (Llamas et al., 2012) and industrial effluent. The internal pollution is due to the geo-*hydrochemical* processes taking place within the formation-bearing groundwater (Kumar, 2000a).

The physical and social parameters within a watershed that determine the degree of these externalities for a local groundwater management regime are (1) watershed characteristics and (2) groundwater basin characteristics. The major watershed characteristics that influence the externality function in a village-level groundwater management action are: (1) the physical characteristics of the watershed: the size of the portion of the watershed falling outside the boundaries of the village and the extent to which that portion of the watershed has been treated; and (2) the social characteristics: number of socio-economic groups within the watershed and the different water use priorities existing (Kumar, 2000a).

The groundwater basin characteristics that influence the externality function are: (1) aquifer characteristics, such as the areal extent and yield characteristics of the aquifer, and the vulnerability of the aquifer to pollution (Llamas et al., 2012); and (2) socio-economic factors affecting groundwater use, such as the presence or absence of urban centers and industrial areas, and water markets. The degree of success in implementing a management decision by a village-level institution will be much higher for an aquifer of a limited areal extent than for an aquifer of a large areal extent. The "degree of stress" imposed by the same amount of pumping on the groundwater regime will be far less for an aquifer with high yields than that with poor yield characteristics (Kumar, 2000a; Llamas et al., 2012). However, such externalities are hardly acknowledged by researchers who work on participatory groundwater management and hence advocate community-based village institutions (see Jadeja et al., 2015; Mukherjee, 2021). Too much faith is put in the self-regulating power of the local communities without recognizing the role of higher authorities, which is essential for making management responses legitimate and effective.

10.5 A framework for the design of groundwater management institutions

The institutional structure for community-based local groundwater management suggested here has three levels in the hierarchy that are vertically integrated, and they operate at the level of groundwater basins or aquifers. These institutions operate under a larger institutional regime provided by a tradable water rights system established by legitimate agencies of the government.

10.5.1 Village-level institutions

The role of groundwater management institution created/formed at the village level will be as follows: (1) evolving groundwater management solutions appropriate for the locality—identification of possible physical management interventions like building of recharge systems and plantation/afforestation, suggesting locations for the same, and identifying potential users of end-use conservation schemes; (2) identification of water use priorities of the communities in the locality; (3) installation of water meters for enforcement of tradable property rights; and (4) investing in infrastructure for transfer of water in order to encourage the development of water markets for water allocation within and across sectors, if needed. The local intuitions—village and watershed institutions—can influence the behavior of all groundwater users, by framing and enforcing volumetric water rights of individual users (Kumar, 2007).

While efficient markets would take care of the transfer of water to economically efficient uses, farmers would need extension inputs to know what kinds of interventions would help them economically and from the point of view of improving water productivity in economic terms. Inputs would be required for the adoption of WSTs and the marketing of high-value products such as fruits, which, though water-efficient, are fast perishable.

Overall, the effectiveness of the institutional arrangement for groundwater management would depend heavily on the robustness of the local village institutions. Apart from implementing watershed management and artificial recharge schemes and monitoring water use to enforce water rights, its functions will include developing efficient water markets between the well owners and the non-well-owning farmers, given the fact that a large percentage of the marginal and small farmers do not have access to individual wells. It has to support these marginal and small farmers to invest in water distribution infrastructure through loan financing. Such investments will happen only if there is security of the water rights. Though the amount of water that can be allocated to individual users as the volumetric use right change from year to year, there has to be a minimum assured quantum of water that is made available in all the years. For the markets to be efficient, it is important that the water withdrawal by the individual rights holders be strictly monitored.

10.5.2 Watershed institutions

As some of the physical activities necessary to augment groundwater supplies are to be carried out at the watershed level, the activities being undertaken in different villages within the same watershed need to be coordinated so that they become effective. An organization is to be formed at the watershed level called a "watershed committee." Such an organization will have representation of village-level institutions from all the villages falling within the watershed. The specific roles envisaged here for such an institution are (1) the setting up of Village Level Institutions in all the villages falling within the watershed; (2) the coordination of various village-level physical activities; and (3) resolving the conflicts between villages (Kumar, 2007).

If one of the villages in the watershed is not cooperative in implementing the management plans for the watershed due to conflicting interests, the watershed committee can intervene and help them resolve the conflict. It is important to note that in the case of a small micro watershed falling completely within the administrative boundaries of a village, there will not be any need for a separate watershed committee. The watershed institutions (WI) can take care of this.

However, the micro-watershed approach that is implemented at the scale of 500–1000 ha in India encounters problems when it comes to up-scaling. Operating at the micro-watershed level does not necessarily aggregate or capture upstream–downstream interactions. It is anticipated that watershed management projects will not only produce local on-site impacts at the micro-watershed level but also induce positive or negative externalities in the form of environmental services or disservices downstream (Syme et al., 2012).

Hence, there are actions that fall outside the mandate of the watershed institutions. For instance, there are several instances in Gujarat, Madhya Pradesh, Rajasthan, and Maharashtra where intensive water harvesting interventions have produced unintended consequences, particularly in the form of reduced downstream flows (Glendenning & Vervoort, 2011; Kumar et al., 2008a; Ray & Mahesh Bijarnia, 2006; Talati & Kumar, 2019) and groundwater over-exploitation (Kale, 2017)[3]. Avoiding such unintended consequences calls for coordinated planning at the level of the river basin or aquifer for optimum benefits across all the watersheds.

10.5.3   Aquifer-level management

The need for aquifer level management of groundwater had earlier been mooted (Burke & Moench, 2000; Kulkarni & Vijay Shankar, 2009; Vijay Shankar et al., 2015; Moench et al., 2013; Shah, 2009; Shen, 2015). However, an appropriate management structure and institutional arrangements, including resource mobilization, for undertaking management interventions have not been suggested. What is generally being echoed is that there is a huge information barrier, which becomes a major constraint for communities trying to get around managing aquifers. Shah (2021) notes, "…Participatory Groundwater Management (PGWM) must form the backbone of our groundwater programmes". Information on aquifer boundaries, water storage capacity, and flows in aquifers should be provided in an accessible, user-friendly manner to primary stakeholders, designated as the custodians of their own aquifers, to enable them to develop protocols for sustainable and equitable management of groundwater, including distance norms, regulation of pumping capacities, crop-water budgeting, matching cropping patterns to agro-ecology, reuse and recycling of water, etc. (Mukherjee, 2021).

---

3. In Maharashtra, the construction of farm ponds by individual farmers was portrayed as a miracle strategy by the state and central governments as well as popular media. But the manner in which it was implemented in the arid and semi-arid regions of Maharashtra had become a cause for serious concern. Farmers extract a huge amount of groundwater to store in large-sized farm ponds (Kale, 2017).

The mandate of the aquifer management committee, being proposed here, should be to facilitate "scaling up" and evolving aquifer management plans. An important task of the aquifer management institution would be to define the water rights of individual users as "tradable property rights" and enforce them through the institutions at the lowest level in the hierarchy. Some of the management decisions are (1) the amount of water that needs to be captured from within and outside the aquifer for augmenting groundwater; (2) the future hydrologic stress for the aquifer, the safe yield and the total allowable abstraction; (3) the identification and demarcation of watersheds for the implementation of local management solutions; (4) the suggestion of suitable cropping patterns for the region with a view to increasing the water use efficiency; and (5) suggesting efficient irrigation water management practices for the region that actually result in real water saving[4].

Evolving management strategies for large groundwater basins will call for reliable information on the physical characteristics of the resource, the current level of groundwater exploitation, the social, economic, and institutional factors affecting the resource use, the socio-economic and ecological impacts of resource exploitation, and the safe yields. Therefore, the aquifer management committee shall have representation from the state groundwater agencies, NGOs, water management specialists, ecologists, and social scientists, apart from the representatives of the watershed committees working in the basin. The role of such scientific inputs is largely under-estimated by a group of researchers and development professionals in India, who wanted to limit the discussions on scientific inputs to mapping the aquifers (see, Kulkarni & Vijaya Shankar, 2009; Shah, 2016). However, as pointed out by Kumar et al. (2017), the outputs of such mapping exercises will only help the rural elite identify areas with good groundwater potential to indulge in "water farming" which in turn will lead to further deprivation of the poor farmers. What is required is that the scientific knowledge generated by the technical agencies about groundwater be carefully used for deciding on the management goals for the aquifer and the specific management strategies.

Conjunctive management of surface water and groundwater in the context involving surplus runoff from Sardar Sarovar is widely discussed in the context of north Gujarat as a groundwater management strategy (Institute of Rural Management Anand/UNICEF, 2001; Paranjape & Joy, 1995; Ranade & Kumar, 2004; Vyas, 2001. There is an ongoing government scheme (known as "*Sujalam Sufalam*"), which involves diversion of surplus water from the Narmada river basin through the main canal of the Sardar Sarovar project and use for decentralized recharge through several of the village ponds in the alluvial areas of north Gujarat.

---

4. See Seckler (1996) for definitions of real and notional water saving in the context of water resources management.

Apart from defining water rights as "property rights," it is also important to introduce a water tax on the basis of volumetric use rights. The idea of a groundwater tax in the Indian context had already been mooted by M. S. Ahluwalia, former deputy chairman of the erstwhile Planning Commission of India. The water tax can be levied annually from the users on the basis of volumetric water pumped from the aquifer or the actual consumptive use in cases where return flows can be quantified. For this, the institutional hierarchy from the aquifer management committee to the village institution could be effective. The imposition of a water tax can create a disincentive among farmers to use up their entire allocation for unproductive purposes and also encourage efficient use and the sale of the saved water at the prevailing market price. On the other hand, as Thobani (1997) argues, it would create incentives for management institutions to define, measure, and enforce water rights. While part of the funds generated from the levying of the water tax could be used for covering the operational costs of the institutions, from aquifer management committees to village institutions, part of the funds would be used for servicing the resource, including large water transfer projects.

The water rights law of Kansas and other parts of the western United States and the Groundwater Management Districts (Peck, 2003), and the property rights reforms introduced by Chile's Water Code and Mexico's water policy[5] (Shah et al., 2004b), provide pointers to India for possible legislative courses of action and institutional interventions. The concerns such as the long-term sustainability of the resource base and safe yield could be taken into consideration in defining use goals for groundwater in a region while establishing a groundwater management regime. One important factor that renders instituting well-defined property rights in water in India almost an impossible task is the presence of millions of tiny users within the jurisdiction of a state or hydrological entity (basin). This problem can be tackled to a great extent by creating management regimes for small geographical units inside a large management area, which could be an aquifer or river basin, and integrating them at the larger management unit level through vertical coordination and linkage.

10.6   Does an enabling environment exist for creating groundwater management institutions?

There are many regions in South Asia where groundwater depletion problems are apparently serious. Groundwater in regions like north Gujarat, western

---

5. Mexico had created tradable property rights in water by: first, declaring water as national property, thereby removing the linkage between land rights and water rights; second, allowing existing users to get their use regularized by obtaining a concession from the CNA; third, setting up a structure for enforcing the concessions; and fourth, levying a volumetric water fee from concession holders. Under the new water law, all diversions other than for direct personal use are allowed only through concessions (Shah et al., 2004b).

Rajasthan, the Rayalaseema region of Andhra Pradesh, Kolar district of Karnataka, the central parts of Indian Punjab, and the Punjab and Baluchistan provinces of Pakistan deserve special attention due to the magnitude of the depletion and quality deterioration problems and their impact on the communities. Several semi-arid regions in India, where surface water resources for meeting the major water demand, i.e., irrigation, are scarce, fall under this category. Some of them are the Deccan Plateau area of Karnataka, Maharashtra, and Madhya Pradesh, crystalline hard rock areas of Tamil Nadu, arid and semi-arid parts of western India, particularly north and central Gujarat, Saurashtra and western Rajasthan. Not only is the rainfall in these regions poor and erratic, but the groundwater potential is very low.

There are several countries across the world, including Iran, Jordan, Saudi Arabia, and Mexico, where the development and use of groundwater have taken place almost entirely in the private sector and legal ownership rights of the resource are not defined as in India. These countries are similar in the sense that there are no established water rights holders; water rights were attached to land ownership rights. In some other countries, it is part of the common heritage (Burke & Moench, 2000, p. 5). Over-development problems similar to those in South Asia (especially India, Pakistan, and Bangladesh) did exist in these countries and exist even today. Institutional interventions of some kind would be critical to achieving sustainable management of the resource in these countries. The reason being that society, at large, faces the negative externalities (social, economic, and environmental) induced by the behavior of individual pumpers, who try to maximize the benefits of groundwater use at the individual level.

In countries where major institutional innovations in groundwater management have been attempted, local management institutions are not common. For instance, the groundwater management districts of Kansas, formed under an enabling Act of the Kansas State Legislature, are centralized entities that perform as regulatory authorities. But those are situations where the number of users within a management unit is very small. On the contrary, in the Indian context, every village has around 200–300 well owners or groundwater users. It would be extremely difficult for a single centralized institution to allocate and enforce water rights.

An exception to this is the farmer organizations in China. Water user associations (WUAs) created at the local level are the basis for self-regulation of groundwater and do not have any distinct management system. Since their introduction to China in the mid-1990s, WUAs have developed quickly and are promoted by the government as the grassroots water management unit. The 2002 Implementation Opinions on Hydraulic Engineering Management Institution Reform by the Office of the State Council require the positive fostering of farmer water user organizations and the development of management institutions based on rural water use cooperative organizations. In 2005, the Ministry of Water Resources, the State Development and Reform Commission, and the Ministry of Civil Affairs issued the Opinions on Strengthening Farmer WUA Development.

The document points out that WUAs solve the construction, management, and maintenance problems in rural water infrastructures and promote farmers' self-regulation. With government support, more than 78,000 WUAs were set up in China (Chen, 2012), managing an irrigation area of 16 m.ha. Almost all irrigation districts in northern China, including groundwater and surface–groundwater joint-use irrigation districts, have set up WUAs (Shen, 2015). Shen (2015) also notes that the groundwater management system that exists at the national level is poorly implemented by these local institutions.

The legal and regulatory measures to be used by the Central Ground Water Authority in India are likely to be ineffective as they deny ownership rights to individual users and do not involve them in management (Kumar, 2000a; Schiffler, 1998). The administrative capability of the agency to take punitive action against violators of regulations is also highly questionable, given the fact that they are armed with very little staff when compared with the millions of users they have to handle (Shah et al., 2003). The Central Ground Water Authority should evolve strategies and guidelines for the joint management of groundwater for "problem areas" and provide legal, financial and technical support for speeding up the process of institutional development from the aquifer to the watershed to the village level. For this, the support of the state governments would be crucial, as groundwater is a state subject in India. The states can pass an enabling Act for the formation of Aquifer Management Institutions.

Experience shows that local user group organizations can emerge in problem areas with support from external agencies such as NGOs if appropriate legal, institutional, and policy regimes exist. Some of these are (1) the recognition of the rights of individuals and communities over groundwater, and (2) the establishment of tradable private property rights. Institutional processes can be initiated to demarcate aquifer basins, evaluate resource availability, and integrate the local institutions with new aquifer/basin-level organizations that are to be created (Jairath, 2001).

In semi-arid and arid regions where groundwater generates millions of micro-economies and forms an important source of rural livelihood (Kumar, 2007), the opportunity cost of not investing in institutional reforms and physical interventions is great. Since farmers, whether resource-rich or poor, across the board, are going to be affected by permanent depletion, the chances of obtaining their cooperation in introducing water rights reforms are likely to be high. The social ingenuity of these farmers would find innovative ways of ensuring equity in access. The groundwater irrigation organizations that have come up in north Gujarat are illustrative of such social ingenuity in designing innovative institutions for water allocation and distribution for irrigation. They have emerged in response to a crisis, as the experience of Manund village in Mehsana district (of Gujarat) amply demonstrates. When the water in the shallow aquifer dried up and the cost of abstraction of groundwater went up rapidly, the resource-poor farmers who used to access groundwater through shallow open wells were deprived of direct access to groundwater. The resource-rich farmers exploited the situation and started selling water to the resource-poor at highly exploitative prices. But many small and marginal farmers got together to invest in deep tube wells instead of buying water from tube well owners at prohibitive prices. These are perfect examples of equitable water allocation, within the limited domain. These irrigation institutions were also found to be economically sound and financially viable (Kumar, 2000b). There are thousands of such tube well partnerships working in north Gujarat, especially in Mehsana and Patan districts.

But the dichotomy is that none of these organizations invest in ensuring sustainable resource use. On the contrary, they ensure that sufficient capital is accrued for investment in a new tube well before the existing one goes out of use. The reason is the lack of well-defined water rights, which creates uncertainty about the negative impact of their interventions on the resource conditions and long-term sustainability of their use. But, once rights are defined, it would increase the incentive among the member farmers to invest in water conservation technologies to enhance water (physical) productivity or switch to crops with inherently higher water productivity in rupee terms, like orchards. This is because water saving gives them direct economic benefits, as they could expand the area under irrigation and generate more income, or reduce the water taxes to be paid. The transaction cost of enforcing water rights would be much lower in areas where these groups exist.

On the other hand, those farmers who have permanently lost access to groundwater would also start investing in appropriating the water rights allocated to them due to the "security of water use tenure" and also for the profits they could make by selling their rights. After Rosegrant and Gazmuri (1994), since the introduction of tradable water right in 1974, two decades of experience in Chile showed that the landless poor gained from the establishment of tradable water rights by selling off their rights to large farm holders.

The norms regarding volumetric water rights can be based on physical and socioeconomic considerations such as the safe yield of the aquifer, overall water availability, the land ownership of the water users, family size, etc. The local institutions can enforce volumetric water rights among the users, apart from implementing a wide range of physical activities related to managing the resource. The property rights in groundwater vested with the users need to be limited to use rights rather than permanent ownership rights over water. The norms for volumetric water rights should also be subject to changes depending on the long-term changes in aquifer conditions with regard to the quantity and quality of water. As discussed in Chapter 8, electricity rationing can be used as a proxy for establishing groundwater use rights in regions where all agro wells are run on electricity. The pre-paid electronic energy meters can be used for restricting electricity access of individual farmers.

A major concern associated with the allocation of tradable water rights is the trade-off between food security and economic efficiency: it encourages farmers to increasingly allocate water for non-agricultural purposes such as industry, with adverse impacts on food grain production, though such a tendency is desirable in a drought year (Kumar, 2007). Results from the study carried out by IFPRI on the Maipo river basin show that even after the reallocation of water from agriculture to urban areas, net profits in irrigated agriculture increased substantially compared to the case of proportional use rights without rights to trade water. Moreover, agricultural production did not decline significantly when water was traded from agriculture. Farmers earned substantial benefits from selling their unused rights during the months with little or no crop production (source: https://www.ifpri.cgiar.org/themes/mp10.htm).

Writing about the impact of rights reforms in Chile, Rosegrant and Gazmuri (1994) note: "The five percent growth rate in Chilean agriculture since the reform of land and water rights also calls into question any presumption of negative area-of-origin effects in agriculture." Two studies that attempted to measure the increase in water use efficiency during 1975–92 found a 26% and a 22% increase in agricultural water use efficiency (source: Rosegrant & Gazmuri, 1994).

There are notable positive experiences with catchment-wide water trading and reallocation of water for the environment in the case of the Murray Darling Basin in Australia, even prior to the implementation of the National Water Act of 2007. Water trading has been progressively introduced in the southern MDB since the late 1980s, along with further reforms in the mid-1990s and again in the mid-2000s. It was based on the premise that trading provides economic benefits to buyers and sellers, and to society as a whole. Trading of water rights enabled increased water use efficiency in irrigated production through the movement of water from low-valued uses to high-valued uses (NWC, 2010). In India, the amount of "real saving" in water that could be achieved through a 5% reduction in irrigated paddy area in Punjab is sufficient to bring about a good groundwater balance in the region. It might be possible to offset the potential reduction in paddy output due to the reallocation of water for the environment through the enhancement in productivity of land and water in paddy.

In the past, there were debates at academic, practitioner, and policy levels on the economic, social, and political viability of introducing a consumption-based power tariff for agricultural pumping (see Dubash, 2007; Gupta, 2012; Narayanamoorthy, 1997; Pfeiffer & Lin, 2014; Saleth, 1997; Shah et al., 2004a). Such debates were not supported by hard empirical data (Kumar, 2018). Dubash (2007) and Shah et al. (2004a) consider energy tariff fixing a political issue and that politicians consider charging for electricity supplied to farmers for groundwater pumping as biting the bullet. The analysis presented in Chapter 8 of this book supports tariff changes, as it shows no negative impacts of such reforms on the economics of irrigated crop production. But a hike in power tariff and change in mode of pricing should be accompanied by improvements in the quality of power supply, which would enable farmers to irrigate the crops at times when it is required and grow water-sensitive and high-value crops (Kumar, 2005; World Bank, 2001). As discussed in Chapter 8, prepaid, electronic energy meters, which operate through scratch cards and can work on satellite and internet technology, can be employed in rural areas to reduce the transaction cost of metering thousands of agro wells. They can help monitor electricity use and groundwater use online from a centralized system.

Introduction of a groundwater tax for the allocated water rights should help the groundwater management institutions invest in management interventions such as the transfer of water from water-abundant regions for artificial recharge, watershed management and village infrastructure for water transfer between buyers and sellers of groundwater, apart from covering their management costs. The water rights and groundwater tax would increase the relevance of the village and watershed institutions further, as monitoring of water use and tax collection would be a regular activity (Kumar, 2007). The management costs of these institutions will be covered by the proceeds from the groundwater tax collection.

The process of institutional development would, however, be arduous and time-consuming given the scientific (resource evaluation), technical (evolving physical options for management with due consideration to micro and macro environments), economic (of water taxes; management costs), legal (issue of who owns water rights), and institutional (norms for allocating water rights) issues involved (Kumar, 2007). As an alternative, several researchers have suggested a greater role for Panchayat Raj Institutions (PRIs) in natural resources management, including water management, to avoid any potential future conflicts between the newly created institutions and the PRIs already existing in the villages (IRMED, 2008). But involving PRI might not yield the desired results. The reason for this is that Panchayat Raj institutions are mostly political entities, not to mention the issue of inadequate human resources and technical capabilities that these institutions are facing. They might use their political power and influence for their vested interests (interests of their own constituencies) in influencing the overall functioning of the village economy, of which access to water rights would be crucial. Hence, thoughts should also be given to short-term interventions to reduce the stress that farmers are experiencing. For instance, they should be aimed at encouraging farmers to use water more efficiently by promoting WSTs to make milk production water efficient, while in the long run, they could move to orchards.

The reason is that, though orchards are more water-efficient[6] and are capable of cutting down groundwater use in the region, many of the small and marginal farmers would hesitate to move out of milk production and adopt it, as it is risky and needs efficient management and proper market support. On the other hand, milk ensures regular cash flow and provides stable returns due to good marketing facilities, and dairy production has good adaptability to water scarcity

---

6. In arid climate, it requires nearly five liters of water per day to irrigate a pomegranate plot with an effective canopy cover of one sq. m, and therefore, the total water requirement for irrigating the orchard farm round the year (for 150 days) would be around 9 cm for an average plot having 1200 plants/ha, against 80–90 cm for a cropping system consisting of wheat and pearl millet (Kumar, 2007).

and drought conditions. For encouraging the adoption of drips, farmers would need subsidies.

In India, the north Gujarat initiative on sustainable groundwater management, launched in 2002 by the International Water Management Institute (IWMI) in partnership with the Sir Ratan Tata Trust (SRTT), began as an action research project to identify the ways to establish local management regimes for addressing north Gujarat's groundwater depletion problems in a few villages. After three years, the initiative had expanded into a full-fledged field project implementing local solutions for tackling groundwater scarcity, based on water demand management in agriculture, involving local NGOs. The technical interventions tried by farmers in the area include: (1) high-valued and water-efficient orchard crops and vegetables replacing conventional cereals; (2) water-saving micro-irrigation technologies for row crops, water-intensive field crops and orchards (pomegranate, lemon, etc.); and (3) vermicomposting and the use of organic manure for all crops replacing chemical fertilizers to ensure enhanced biomass utilization efficiencies and improved primary productivity and water retention capacity of degraded soils. A study of the impact of the water-saving irrigation devices (drips and sprinklers) in the region showed a 27% reduction in water use at the farm level, along with improvements in the water productivity of several crops (Singh & Kumar, 2012).

The estimates of average farm-level water use for different crops before and after adoption of the micro-irrigation (MI) system showed a reduction from 34,870 m^3 to 27,343 m^3 owing to the adoption of the technologies. The total reduction in groundwater use was 7527 m^3, which was considered as real water saving due to a reduction in evaporation from the soil and non-recoverable deep percolation. Based on the data available at the macro level that around 24,285 ha of land had been brought under MI in the three districts of north Gujarat, consisting of Banaskantha, Mehsana, and Patan, during a period of 4–5 years, the annual saving in groundwater for irrigation was estimated to be 56.90 MCM per annum (Singh & Kumar, 2012). Today, there is hardly any farm in the area that does not have any of the micro irrigation devices. Projects like this could be conceptualized by the government in partnership with NGOs and donors.

On the other hand, there are other reasons why regions like north Gujarat, western Rajasthan, central Punjab and the Rayalaseema region (of Andhra Pradesh) in India should attract investments for their management of ground water resources. Many semi-arid and water-scarce regions in India, including the Indus basin and Pennar basin, are net exporters of crops to states, including those which are physically water surplus but face economic water scarcity, such as Bihar, Uttar Pradesh, Madhya Pradesh, and parts of Peninsular India (source: based on Amarasinghe et al., 2004: Fig. 3, p. 14). Gujarat is largely a food-deficit region. But within this state, water-scarce areas such as Saurashtra and north Gujarat export virtual water in the form of milk (Singh et al., 2004). The amount of water it exports in the form of milk and dairy products from two districts alone comes to around 2116 MCM annually, which is far more than the groundwater imbalance of the region[7]. From that perspective, the region's demand for imported water for recharging its depleted aquifers is legitimate. Over and above that, recharge interventions would be crucial for addressing groundwater quality problems such as high levels of fluorides and TDS causing health hazards.

The *"Sujalam Sufalam"* scheme of the government of Gujarat takes water from the Sardar Sarovar Project (SSP) to use for recharging the alluvial areas of Mehsana and Banaskantha districts through a network of canals and rivers (Kumar, 2007). But the analysis of the economic surplus generated from the use of groundwater for irrigation showed that the investment can be justified only if the existing ponds and tanks in the proposed command areas of Sardar Sarovar are used for recharging, in which case the study estimated the total recharge potential to be 200 MCM per annum. Further analysis showed that recharging in the non-command area with a total recharge volume of 1000 MCM/year would involve lifting water and creating new canals for diversion to recharge sites spread over a large area and therefore would be economically unviable (for details, see Ranade & Kumar, 2004).

The Kaleshwaram project in Telangana, which involves lifting 215 TMC of water from the Godavari River for transfer to agricultural areas in the dry regions of the state (to irrigate 1.82 million acres) and cities, will also divert water to thousands of irrigation tanks *en route*. This measure of replenishing the tanks would also eventually recharge the stressed aquifers and rejuvenate the unproductive farm wells, through return flows from irrigated paddy fields and, to a limited extent, direct percolation from the tanks, apart from directly benefiting the farmers in their command area. However, such publicly funded projects are prohibitively expensive[8], with no major revenue to the states through irrigation water charge. The only way to make such large projects viable is to do rationing of the volume of water supplied to the farmers and efficient pricing of groundwater, which would lead to enhanced water productivity and economic surplus.

In the Indus basin area of Pakistan, a lot can be achieved through the redistribution of surface water from canals, with more water allocated to the tail reaches that are facing groundwater depletion and mineralization of groundwater and less water to the head reaches that are receiving excess amounts of seepage instead of allocating equal amounts of water across the system. Hence, possibilities exist to arrest further depletion of and even rejuvenate the over-exploited aquifers in different regions of South Asia. But the initiatives to do so should be part

---

7. Estimates by Singh et al. (2004) show that north Gujarat is exporting a net volume of about 2116.05 MCM of virtual water annually through the dairy business.
8. The Kaleshwaram project of Telangana is expected to cost a total of Rs. 1.2 lac crore or 16 billion US dollars at current prices. The Sardar Sarovar Project is expected to cost nearly Rs. 68,000 crore or around 9 billion US dollars, on completion.

of a proper groundwater management plan, evolved through a proper analysis of the various options available for rescuing the aquifer under investigation—such as demand side management, long-distance water import and conjunctive management, and groundwater banking—and their economic viability. Over and above, developing such a management plan should be the responsibility of the groundwater management institutions. However, the effectiveness of the management institutions in achieving the management objectives (cutting down overall groundwater abstraction, improving equity in access to groundwater, and improving the efficiency of use of water, etc.), would depend on the institutional and policy support from the state-level institutions for applying the instruments such as electricity metering, pro-rata energy pricing and energy rationing.

10.7 The future steps

The institutional model for groundwater management proposed in this chapter has some key aspects: (1) the establishment of tradable water rights among all the water users; (2) promoting formal water markets; and (3) the establishment of formal institutions at various levels, starting from village institutions to institutions at the watershed level and aquifer level. The real challenge is in identifying regions where these concepts can be introduced and operationalized with the least transaction cost. The most ideal regions are those where the negative externalities of over-exploitation problems on the society are quite perceptible and significant—in the form of well failures, very high cost of abstraction of groundwater, salinization of aquifers and soils, etc.—and where well density is low, yet the intensity of groundwater use is high. The regions where informal water markets already exist for trading water within the agriculture sector and between agriculture and urban uses are the most ideal, as the selection of such regions would reduce the infrastructure requirements for developing water markets. The areas where demand management interventions in agriculture to save water in that sector are feasible should be preferred. However, the first and foremost thing is that the concerned state government is willing to experiment with water rights systems with well-defined criteria for water allocation amongst the uses and users within them, which would require legal reforms that recognize property rights systems in groundwater and local groundwater management institutions as agencies for enforcing them. Such institutional models should be piloted in both alluvial areas with groundwater stock and also hard rock areas. The purpose of doing this is to deal with the challenges of allocating water in two different situations. The first one is where there will be no significant variations in allocation between years owing to the availability of stock that is not dependent on the annual recharge. The second one is where there can be substantial variation in the amount of water that can be allocated due to the inter-annual variability in the replenishable groundwater resource.

However, the allocation of water use rights should also include rights to use surface water in regions where it can play a critical role in improving the sustainability of groundwater use, such as the Indus basin area in Pakistan. In the Indus basin irrigation system, the total surface water allocation for irrigation is around 160 billion cubic meters (Bakshi & Trivedi, 2011; Basharat et al., 2014), spread over an area of 16 m.ha. (Government of Pakistan, 2019). In the Indus basin areas, the intensive use of groundwater for irrigation in areas with limited access to canal water (especially in the tail reaches of canals) is causing salinization of the groundwater due to the intrusion of water from saline formations into freshwater layers. Excessive use of this saline groundwater for irrigation in the absence of sufficient freshwater for blending is also causing soil salinization. On the other hand, increased availability of cheap surface water in some areas (head reaches of the canal network) is restricting the use of groundwater (Qureshi, 2020; Qureshi et al., 2010). This, combined with excessive return flows from canal irrigation and canal seepage, is causing waterlogging problems in many areas. Balanced development and use of groundwater in both areas can be achieved through the reallocation of surface water.

References

Amarasinghe, U., Sharma, B. R., Aloysius, N., Scott, C. A., Smakhtin, V., de Fraiture, C., Sinha, A. K., & Shukla, A. K. (2004). *Spatial variation in water supply and demand across river basins of India* [Research Report 83]. IWMI.

Basharat, M., Hassan, D., Bajkani, A., & Sultan, S. (2014). *Surface water and groundwater nexus: Groundwater management options for Indus Basin irrigation system.* Water and Power Development Authority (WAPDA).

Bakshi, G., & Trivedi, S. (2011). *The Indus equation.* Strategic Forsight Group https://www.strategicforesight.com/publication_pdf/10345110617.pdf.

Bauer, C. J. (2012). *Against the current: Privatization, water markets, and the state in Chile (Natural resource management and policy, 14).* Springer.

Bhanja, S. N., Mukherjee, A., Rodell, M., Wada, Y., Chattopadhyay, S., Velicogna, I., Pangaluru, K., & Famiglietti, J. S. (2017). Groundwater rejuvenation in parts of India influenced by water-policy change implementation. *Scientific Reports, 7,* 7453. https://doi.org/10.1038/s41598-017-07058-2.

Brozović, N., & Young, R. (2014). Design and implementation of markets for groundwater pumping rights. In K. Easter, & Q. Huang (Eds.), *Water markets for the 21st century: What have we learned? (Global issues in water policy, 11)* (pp. 283–303). Springer.

Burke, J., & Moench, M. (2000). *Groundwater and society: Resources, tensions and opportunities.* New York, NY: United Nations.

Central Ground Water Board. (2007). *Manual on Artificial Recharge of Ground Water, Central Ground Water Board, Ministry of Water Resources.* Faridabad: Govt. of India, September.

Central Ground Water Board. (2019). *National compilation on dynamic ground water resources of India 2017.* Central Ground Water Board, Dept. of Water Resources, RD & GR, Ministry of Jal Shakti.

Chen, L. (2012). The State Council report on rural water resources development. In *On the 26th Meeting of the Standing Committee of 11th National People's Congress, 25 April 2012.*

Cullet, P. (2014). Groundwater law in India: Towards a framework ensuring equitable access and aquifer protection. *Journal of Environmental Law, 26*(1), 55–81. https://doi.org/10.1093/jel/eqt031.

Dubash, N. K. (2007). The electricity-groundwater conundrum: Case for a political solution to a political problem. *Economic and Political Weekly, 42*(52), 45–55.

Easter, K. W., & Huang, Q (Eds.). (2014). *Water markets for the 21st century: What have we learned?.* Springer.

Easter, K. W., Rosegrant, M. W., & Dinar, A. (1999). Formal and informal markets for water: Institutions, performance, and constraints. *The World Bank Research Observer, 14*(1), 99–116. https://doi.org/10.1093/wbro/14.1.99.

Frederick, K. D. (1993). *Balancing water demands with supplies: The role of management in a world of increasing scarcity* [Technical Paper Number 189]. World Bank.

Frederiksen, H. D. (1998). Institutional principles for sound management of water and related environmental resources. In Asit K. Biswas (Ed.), *Water resources: Environmental planning, management and development.* Tata McGraw-Hill Publishing Company Limited.

Garduño, H. (2005, January 26–28). *Making water rights administration work* [Paper presentation]. International workshop on African Water Laws: Plural Legislative Frameworks for Rural Water Management in Africa.

Glendenning, C. J, & Vervoort, R. W. (2011). Hydrological impacts of rainwater harvesting (RWH) in a case study catchment: The Arvari River, Rajasthan, India: Part 2. Catchment-scale impacts. *Agricultural Water Management, 98*(4), 715–730.

Government of India. (2014). *Report of 4th census of minor irrigation schemes.* Minor Irrigation Statistics Wing, Ministry of Water Resources, River Development and Ganga Rejuvenation, Government of India.

Government of India. (2017). *Report of 5th census of minor irrigation schemes.* Minor Irrigation Statistics Wing, Ministry of Water Resources, River Development and Ganga Rejuvenation, Government of India.

Government of Pakistan. (2019). *Pakistan Economic Survey 2018-19; Economic Adviser's Wing, Finance Division.* Islamabad, Pakistan: Government of Pakistan.

Gupta, R. K. (2012). The role of water technology in development: A case study in Gujarat, India. In R. Ardakanian, & D. Jaeger (Eds.), *Water and the green economy: Capacity development aspects.* UNW-DPA.

Howe, Charles W., Schurmeier, Dennis R., & Douglas Shaw, W., Jr. (1986). Innovative approaches to water allocation: The potential for water markets. *Water Resources Research, 22*(4), 439–445 April.

Hérivaux, C., Rinaudo, J., & Montginoul, M. (2020). Exploring the potential of groundwater markets in agriculture: Results of a participatory evaluation in Five French case studies. *Water Economics and Policy, 6*(1), 1950009. https://doi.org/10.1142/s2382624x19500097.

Institute for Resource Management and Economic Development (IRMED). (2008). *Institutional framework for regulating use of ground water in India: Final report submitted to the Ministry of Water Resources, Government of India.* Institute for Resource Management and Economic Development.

Institute of Rural Management Anand/UNICEF. (2001). *White paper on water in Gujarat: Report submitted to the Government of Gujarat, Gandhinagar.*

Jadeja, Y., Maheshwari, B., & Packham, R.others. (2015). Participatory groundwater management at village level in India–Empowering communities with science for effective decision making *[Paper presentation]* November 3–5. Australian Groundwater Conference 2015.

Jairath, J. (2001). *Water user associations in Andhra Pradesh: Initial feedback.* Published for Centre for Economic and Social Studies, Hyderabad by Concept Pub. Co.

Kale, E. (2017). Problematic uses and practices of farm ponds in Maharashtra. *Economic and Political Weekly, 52*(3), 20–22.

Kay, M., Franks, T., & Smith, L. E. D. (1997). *Water: Economics, management and demand.* Chapman & Hall (E & FN Spon).

Kemper, K. E (2007). Instruments and Institutions for Groundwater Management. In M. Giordano, & K. G. Villholth (Eds.), *Agricultural groundwater revolution: Opportunities and threats to development.* CAB International.

Kulkarni, H., & Vijay Shankar, P. S. (2009b). Groundwater: Towards an aquifer management framework. *Economic and Political Weekly, 44*(6), 13–17. https://www.jstor.org/stable/pdfplus/40278472.

Kumar, M. D. (1996). *Tube well turn over: a study of groundwater management organizations in Gujarat.* Ahmedabad: Monograph, VIKSAT.

Kumar, M. D., Chopde, S., Mudrakartha, S., & Prakash, A. (1999). Addressing water scarcity and pollution problems in Sabarmati Basin: Local strategies for water supply and conservation management. In M. Moench, E. Caspari, & A. Dixit (Eds.), *Rethinking the Mosaic: Investigations into local water management.* Nepal Water Conservation Foundation, Kathmandu and Institute for Social and Environmental Transition.

Kumar, M. D. (2000a). Institutional framework for groundwater management: A case study of community organizations in Gujarat, India. *Water Policy, 2*(6), 423–432.

Kumar, M. D. (2000b). Institutions for efficient and equitable use of groundwater: Irrigation management institutions and water markets in Gujarat, Western India. *Asia-Pacific Journal of Rural Development, 10*(1), 52–65.

Kumar, M. D., & Singh, O. P. (2001). Market instruments for demand management in the face of scarcity and overuse of water in Gujarat, Western India. *Water Policy, 3*(5), 387–403.

Kumar, M. D. (2002). *Reconciling water use and environment: Water resource management in Gujarat, resource, problems, issues, strategies and framework for action: Report prepared for State Environmental Action Project supported by the World Bank.* Gujarat Ecology Commission.

Kumar, M. D. (2003). Demand management in the face of growing water scarcity and conflicts in India: Institutional and policy alternatives for future. In K. Chopra, C. H. H. Rao, & R. P. Sengupta (Eds.), *Water resources, sustainable livelihoods and ecosystem services.* Indian Society for Ecological Economics, & Concept Publishing Company.

Kumar, M. (2005). Impact of electricity prices and volumetric water allocation on energy and groundwater demand management:. *Energy Policy, 33*(1), 39–51. https://doi.org/10.1016/s0301-4215(03)00196-4.

Kumar, M. D., Ghosh, S., Patel, A., Sundar, S., & Ravindranath, R. (2006). Rainwater harvesting in India: Some critical issues for basin planning and research. *Land Use and Water Resources Research, 6,* 1–17.

Kumar, M. D. (2007). *Groundwater management in India: physical, institutional and policy alternatives.* New Delhi: SAGE Publications Pvt. Ltd.

Kumar, M. D. (2018). Institutions and policies governing groundwater development, use and management in the Indo-Gangetic Plains of India. In K. G. Vilholth, E. L. Gunn, K. Conti, A. Garrido, & J. van der Gun (Eds.), *Advances in groundwater governance.* CRS Press/Balkema.

Kumar, M. D., Patel, A., Ravindranath, R., & Singh, O. P. (2008). Chasing a mirage: Water harvesting and artificial recharge in naturally water-scarce regions. *Economic and Political Weekly, 43*(35), 61–71.

Kumar, M. D., & Bassi, N. (2011). Maximizing the Social and Economic Returns from Sardar Sarovar Project: Thinking beyond the Convention. In Parthasarathy, R., & Dholakia, R. (Eds.), *Sardar Sarovar Project on the River Narmada, Impacts so far and ways forward* (pp. 747–776). Delhi: Concept Publishing.

Kumar, M. D., Scott, C. A., & Singh, O. P. (2011). Inducing the shift from flat-rate or free agricultural power to metered supply: Implications for groundwater depletion and power sector viability in India. *Journal of Hydrology, 409*(1–2), 382–394.

Kumar, M. D., Scott, C. A., & Singh, O. (2013). Can India raise agricultural productivity while reducing groundwater and energy use? *International Journal of Water Resources Development, 29*(4), 557–573.

Kumar, M. D., & Perry, C. J. (2019). What can explain groundwater rejuvenation in Gujarat in recent years? *International Journal of Water Resources Development, 35*(5), 891–906. https://doi.org/ 10.1080/07900627.2018.1501350.

Kumar, M. D., Prathapar, S. A., Saleth, R. M., Reddy, V. R., Narayanamoorthy, A., & Bassi, N. (2017). New water management paradigm: outdated concepts? *Economic and Political Weekly, 52*(49), 89–94.

Lawson, S., Buckingham, A., Hamstead', M., & O'Keefe, V (2010). Overcoming impediments to groundwater trading in Australia. In *Australian National Groundwater Conference, National Convention Centre, Canberra.*

Llamas, M. R., De Stefano, L., Aldaya, M., Custodio, E., Garrido, A., López-Gunn, E., & Willaarts, B. (2012). Introduction. In L. De Stefano, & M. R. Llamas (Eds.), *Water, agriculture and environment: Can we square the circle?*. Taylor & Francis.

Lopez-Gunn, E., & Cortina, L. M. (2006). Is self-regulation a myth? Case study on Spanish groundwater user associations and the role of higher-level authorities. *Hydrogeology Journal, 14*(3), 361–379. https://doi.org/10.1007/s10040-005-0014-z.

Moench, M. (1995). Approaches to groundwater management in India: To control or enable? In M. Moench (Ed.), *Groundwater law: The growing debate*. VIKSAT-Natural Heritage Institute Monograph, VIKSAT.

Moench, M., Kulkarni, H., & Burke, J. (2013). *Trends in local groundwater management institutions* [Technical Report 7]. *Groundwater Governance: A Global Framework for Action* www.groundwatergovernance.org.

Mukherjee, M. (2021, October 19). Nationalisation of water is not an option for India: Mihir Shah [Interview]. *The Wire*. https://thewire.in/environment/interview-national-water-policy-management-mihir-shah.

Narayanamoorthy, A. (1997). Impact of electricity tariff policies on the use of electricity and groundwater: arguments and facts. *Artha Vijnana, 39*(3), 323–340.

National Water Commission. (2010). *The impacts of water trading in the southern Murray–Darling Basin: An economic, social and environmental assessment*. NWC, Canberra. https://webarchive. nla.gov.au/awa/20100619154046/http://www.nwc.gov.au/www/html/2816-impacts-of-water-trading-in-the-southern-murraydarling-basin.asp?intSiteID=1.

Ostrom, E. (1992). *Crafting institutions for self governing irrigation systems*. Institute of Contemporary Studies.

Palanisami, K. (1994). Evolution of agricultural and urban water markets in Tamil Nadu, India. *Arlington, Virginia: Irrigation Support Project for Asia and the Near East (ISPAN)*. United States Agency for International Development.

Paranjape, S., & Joy, K. J. (1995). *Sustainable technology making the Sardar Sarovar Project viable: A comprehensive proposal to modify the project for greater equity and ecological sustainability*. Centre for Environment Education.

Pearce, D. W., & Warford, J. (1993). *World without end: Economics, environment and sustainable development*. Oxford University Press.

Peck, J. C. (2003). *Property rights in groundwater: Lessons from Kansas experience* [Paper presentation] March 16–25. 3rd World Water Forum.

Pfeiffer, L., & Lin, C. Y. C. (2014). The effects of energy prices on agricultural groundwater extraction from the high plains aquifer. *American Journal of Agricultural Economics, 96*(5), 1349–1362. https://doi.org/10.1093/ajae/aau020.

Qureshi, A. S. (2020). Groundwater governance in Pakistan: From colossal development to neglected management. *Water, 12*(11), 3017. https://doi.org/10.3390/w12113017.

Qureshi, A. S., McCornick, P. G., Sarwar, A., & Sharma, B. R. (2010). Challenges and prospects of sustainable groundwater management in the Indus Basin, Pakistan. *Water Resources Management, 24*(8), 1551–1569. https://doi.org/10.1007/s11269-009-9513-3.

Ranade, R., & Kumar, M. D. (2004). Narmada water for groundwater recharge in north Gujarat conjunctive management in large irrigation projects. *Economic and Political Weekly, 39*(31), 3498–3503.

Ray, S., & Bijarnia, M. (2006): Upstream vs Downstream: Groundwater Management and Rainwater Harvesting, Economic & Political Weekly, 41 (23): 2375–2384.

Rosegrant, M. W., & Gazmuri, R. S. (1994). *Reforming water allocation policy through markets in tradable water rights: Lessons from Chile, Mexico, and California* [EPTD Discussion paper 2]. International Food Policy Research Institute.

Rosegrant, M. W., & Ringler, C. (2000). Impact on food security and rural development of transferring water out of agriculture. *Water Policy, 1*(6), 567–586. https://doi.org/10.1016/s1366-7017(99)00018-5.

Saleth, R. M. (1996). *Water institutions in India: Economics, law and policy.* Commonwealth Publishers.

Saleth, R. M. (1997). Power tariff policy for groundwater regulation: Efficiency, equity and sustainability. *Artha Vijnana, 39*(3), 312–322.

Saleth, R. M., & Dinar, A. (1999). *Water challenge and institutional responses (A cross country perspective)* [Policy research working paper series 2045]. World Bank.

Schiffler, M. (1998). *The economics of groundwater management in arid countries: Theory, international experience and a case study of Jordan (GDI Book Series No. 11).* Routledge.

Shah, T. (1993). *Water markets and irrigation development: Political economy and practical policy.* Oxford University Press.

Shah, T., & Bhattacharya, S. (1993). *Farmers organizations for lift irrigation: Companies and tubewell co-operatives of Gujarat [ODI irrigation management network paper 26].*

Shah, T., Roy, A. D., Qureshi, A. S., & Wang, J. (2003). Sustaining Asia's groundwater boom: An overview of issues and evidence. *Natural Resources Forum, 27*(2), 130–141. https://doi.org/10.1111/1477-8947.00048.

Shah, T., Scott, C. A., Kishore, A., & Sharma, A. (2004a). *Energy-irrigation nexus in South Asia: Improving groundwater conservation and power sector viability* [Research report 70]. International Water Management Institute.

Shah, T., Scott, C. A., & Buechler, S. (2004b). Water sector reforms in Mexico: Lessons for India's water policy. *Economic and Political Weekly, 39*(4), 361–370.

Shah, T. (2009). *Taming the anarchy: Groundwater governance in South Asia.* Resources for the Future.

Shah, M. (2016). *A 21st century institutional architecture for water reforms in India: Final Report submitted to the Ministry of Water Resources. River Development & Ganga Rejuvenation.* Government of India.

Shah, M. (2021). Nationalisation of water is not an option for India. *Wire, 19* October, 2021.

Sharma, H. P. (2005). Integrated water resources management and water efficiency plans: With special reference to Mauritius. In G. N. Mathur (Ed.), *Proceedings of the XII World Water*

Congress "Water for Sustainable Development, Towards Innovative Solutions" Vol. 1 (pp. 287–295). Central Board of Irrigation and Power.

Shah, T., & Verma, S. (2008). Co-management of electricity and groundwater: an assessment of Gujarat's Jyotirgram scheme. *Economic and Political Weekly, 43*(7), 59–66.

Sharma, S. C. (1995). Regulation of groundwater development in India: Existing provisions and future options. In M. Moench (Ed.), *Groundwater law: The growing debate.* Vikram Sarabhai Centre for Development Interaction (VIKSAT), Natural Heritage Institute Monograph.

Shen, D. (2015). Groundwater management in China. *Water Policy, 17*(1), 61–82. https://doi.org/10.2166/wp.2014.135.

Singh, O. P., Sharma, A., Singh, R., & Shah, T. (2004). Virtual water trade in the dairy economy: Analyses of irrigation water productivity in dairy production in Gujarat, India. *Economic and Political Weekly, 39*(31), 3492–3497.

Singh, O. P., & Kumar, M. D. (2012). Hydrological and farming system impacts of agricultural water management interventions in North Gujarat, India. In M. D. Kumar, M. V. K. Sivamohan, & N. Bassi (Eds.), *Water management, food security and sustainable agriculture in developing economies.* Routledge/Earthscan.

Skurray, J. H., Roberts, E., & Pannell, D. J. (2012). Hydrological challenges to groundwater trading: Lessons from south-west Western Australia. *Journal of Hydrology, 412–413*, 256–268. https://doi.org/10.1016/j.jhydrol.2011.05.034.

Syme, G. J., Reddy, V. R., Pavelic, P., Croke, B., & Ranjan, R. (2012). Confronting scale in watershed development in India. *Hydrogeology Journal, 20*(5), 985–993. https://doi.org/10.1007/s10040-011-0824-0.

Talati, J., Kumar, M. D., Kumar, M. D., Reddy, V. R., & James, A. J. (2019). From watershed to sub-basins: Analyzing the scale effects of watershed development in Narmada river basin, central India. *From catchment management to managing river Basins: Science, technology choices, institutions and policy (Current directions in water scarcity research,* Vol. 1). Elsevier Science.

Thobanl, M. (1997). Formal water markets: Why, when, and how to introduce tradable water rights. *The World Bank Research Observer, 12*(2), 161–179. https://doi.org/10.1093/wbro/12.2.161.

Thompson, C. L., Supalla, R. J., Martin, D. L., & McMullen, B. P. (2009). Evidence supporting cap and trade as a groundwater policy option for reducing irrigation consumptive use. *Journal of the American Water Resources Association, 45*(6), 1508–1518. https://doi.org/10.1111/j.1752-1688.2009.00384.x.

Vijay Shankar, P. S., Kulkarni, H., & Krishnan, S. (2015). India's groundwater challenge and the way forward. *Economic and Political Weekly, 46*(2), 37–45.

Vyas, J. N. (2001). Water and Energy for Development in Gujarat with Special Focus on the Sardar Sarovar project. *International Journal of Water Resources Development, 17*(1), 37–54.

Wester, P., Sandoval Minero, R., & Hoogesteger, J. (2011). Assessment of the development of aquifer management councils (COTAS) for sustainable groundwater management in Guanajuato, Mexico. *Hydrogeology Journal, 19*(4), 889–899. https://doi.org/10.1007/s10040-011-0733-2.

Wheeler, S. A., Schoengold, K., & Bjornlund, H. (2016). Lessons to be learned from groundwater: Trading in Australia and the United States. *Integrated Groundwater Management*, 493–517.

World Bank. (2001). *India power supply to agriculture, volume 1: Summary report* [Report no. 22171-IN]. Energy Sector Unit, South Asia Regional Office. www.ifpri.cgiar.org/themes/mp10.html.

Zekri, S. (2008). Using economic incentives and regulations to reduce seawater intrusion in the Batinah coastal area of Oman. *Agricultural Water Management, 95*(3), 243–252. https://doi.org/10.1016/j.agwat.2007.10.006.

Chapter 11

Managing groundwater in under-developed regions: Policy lessons

11.1 Synthesis of findings

As an opening remark, we would like to state that too many myths eclipse attempts by scholars to really investigate the factors driving and the solutions for South Asia's groundwater problems. These myths were created over the past three decades, mostly by researchers and field professionals working on groundwater, and at times by some policymakers. They were part of a larger narrative pushed by vested interests to create the space for pushing certain unscientific solutions with no evidence of effectiveness.

Some of the myths that prevail in the groundwater sector concerning its use are: (1) the wells are where people are; intensity of groundwater use is largely driven by population pressures; (2) groundwater is a democratic resource, and ownership of and access to the resource are highly equitable, with a large proportion of them being owned by small and marginal farmers; (3) the performance of surface irrigation is very poor, with negative returns on investment and has no future; (4) groundwater is a natural oligopoly and groundwater markets (or pump rental markets) actually promote access equity in groundwater by making the resource accessible to the poor farmers at affordable prices; (5) electricity metering and its pro-rata pricing, though good options, are extremely difficult options to improve the efficiency and sustainability of groundwater use because of their politically sensitive nature; and (6) market instruments (such as water rights systems, water entitlements, water resource tax/fee, pollution tax, etc.) and the formal institutional arrangements that were attempted by developed countries (Kemper, 2007), whose water economies are more formal, will not work in the Indian context (Shah & van Koppen, 2006). All these propositions basically imply that groundwater use in South Asia is a case of colossal anarchy and there are no textbook solutions.

Thereafter, a few "homegrown" solutions were prescribed by those who perpetuate these myths. One of them is that decentralized water harvesting and "local recharge" are viable options to arrest depletion and improve the

Groundwater Economics and Policy in South Asia. DOI: https://doi.org/10.1016/B978-0-443-14011-2.00009-7

agricultural economy in areas witnessing over-exploitation and there is sufficient runoff water available in these areas to capture through water harvesting—with many variants of it such as farm ponds, check dams, dug well recharging, managed aquifer recharge, etc. being tried in different parts of the country. Another one is that micro irrigation systems are a viable option to improve crop productivity and also save water in agriculture, enabling the reallocation of water to other sectors, especially the environment, and have huge potential. The third one is the provision of a separate feeder for agricultural power supply and restricted hours of supply (8 hours) to prevent over-pumping and ensure judicious use of electricity and groundwater. The other options include: delayed sowing of paddy (as tried in Punjab) and direct delivery of power subsidies to agriculture to incentivize farmers to conserve electricity and groundwater.

Such myths have dominated the academic debates in the sector for a long time because of the limited ability of the majority of researchers and academicians to appreciate the problems and issues from a multi-disciplinary and inter-disciplinary perspective, which had tempted them to embrace the quick fixes and piecemeal solutions that were churned out by influential academic/research groups. The irrigation economists hardly understood hydrology—which, for instance, mattered when it came to understanding the importance of surface irrigation in improving the sustainability of well irrigation, barring the exception of scholars like the late Prof. B.D. Dhawan was vociferous about the importance of return flows from gravity irrigation in sustaining well irrigation. Very few economists from India ever understood how the charges for electricity could positively affect the returns from farming, through adjustments in cropping patterns and the efficiency of the use of inputs, including irrigation water. A large majority of them believed that pricing would adversely impact the net returns from crop production due to a rise in the cost of irrigation. The geo-hydrologists hardly understood the socio-economic and livelihood dimensions of groundwater use—which, for instance, mattered when it came to recognizing the fact that top-down regulations to check groundwater draft are unlikely to work. They hardly ever appreciate the fact that it is hard to find surplus water (runoff) for recharging the aquifers without causing negative effects in regions witnessing depletion because of the very characteristics of the river basins there.

The academic books on groundwater largely dealt with the theoretical aspects of the physical processes–groundwater flow dynamics, recharge process, well hydraulics and the design of pumping wells, groundwater-surface water interactions, seawater intrusion, artificial recharge, aquifer modeling, etc. They hardly examined the practical issues associated with the design and construction of wells, artificial recharge structures, the economics of groundwater abstraction, modeling complex aquifers, etc. in complex real-life situations. On the socio-economic side, academic literature that deals with the economic valuation of groundwater, the effectiveness of various technical interventions, policy instruments, and institutions in regulating groundwater use or managing groundwater, and the crafting of groundwater management institutions in different socio-economic settings are rather scanty.

This is the general scenario, the infamous nexus that exists between the academicians, bureaucrats, and politicians in the water sector (Kumar, 2018), and the orchestration by a large majority of the researchers and academicians to parade these myths and cheer the quick fix solutions advocated by these influential groups had prompted government agencies to try them, sometimes as pilot projects and sometimes as large, publicly-funded schemes (to name a few, there is the farm pond scheme in Maharashtra, decentralized water conservation through check dam construction in Gujarat, agricultural feeder line separation in Haryana, following the Gujarat model, dug well recharging in 100 hard rock districts of India, and direct delivery of power subsidy to farmers in Punjab).

Through several chapters of this book, we demonstrated that none of these solutions can actually make a difference in improving the groundwater conditions and instead would create more problems. To begin, we have shown how inequitable the access to groundwater is in South Asian countries such as India, Pakistan, and Bangladesh, with only a small fraction of the marginal and small farmers owning them even after five decades of the advent of drilling and pumping technologies. What is most worrisome is the fact that the inequity is growing, as shown by a comparison between the data available from the 3rd and 5th Minor Irrigation Censuses, with a time gap of 14 years (Chapter 2). In the Punjab province of Pakistan, 99.4% of those farmers who were buying water from well owners were marginal and small farmers, and all those who were selling water were large farmers. Moreover, nearly 46% of the large farmers were found to be selling water. In India, the growth in well irrigation is declining as a result of over-exploitation of the regions where there is still a lot of arable land that can be brought under crop cultivation. In Bangladesh, the actual area irrigated by wells had declined in the recent past as a result of an increase in the cost of pumping groundwater owing to a decline in water levels. Continued depletion of groundwater is likely to affect small farmers as the cost of abstraction of groundwater further increases.

We have also shown that in India, the connected load-based tariffs for electricity actually benefit large farmers by keeping their implicit cost of pumping groundwater very low. Our analysis shows that it also increases the monopoly power of large well owners, who dictate the price of water sold in the market, by discouraging small holders from investing in wells and pump sets and making it a sellers' market. In Pakistan, the electricity subsidies for groundwater irrigation are appropriated by less than 0.30% of the population, which is essentially constituted by large farm households owning deep tube wells.

In Chapter 3, we have discussed the complex concept of groundwater over-exploitation, and how the official methodologies are still inadequate to assess the over-exploitation problems from physical, social, economic, environmental, and ethical points of view. We have shown how, under different geological settings, groundwater behaves differently—with a sharp seasonal decline in water levels in hard rock areas and much lower seasonal drops in alluvial areas—and therefore how important it is to factor in the geological conditions of the area while analyzing groundwater level trends and interpreting them. We have also shown how many hard rock areas, which record a very high incidence of temporary (reduction in well yields) and permanent well failures, are still being categorized as "safe" as per the official assessments. Many areas in central India, which are facing problems of yield decline, are still categorized as under-developed areas. Such anomalies between official estimates and the ground situation need to be rectified through methodological refinements.

Using selected illustrative cases, we have demonstrated how combining official statistics of the status of groundwater development with information on detailed water balance, geology, water level fluctuations, and socio-economic, ecological, and ethical aspects would cast an altogether different scenario of the degree of over-exploitation problems. To get an assessment of aquifer over-exploitation that reflects the concerns of the stakeholders, two steps are important. The first step is improving the reliability of groundwater recharge and withdrawal estimates. The most important challenge is the accurate estimation of groundwater drafts and natural outflows from and lateral flows in the groundwater system. With the easy availability of good quality data on water level fluctuations from observation wells for considerably long time periods, the accuracy in the estimation of groundwater recharge would depend much on the availability of reliable data on specific yield values for the aquifer under consideration. The second step is broadening the criteria for assessing aquifer over-exploitation to capture the complex hydrologic, hydro-dynamic, economic, social, and ecological variables that reflect the negative consequences of over-development.

Besides this, our analyses of data from Ahmednagar, Ludhiana, and the Narmada river basin have shown why the interpretation of long-term data on water levels collected by the official agencies in India needs to factor in the differences in characteristics of the aquifers (i.e., whether consolidated, unconsolidated, or semi consolidated, and whether there is static groundwater or not, and whether the terrain is undulating or plain). The reason is that water-level trends are the net effect of the hydrologic stresses acting on aquifers and their effect on the processes of groundwater recharge, storage, and discharge, and the conditions that affect these processes (such as hydrology, geology, geohydrology, and geomorphology) change from region to region. We have also indicated how the important findings from the review of the groundwater assessment methodology followed by the Indian agencies could be used in developing comprehensive criteria for the assessment of groundwater in Pakistan and Bangladesh that are capable of capturing the effect of water level decline on the cost of well irrigation and the ecological effects of groundwater irrigation.

Chapter 4 provides a good comparison between surface gravity irrigation and well irrigation using examples from India with a strong empirical basis. It showed how a myopic view among some researchers, favoring well irrigation over surface irrigation, is doing damage to both sectors. The chapter shows how the criteria being used for assessing the performance of surface irrigation schemes based on gravity, which are simplistic, poor, and outdated, cast such systems in a poor light, and discourage the much-needed public investments in such schemes. The criteria only lead to underestimating the performance of gravity irrigation systems, while ignoring the huge positive externalities they induce on well irrigation. While advocating an integrated approach to the development of groundwater and surface water resources, it shows the fallacy behind arguing for using surface water from reservoirs for recharging aquifers. Instead, what the chapter shows us is that surface irrigation through gravity that favors crops like paddy and sugarcane augments groundwater recharge reduces the energy footprint for groundwater irrigation, and thus sustains well irrigation.

In Chapter 4, we also illustrate how such a myopic view favoring well irrigation is quite dominant even in Pakistan, a country with the world's largest gravity irrigation network, in spite of the wide recognition arising from the data available from technical studies that a large proportion of the groundwater recharge in Pakistan is due to return flows from canal irrigated areas.

In Chapter 5, we showed how schemes that are designed on the basis of poor concepts in hydrology and economics can produce unintended developmental outcomes when implemented. It showed that the highly touted farm pond scheme for reducing groundwater stress in the drought-prone regions of Maharashtra helped the resource-rich farmers to over-exploit the groundwater and store it in the surface storage ponds, which in turn enabled them to grow cash crops and perennial crops (vegetables, pomegranate) throughout the year and irrigate them more intensively to obtain a higher yield and income per unit of land. It resulted in a much greater use of groundwater, but with a very small improvement in crop water productivity. The chapter also showed how the heavy capital subsidies for solar-powered irrigation are benefiting the resource-rich farmers in Gujarat by allowing them to extract more groundwater with the help of the excessive power available to them and sell it to their neighbors for increasing their profit, in spite of having a feed-in tariff system that enabled them to sell the excess power to the electricity utility. Both the schemes, which were designed to create social welfare through the conservation of groundwater (in both cases) and electricity (in the second case), resulted in unintended consequences in the form of increased groundwater use in agriculture and private benefits at a large cost to the public.

Chapter 6 looks at the impacts of large irrigation systems and monsoon precipitation on reducing groundwater imbalance and arresting depletion at the regional scale. While bursting the myth about the reasons for groundwater reju-venation in Gujarat witnessed in recent years, we have shown in Chapter 6 that abundant rainfall received in many years during the period, 2003–14, followed by a gradual increase in surface water irrigation through gravity in large areas (from 2002 onwards), substantially reduced the pressure on groundwater resources and reversed the downward trend in water levels in regions that once witnessed groundwater mining problems. In the large command area of SSP, a consistent increase in surface irrigation through gravity not only reduced the dependence of farmers on groundwater for irrigation but also improved the recharge through irrigation return flows and canal seepage. This, along with increased rainfall that resulted in better natural recharge and improved availability of soil moisture, improved the groundwater balance. We provided empirical data to show that neither decentralized water harvesting (check dams) nor the *Jyotigram Yojna* played any role in reversing the downward trend of regional groundwater.

We have also shown how the important findings from the analysis of the Gujarat groundwater rejuvenation phenomenon can be used to explore options for improving the groundwater condition in other semi-arid and arid regions of South Asia in the future. The strategy would revolve around the intelligent reallocation of water from large irrigation systems to the command areas and extending the canal networks of such large systems to areas of groundwater mining to be used as replacement water.

Another myth that is created is that paddy is a water-intensive crop and that groundwater depletion in many semi-arid regions of India (like in erstwhile AP as well as Punjab) is because of the cultivation of irrigated paddy (Chapters 7 and 9). By estimating the return flows from irrigated paddy (presented in Chapter 4), we have shown that the solutions advocated by some practitioners and researchers, such as restricting irrigation for paddy to save water (Chaba, 2019; Prasad, 2018) and replacement of paddy by low water-intensive cereals such as sorghum and pearl millet (Shah et al., 2021), are not going to be valid everywhere.

On the other hand, we have also shown that for erstwhile Andhra Pradesh, micro irrigation systems have very little potential to save groundwater because the region's cropping system is dominated by cereal crops like paddy. The region is left with no other option than to go for energy metering, consumption-based tariffs, and energy rationing. In the Rayalaseema region, which experiences problems of rampant well failures, such an arrangement would motivate farmers to allocate the costly groundwater to highly water-efficient crops and introduce water-saving irrigation technologies such as drips and mulching. Another viable option to address the problem of groundwater overdraft in the region, as discussed in Chapter 7, is the inter-basin transfer of water from the 'water-surplus' Godavari basin to the part of the region falling inside the Krishna basin, to be used for irrigation. Gravity irrigation for paddy can quickly recharge the shallow aquifers of the region through return flows.

Chapter 8 discussed the institutional setup for groundwater development in the Indo-Gangetic plain, and the policies relating to water, energy, and food, which have an influence on groundwater development and use in the region. The institutions that controlled groundwater development were: interventions by state agencies for irrigation development through public tube wells, water markets of small and marginal farmers, legal and regulatory instruments for checking groundwater over-development, and minimum support price for the procurement of cereals. The key policies were: the provision of subsidized diesel pumps, electricity and diesel pricing policies for groundwater pumping, policies for providing power connections to the farm sector, electricity supply policies for the farm sector, and procurement policies for cereals.

The chapter also systematically analyzed the impacts of these institutional and policy regimes on the sustainability of groundwater use—such as the drying up of wetlands, groundwater contamination, and resource depletion in the western Gangetic plains. In the chapter, we have shown how two different models of groundwater management are important for the region, one for the western Indo-Gangetic plain (which have lower rainfall and higher aridity) and the other for the eastern Gangetic plains. For the former, sustainable resource use along with equity and efficiency are the major concerns as resources are depleting fast. But for the eastern Gangetic plains, with abundant groundwater, equity in access to the resource is important. Pro rata pricing of electricity would allow small and marginal farmers in the eastern IGP to invest in wells instead of depending on water purchased at prohibitive prices, provided obtaining electricity connections became easy. This will also reduce the monopoly price of water traded in the market as the number of potential sellers of water increases. Here, cheap diesel engines will also have to be introduced. In the western Gangetic plains, subsidy removal, metering, and pro-rata pricing, combined with energy rationing as a proxy for a functional water rights system, need to be introduced with a minimum transaction cost. The unit charge for electricity, however, has to be fixed cautiously with due consideration to the water table depth, as depth to the water table varies widely across the region with a consequent effect on the amount of electricity required to pump a unit volume of water.

These apart, on the governance front, refining the resource assessment methodologies to suit the conditions of these two distinct regions is important. In the case of the western IGP, the methodology needs to be sound enough to assess the impact of imported surface water. In the case of the eastern IGP, understanding groundwater-wetland interactions is important to ascertain the extent to which groundwater use in the region could be intensified. Also, framing appropriate rules for groundwater abstraction in the eastern Gangetic plains based on sound environmental criteria is important to ensure that groundwater use does not affect the functioning of the wetlands.

The chapter concludes by drawing important inferences for alluvial areas of Pakistan and Bangladesh that are part of the Indo-Gangetic plains. We have argued that in Pakistan, energy metering, pro-rata pricing, and energy rationing in the agriculture sector can be explored like in the western IGP of India, but with a focus on the deep tube well-owning farmers who pump large volumes of water. We also argued that in the case of the alluvial plains of Bangladesh, the focus of the strategy to arrest groundwater depletion (in the north-central and north-western regions) and to protect the wetlands (that provide water for surface irrigation) should be to reduce the consumptive use of water for crop production through interventions that can help reduce evapotranspiration per unit cropped area.

On the use of economic incentives for regulating groundwater use, our analysis presented in Chapter 9 shows that the schemes that restrict electricity quota for groundwater pumping on the basis of the "connected load" of the pump owners (as attempted in Indian Punjab) favor rich well owners, whose pump capacity is disproportionately higher than the land holding they have, by allowing them to use more electricity and water and yet claim cash incentives. The analysis also shows that consumptive water use in paddy in Punjab is not as high as normally believed and the root cause of groundwater depletion in Punjab is the intensive cultivation of two crops, paddy and wheat, that increases the evapotranspiration per unit of land.

In the same chapter, while discussing an alternative to direct delivery of power subsidy to farmers, we showed how a well-designed model of energy supply to individual farms that are fixed on the basis of the land holding size, water requirement of the prevailing cropping system and the sustainable yield of the regional aquifers, with electricity metering at the farm level, could incentivize farmers to efficiently use energy and groundwater and achieve sustainable groundwater use. We proposed the use of smart meters to be installed in pumps under a prepaid system, and energy quotas can be decided on the basis of socio-economic parameters such as land holdings to be irrigated, the water quota per unit area of land (based on crop water requirements) and the energy required to lift a unit volume of water. We expect that the establishment of water/energy quotas will promote the allocation of water to economically efficient crops and extensive water trading.

In Chapter 10, we have shown how the idea of participatory, community-based management of groundwater at the local (village) level, widely talked about by many researchers, practitioners, and policymakers, remained a fallacy even after three decades since the idea was first coined. While the researchers and practitioners who advocated the idea of 'local management' failed to recognize the externalities in local management, they focused their attention only on the capacity building of the village institutions of the local communities. While the debate was mainly on how the 'scale' of management responses should be decided on the basis of the aquifer characteristics—i.e., karstic, alluvial, hard rock, etc. (see, Moench et al., 2013), the consensus was strong on the view that lack of adequate knowledge of the aquifers is the biggest barrier for organizing communities around the resource to manage it (Jadeja et al., 2015; Kulkarni & Vijay Shankar, 2009; Kulkarni et al., 2015). Participatory aquifer mapping was touted as a panacea for all the "information" barriers limiting the community's ability to manage groundwater locally (Shah, 2016). A few have also mooted the idea of participatory aquifer management, but without any clarity on the management structure and the institutional arrangements (see Kulkarni & Vijay Shankar, 2009; Kulkarni et al., 2015; Shah, 2021).

The chapter first discussed the reasons for the failure of "top-down regulatory approaches" to manage groundwater attempted to date in India and then went on to discuss alternative institutional approaches to manage groundwater, with the establishment of tradable water uses rights and promotion of formal water markets to affect groundwater demand management. After discussing the externalities in local management at the village level, we presented a three-tier hierarchy of institutions from aquifer-level management committees to watershed institutions to village institutions. The chapter also discussed the functions of institutions at every level and also moot the idea of a groundwater tax (linked to property rights) as a major avenue to generate the resources required for implementing the resource management activities and for covering the management costs of the institutions. In the chapter, we also pointed out that the process of institutional development would be arduous and time-consuming given the scientific, technical, economic, legal, and institutional issues involved and also discussed the various opportunities that are available for investing in institutional development by citing examples from within India and other parts of the world.

In the concluding section, we have also discussed how surface water allocation from public irrigation systems can be used as a powerful instrument for regulating groundwater use and therefore why rights to use surface water from canals should be brought within the purview of water rights for improving the sustainability of groundwater use in regions such as the Indus basin area in Pakistan, where farmers use both groundwater and surface water conjunctively.

11.2 What can other semi-arid countries learn from the South Asian experience?

There are many important lessons that other developing countries (especially from Africa) can learn from India regarding the approaches to developing and managing groundwater resources. In South Asia as a whole and especially in India, Pakistan, and Bangladesh, the pace of development of groundwater was much higher than the pace at which the science of managing groundwater evolved. As the population grew and with the growing pressure to intensify cropping and improve the yields to achieve food security, India went all out to increase irrigated areas. Subsidies were offered to farmers to dig or drill wells and install pump sets, and the subsidy was also offered for electricity supplies for running engines. For those who had diesel engines, subsidized diesel was available. These factors, compounded by the government's thrust on massive rural electrification, had a significant impact on well irrigation. In India, groundwater investigations by the Geological Survey of India and later by the Central Ground Water Board generated a good amount of knowledge about the aquifer systems and their yield potential. In Pakistan and Bangladesh, the presence of rich alluvial aquifers with shallow water table conditions made well development rather easy except in areas where groundwater is inherently saline, though, in Bangladesh, groundwater development started picking up momentum only in the 1970s.

While India lost a large amount of irrigated area after the partition in 1947 to the newly formed nation of Pakistan, the subsequent government investments in public irrigation could not keep pace with the growing population and increasing food demand. Many regions of India (with low to medium rainfalls and where cultivation of the second crop after kharif was not possible without irrigation facilities), such as peninsular India (excluding Kerala), the central Indian plateau (Madhya Pradesh, Chhattisgarh), and western India (Gujarat, Maharashtra, and Rajasthan), were left with very little area under public (canal) irrigation. In some areas, even kharif crop cultivation wasn't possible without irrigation for many years, given the extreme variability in rainfall conditions and aridity in these low-rainfall regions. This has forced farmers in these regions to go digging open wells and, later, bore wells and tube wells, depending on the aquifer characteristics. While open wells were preferred in the initial years in both alluvial and hard rock areas, with the lowering of water levels and increasing incidence of well failures (permanent failure and reduction in well yields) due to overdraft, farmers had to shift to bore wells in hard rock areas and deep tube wells in alluvial and sedimentary areas.

Borrowing from the West (especially from western United States and Australia), the focus of management responses in India was on the artificial recharge of groundwater to augment supplies, as the subject was largely handled by groundwater scientists at the national and state levels[1]. The mainstream groundwater scientists from the agencies had very little appreciation of the hydrology of the regions concerned, which mattered a lot when it comes to determining the feasibility of artificial recharge, apart from the knowledge of geology and geohydrology. The regional hydrology and the hydrology of the local catchment decide the amount of surface runoff that would be available for recharge, the flow regimes (peak discharge and duration of flow), sediment yield, etc. These hydrological variables have serious implications for the scope of artificial recharge and the design of the recharge structures. The limited success of one or two NGOs in groundwater recharge (which was highly localized) also encouraged the government agencies (such as CAPART and NABARD) to push the idea through the NGOs they supported.

A large number of NGOs also started working in many water-scarce regions without taking into consideration the hydrology and geohydrology of the areas. Listening to the "clarion call" for artificial recharge of groundwater from the central government, many schemes for water harvesting and recharge were designed and implemented by the state governments in western, central, and southern India. Unfortunately, the condition in India is such that in the semi-arid and arid regions where dependence on groundwater for agriculture is high

---

1. For instance, Australia uses techniques such as aquifer storage and recovery; aquifer storage, transfer and recovery; infiltration gallery; pond infiltration; recharge weir, soil aquifer treatment and reservoir release. The sources of water used are surface water/storm water, reclaimed water and groundwater (source: https://research.csiro.au/mar/using-managed-aquifer-recharge/#map).

and where groundwater resources are over-exploited, the river basins are already over-developed (Kumar et al., 2008a, 2012). Moreover, the high inter-annual variability of rainfall in these semi-arid and arid regions means that when the crops require irrigation water more (as a result of monsoon failure, which also increases the aridity and causes fast depletion of soil moisture), the runoff available from the catchments is very low (Kumar et al., 2006). Hence, the artificial recharge systems fail when water from such storage is badly needed.

Hence, most of the initiatives from the government and NGOs failed to make any positive impact on the groundwater situation. On the contrary, several ill effects of indiscriminate water harvesting were widely seen (Glendenning & Vervoort, 2011; Kumar et al., 2008a; Ray & Bijarnia, 2006; Talati & Kumar, 2019). While critical analysis of these initiatives brought out the fallacies in the claims made by the proponents of water harvesting and artificial recharge, the motivated writing of some of the research groups from within India and outside, like evangelists, helped keep the activities afloat for some more time. However, over a period of 20 years or so, the enthusiasm has now died down almost everywhere.

However, such a push for water harvesting was not evident in Pakistan, probably because of four reasons. *First:* groundwater availability as such is not an issue in the Indus Basin alluvial area, which has a large stock of groundwater, though the salinity of groundwater is an issue in many areas. *Second:* the areas facing groundwater over-exploitation (and the associated problems of rising salinity) were receiving canal water and therefore there was a demand for increased allocation of surface water for irrigation, which would enable farmers to use more poor-quality groundwater after blending. *Third:* rainfall is very low and aridity is very high in most parts of the alluvial areas of the Indus basin, with very little surface runoff generated in normal years. *Fourth:* the flat topography and poor drainage density of the region mean that the construction of dams for impounding surface runoff is not feasible.

With a big government push, the new fad is micro irrigation to save water in agriculture and produce more crops per drop of water. Subsidies are offered to farmers by both the Union government and the state governments to adopt micro irrigation. But hardly any attention is being paid to factors such as where and under what conditions micro irrigation systems can save water in real terms, and what the aggregate impact of large-scale adoption of such systems on groundwater would be. The critical questions are: (1) to what extent the adoption of MI leads to a reduction in the consumptive use of water in the field (which actually determines the extent of real water saving per unit area); and (2) what do the farmers do with the saved water, i.e., whether they expand the area or reduce pumping and leave water in the aquifers (Kumar, 2016). Most of the studies sponsored by the Ministries and the financial institutions (like NABARD) in this regard basically looked at the data collected from farmers on water application to crops and yields, which were totally inadequate to address the first question.

While the objective of offering subsidies was to reduce water consumption in crop production, this question of real water saving was never addressed by such studies. Long-term water resource management studies in regions that have witnessed large-scale adoption of micro irrigation (such as the Marathwada region of Maharashtra, north Gujarat and western Rajasthan) that specifically look at the impact on groundwater are absent. The government agencies concerned are not oriented to such thinking. However, there are several studies from around the world that looked at these issues more critically, and concerns were raised by some scholars about a potential increase in water depletion as a result of farmers increasing the area under cultivation following the adoption of MI systems.

Kumar et al. (2008b) and Kumar (2016) hold the view that with a reduction in water applied per unit area of land, the farmers would divert the saved water for expanding the area, subject to favorable conditions with respect to water and equipment availability and power supplies for pumping water, and under such circumstances, the net effect of adoption of MI systems on water use could be insignificant at the aggregate level unless the cropping pattern is changed after adoption.

Others argue that with the adoption of WSTs, there is a greater threat of resource depletion (Molle et al., 2004; Perry, 2007; Ward & Pulido-Velazquez, 2008), because in the long run, either the return flows from irrigated fields would decline and area under irrigation would increase if conditions are favorable for cropped area expansion (Molle et al., 2004; Perry, 2007) or crop consumptive use (ET) would increase with drip-irrigated area replacing area under the traditional method of irrigation (Ward & Pulido-Velazquez, 2008).

However, both views fail to make a distinction between consumptive water use in crop production (depletion) and crop consumptive use (ET) (see Allen et al., 1997 for definitions) and how they could be vastly different in different contexts (Kumar, 2016; Kumar and van Dam, 2013). Nevertheless, some do argue that with the adoption of efficient irrigation technologies, aggregate water use may decrease or increase depending on a variety of conditions such as agricultural system characteristics, aquifer characteristics, previous conditions of irrigation infrastructure, and induced changes in crop rotations (Berbel et al., 2014; Sanchis-Ibor et al., 2015).

The key message for other semi-arid and arid countries, especially those from Africa and a few countries from South Asia (such as Afghanistan, Nepal, and Sri Lanka), which are now embarking on large-scale groundwater development for irrigation and drinking water supply, is that they should tread carefully. The following are the specific recommendations for such countries.

1. **Scientific resource assessment**: Map different aquifer systems in the country/province and for each aquifer system, determine the aquifer properties through pumping tests, and ascertain the sustainable yield of the aquifers and the amount of annual groundwater draft that the aquifers can support through proper modeling studies based on real-time data on recharge from

precipitation and other sources. Such assessments can reduce the bad investments in the indiscriminate sinking of wells with no proportional increase in irrigated area, resulting in poor economic efficiency in water abstraction. Subsequently, the types of wells that can be drilled in a cost-effective manner can be decided. Predicting the future on the basis of the current trend that the dependence on groundwater for agriculture would remain low could lead to serious negative outcomes. Indian experience suggests that the drastic reduction in the cost of well drilling and the price of pump sets and fuel in the market could dramatically change the scenario for many countries in Africa in the near future.

2. **Mandatory registration of all the wells tapping the aquifers**: It is better to start registering all the wells (with their features and characteristics) through legislation enacted with proper enforcement mechanisms in place at the very beginning itself before the numbers explode and new registrations become almost impossible. Such processes will help the water management agencies in these countries design instruments for regulating groundwater use, including the fixing of the volumetric water rights/water quota under the water rights system or well spacing regulations at a later stage if required. Proper registration of wells would also provide hints on when to ban the construction of new groundwater abstraction structures in an area or take a relook at the allocated water rights. All the newly constructed wells should be geo-tagged.

3. **Encourage balanced development of groundwater and surface water**: The experience of India, Pakistan, and Bangladesh shows that well irrigation is sustainable in areas that are also served by gravity irrigation. A decline in well irrigation has been noticed in the case of Bangladesh. This is because of the return flows from gravity irrigation that benefits shallow aquifers by augmenting recharge and raising water levels. It is therefore necessary that, wherever hydrology and topography permit, surface irrigation systems be planned in conjunction with well irrigation. Conversely, groundwater irrigation should be promoted wherever canal irrigation results in excessive seepage (from canal networks) and return flows from the irrigated fields.

4. **Framing resource development policies with a long-term perspective**: The governments should frame policies to promote groundwater development more carefully and with a long-term vision. With changes in the trajectory of development, the policies need to be reviewed on the basis of impact studies and changed if required. Sticking to the same policy for many decades can turn out to be disastrous for any region. In areas where groundwater is already over-exploited or is on the verge of being over-exploited, continuing with the subsidies for wells and pump sets and free electricity, which were probably required when people were not using water from wells, cannot be justified as they will produce negative welfare outcomes. However, restricting the access of resource-poor farmers to groundwater can have (intragenerational) equity implications. Therefore, pump subsidies and energy

subsidies may be offered to the farmers who are willing to install energy meters, and water-saving irrigation equipment and restrict their energy use to a pre-determined level fixed by the utility based on an assessment of crop water requirement, land holdings, and safe yield of the aquifer.

5. **Align private interests with societal goals through policies and institutions**: Policies relating to water and energy pricing and the institutional regime governing access to underground water—water entitlements, water rights, water quota, etc.—with due consideration for the laws existing in those countries, need to be carefully thought of and designed in such a way that the private interests of the farmers are well aligned with the societal goals of water conservation and environmental protection.

If the farmer has to pay for every unit of electricity used for pumping groundwater, it would lead to optimal use of water and electricity, which would benefit him/her in terms of reduced cost of inputs. Similarly, even if the government is not in a position to change the existing water rights through a new law, if there are restrictions on the amount of energy that can be used from a power grid (which is normally run by a government agency), it would be equivalent to a functional water rights system. The farmers would then be cautious in choosing crops in a way that yields high allocative efficiency (in $ per m^3 of water allocated to the farm). Such actions would also simultaneously benefit society with reduced water consumption and resource depletion. If this is not done, the farmers might use the technologies to get more water from underground to intensively irrigate their crops without caring for the resource (like in the case of farm ponds) or for more yield (like in the case of micro irrigation) but they may not reduce water and energy use as both actions do not get translated into cost saving for him/her.

While subsidies for energy would be required in the initial phases of groundwater development to promote well irrigation (to reduce the cost of irrigation), what is important is that the utility starts charging on the basis of actual electricity consumption, thereby making the farmers confront the positive marginal cost of using electricity for irrigating crops. Energy metering of all agro-wells is a must to know how much water is being pumped (roughly) and how it is changing over time. Slowly, the subsidy rate for energy will have to be reduced. Similarly, the water entitlement or the water quota or the water right per ha of cropland can be fixed at a higher level initially and then gradually reduced as more farmers start using the water from the aquifers.

6. **Careful design of capital subsidies**: Government interventions for promoting well irrigation (for well digging/drilling), water conservation (and recharge) works, and water-saving technologies (micro irrigation systems) at the farm level in the form of capital subsidies should be based on a thorough scientific assessment of the likely impacts of such technologies on crop production and resource use. Such impact studies should be done under different scenarios of water use, with different policies with respect

to water price, energy price, and water supply regulations. In the absence of unit charges for electricity or water, it is quite unlikely that the farmers will reduce their water consumption (per unit irrigated area) after the adoption of efficient irrigation equipment such as drip systems if such equipment comes with heavy capital subsidies.

7. **Follow a catchment/basin approach for planning water harvesting and artificial recharge schemes**: Promotion of decentralized water harvesting and recharge works at the local level (at the level of villages and micro watersheds) should be based on sound hydrological planning that helps assess basin/sub-basin yields and the amount of un-utilized water in the basins and their sub-watersheds. Micro-level planning of such activities being undertaken by the local communities at the village level should be integrated at the macro level for the sub-basins and river basin to make sure that there is no over-appropriation of the runoff (Kumar et al., 2019). The work should be coordinated by one centralized agency, and periodic reviews of such work should be undertaken to make sure that there are no major adverse effects, especially in the downstream areas.

References

Allen, R. G., Willardson, L. S., & Frederiksen, H. (1997). Water use definitions and their use for assessing the impacts of water conservation. In J. M. deJager, L. P. Vermes, & R. Rageb (Eds.), *Proceedings of the ICID Workshop on Sustainable Irrigation in Areas of Water Scarcity and Drought,* September 11–12, Oxford, England (pp. 72–82).

Berbel, J., Gutiérrez-Martín, C., Rodríguez-Díaz, J. A., Camacho, E., & Montesinos, P. (2014). Literature review on rebound effect of water saving measures and analysis of a Spanish case study. *Water Resources Management, 29*(3), 663–678. https://doi.org/10.1007/s11269-014-0839-0.

Chaba, A. A. (2019, January 4). Reforming agriculture: When power is worth saving. The Indian Express. https://indianexpress.com/article/india/reforming-agriculture-when-power-is-worth-saving-5520945/. (Accessed: 30 July 2019).

Glendenning, C., & Vervoort, R. (2011). Hydrological impacts of rainwater harvesting (RWH) in a case study catchment: The Arvari River, Rajasthan, India. *Agricultural Water Management, 98*(4), 715–730. https://doi.org/10.1016/j.agwat.2010.11.010.

Jadeja, Y., Maheshwari, B., Packham, R., Hakimuddin, Purohit, R., Thaker, B., Goradiya, V., Oza, S., Dave, S., Soni, P., Dashora, Y., Dashora, R., Shah, T., Gorsiya, J., Katara, P., Ward, J., Kookana, R., Dillon, P., Prathapar, S., Chinnasamy, P., & Varua, M. (2015). Participatory groundwater management at village level in India: Empowering communities with science for effective decision making [Paper presentation]. In *Australian Groundwater Conference 2015.*

Kemper, K. E. (2007). Instruments and institutions for groundwater management. In M. Giordano, & K. G. Villholth (Eds.), *Agricultural groundwater revolution: Opportunities and threats to development.* CAB International.

Kulkarni, H., Shah, M., & Vijay Shankar, P. S. (2015). Shaping the contours of groundwater governance in India. *Journal of Hydrology: Regional Studies, 4,* 172–192. https://doi.org/10.1016/j.ejrh.2014.11.004.

Kulkarni, H., & Vijay Shankar, P. S. (2009). *Groundwater: Towards an aquifer management framework. Economic and Political Weekly, 44*(6), 13–17.

Kumar, M. D. (2016). Water saving and yield enhancing micro irrigation technologies: Theory and practice. In P. K. Viswanathan, M. D. Kumar, & A. Narayanamoorthy (Eds.), *Micro irrigation systems in India: Emergence, status and impacts* (pp. 13–26). Springer.

Kumar, M. D. (2018). Water Policy, Science and Politics: An Indian Perspective. Amsterdam: Elsevier Science.

Kumar, M. D., Batchelor, C., & James, A. J. (2019). Operationalizing IWRM concepts at the basin level: From theory to practice. In M. D. Kumar, V. R. Reddy, & A. J. James (Eds.), *From catchment management to managing river basins: science, technology choices, institutions and policy* (pp. 299–329). Amsterdam: Elsevier Science.

Kumar, M. D., Ghosh, S., Singh, O. P., Patel, A., & Ravindranath, R. (2006). Water harvesting and groundwater recharge in India: Critical issues for basin planning and research. *Land Use and Water Resource Research, 6*(1), 1–17.

Kumar, M. D., Patel, A., Ravindranath, R., & Singh, O. P. (2008a). Chasing a mirage: Water harvesting and artificial recharge in naturally water-scarce regions. *Economic and Political Weekly, 43*(35), 61–71.

Kumar, M. D., Sivamohan, M. V. K., & Narayanamoorthy, A. (2012). *The food security challenge of the food-land-water nexus. Food Security Journal, 4*(4), 539–556.

Kumar, M. D., Turral, H., Sharma, B. R., Amarasinghe, U., & Singh, O. P. (2008b). Water saving and yield enhancing micro irrigation technologies in India: When do they become best bet technologies? In M. D. Kumar (Ed.), *Managing water in the face of growing scarcity, inequity and declining returns: Exploring fresh approaches, Vol. 1. Proceedings of the 7th Annual Partners' Meet of IWMI-Tata Water Policy Research program*, ICRISAT, Hyderabad (pp. 13–36).

Kumar, M. D., & van Dam, J. C. (2013). Drivers of change in agricultural water productivity and its improvement at basin scale in developing economies. *Water International, 38*(3), 312–325.

Moench, M., Kulkarni, H., & Burke, J. (2013). Trends in local groundwater management institutions [Technical report 7]. *Groundwater Governance: A Global Framework for Action.* www.groundwatergovernance.org.

Molle, F., Mamanpoush, A., & Miranzadeh, M. (2004). *Robbing Yadullah's water to irrigate Saeid's garden: Hydrology and water rights in a village of Central Iran* [Research report 80]. International Water Management Institute.

Perry, C. J. (2007). Efficient irrigation; inefficient communication; flawed recommendations. *Irrigation and Drainage, 56*(4), 367–378. https://doi.org/10.1002/ird.323.

Prasad, A. V. (2018). India: Direct Delivery of Electricity Subsidy for Agriculture, An Innovative Pilot in Punjab. Presented at Geneva. *October, 30,* 2018. http://esmap.org/sites/default/files/KEF%20Geneva%202018/Energy%20Subsidy%20Reform%20Forum%20%202018_Punjab_Final%20-%2025.10.2018.pdf.

Ray, S., & Bijarnia, M. (2006). Upstream vs downstream: Groundwater management and rainwater harvesting. *Economic and Political Weekly, 41*(23), 2375–2383.

Sanchis-Ibor, C., Macian-Sorribes, H., García-Mollá, M., & Pulido-Velazquez, M. (2015). Effects of drip irrigation on water consumption at Basin Scale (MIJARES RIVER, SPAIN). In *26th Euro-Mediterranean Regional Conference and Workshops, "Innovate to improve Irrigation performances"*. https://icid2015.sciencesconf.org/75019/ICID2015_IMPADAPT_fnl.pdf.

Shah, M. (2016). *A 21st century institutional architecture for water reforms in India. Final Report submitted to the Ministry of Water Resources.* River Development & Ganga Rejuvenation, Government of India.

Shah, M. (2021, October 19). Nationalisation of water is not an option for India [Interview]. Wire, https://thewire.in/environment/interview-national-water-policy-management-mihir-shah.

Shah, M., Vijay Shankar, P. S., & Harris, F. (2021). Water and agricultural transformation: a symbiotic relationship I, Economic and Political Weekly, *56*(29): 43–55.

Shah, T., & van Koppen, B. (2006). Is India ripe for integrated water resources management? Fitting water policy to national development context. *Economic and Political Weekly, 41*(31), 3413–3421.

Talati, J., & Kumar, M. D. (2019). From watershed to sub-basins: Analyzing the scale effects of watershed development in Narmada river basin, central India. In M. D. Kumar, V. R. Reddy, & A. J. James (Eds.), *From catchment management to managing river basins: Science, technology choices, institutions and policy (Water scarcity research Vol. 1).* Elsevier Science.

Ward, F. A., & Pulido-Velazquez, M. (2008). Water conservation in irrigation can increase water use. *Proceedings of the National Academy of Sciences, 105*(47), 18215–18220. https://doi.org/10.1073/pnas.0805554105.

Index

Printed in the United States
by Baker & Taylor Publisher Services